Crossing Borders

Michelle Ann Miller
Michael Douglass • Matthias Garschagen
Editors

Crossing Borders

Governing Environmental Disasters in a Global Urban Age in Asia and the Pacific

Springer

Editors
Michelle Ann Miller
Asia Research Institute
National University of Singapore
Singapore, Singapore

Michael Douglass
Asia Research Institute
National University of Singapore
Singapore, Singapore

Matthias Garschagen
Institute for Environment and Human Security (UNU-EHS)
United Nations University Bonn
Bonn, Nordrhein-Westfalen, Germany

ISBN 978-981-10-6125-7 ISBN 978-981-10-6126-4 (eBook)
DOI 10.1007/978-981-10-6126-4

Library of Congress Control Number: 2017949704

Printed on acid-free paper

This Springer imprint is published by Springer Nature
The registered company is Springer Nature Singapore Pte Ltd.
The registered company address is: 152 Beach Road, #21-01/04 Gateway East, Singapore 189721, Singapore

Acknowledgements

Four funding sources enabled the production of this book. The Asia Research Institute of the National University of Singapore, in collaboration with the United Nations University Institute for Environment and Human Security (UNU-EHS), and the Urban Knowledge Network Asia, International Institute of Asian Studies (UKNA-IIAS), made possible the *International Conference on Crossing Borders: Governing Environmental Disasters in a Global Urban Age in Asia and the Pacific* (5–6 November 2015), which provided the initial impetus for producing this edited collection through the conversations generated by the event as well as through the written contributions of selected presenters. The project also benefited from the financial support of a Singapore Ministry of Education Academic Research Fund Tier 2 grant entitled "Governing Compound Disasters in Urbanising Asia" (MOE2014-T2-1-017). For their overall administration of this project, we thank Marcel Bandur, Sharon Ong, Valerie Yeo, Tay Minghua and Henry Kwan. For their constructive comments on earlier iterations of the chapters in this book, we are grateful to Gregory Clancey, Rita Padawangi, Jerome Whitington, Tyson Vaughan, Jonathan Rigg, Eli Elinoff, Christopher Courtney and Eric Kerr. Thank you also to the anonymous reviewers, who helped to improve the quality of this manuscript, and to Jayanthie Krishnan, Ameena Jaafar and the team at Springer Publishing for seeing this volume through to fruition. For her invaluable editorial support, we are indebted to Dr. Sarah Starkweather.

Contents

Contributors

Sarfaraz Alam Department of Geography, Institute of Science, Banaras Hindu University, Varanasi, UP, India

Kristoffer B. Berse National College of Public Administration Governance, University of the Philippines, Quezon City (NCR), Philippines

Michael Douglass Asia Research Institute, National University of Singapore, Singapore, Singapore

Konia Freitas Kamakakūokalani Center for Hawaiian Studies, University of Hawai'i at Mānoa, Honolulu, HI, USA

Richard Friend Environment Department, University of York, York, North Yorkshire, UK

Matthias Garschagen United Nations University – Institute for Environment and Human Security (UNU-EHS), Bonn, Nordrhein-Westfalen, Germany

Alain Guilloux Independent Scholar, Lamma Island, Hong Kong

Andrea Haefner Griffith Asia Institute, Brisbane, QLD, Australia

John Hannigan Department of Sociology, University of Toronto, Toronto, ON, Canada

Karl Kim Department of Urban & Regional Planning, University of Hawaii at Mānoa, Honolulu, HI, USA

Michelle Ann Miller Asia Research Institute, National University of Singapore, Singapore, Singapore

Yenny Rahmayati Centre for Design Innovation (CDI), Swinburne University of Technology, Melbourne, VIC, Australia

Anthony Reid The Australian National University, Canberra, ACT, Australia

Matthew A. Shapiro Department of Political Science, Illinois Institute of Technology, Chicago, IL, USA

Pakamas Thinphanga Thailand Environment Institute Foundation (TEI), Bangkok, Bangkok, Thailand

Robert James Wasson Institute of Water Policy, Lee Kuan Yew School of Public Policy, National University of Singapore, Singapore, Singapore

Fiona Williamson Asia Research Institute, National University of Singapore, Singapore, Singapore

About the Authors

Sarfaraz Alam is associate professor in the Department of Geography, Institute of Science, Banaras Hindu University, Varanasi. He has briefly served in IGNOU as a reader in geography (School of Sciences). Dr Alam also previously worked in the Institute for Defence Studies and Analyses (IDSA), New Delhi, and at JP University, Chapra (Bihar). His research interests include environmental conflict and security in South Asia, geography education and social geography. Presently, he is writing a book on the theme of geography, environment and the national security of Bangladesh to be published by Cambridge Scholars Publishing, UK. Dr Alam is also involved in a sponsored research project on the theme of neo-urban processes and emerging social geographies of Varanasi City.

Kristoffer B. Berse is assistant professor at the University of the Philippines' National College of Public Administration Governance in Quezon City, Philippines. Outside the academe, he serves as a consultant for the government, civil society and international development organisations in areas broadly criss-crossing the intersections of urban development, governance and environment. His recent publications include "From Yokohama with love: Transferring best practices through international municipal cooperation" (in Y. Nishimura & C. Dimmer, eds., Planning for Sustainable Asian Cities, APSA 2011 Selected Papers, 2012) and "Building local government resilience through city-to-city cooperation" (in R. Shaw & A. Sharma, eds., Climate and Disaster Resilience in Cities, Emerald, 2011). Dr. Berse earned his PhD (urban engineering) and master's degrees (environmental studies) from the University of Tokyo, as a Japanese government and Asian Development Bank scholar, and his bachelor's degree in public administration from the University of the Philippines.

Michael Douglass is professor at the Asia Research Institute, where he is leader of the Asian Urbanisms Cluster, and professor at the Lee Kuan Yew School of Public Policy, National University of Singapore. He is emeritus professor and former chair of the Department of Urban and Regional Planning at the University of Hawaii where he was also the director of the Globalization Research Center. He received his PhD in urban planning from UCLA. He previously held positions in the United Nations and taught at the Institute of Social Studies (the Netherlands) and the University of East Anglia (UK).

Konia Freitas was born and raised in Hawai'i on O'ahu Island. She is a specialist faculty member at the Kamakakūokalani Center for Hawaiian Studies at the University of Hawai'i at Mānoa. As a special projects coordinator, she works in the area of programme planning, curriculum development and programme assessment and evaluation. She has worked for several years in the area of community engagement and engaged scholarship. Projects funded under this initiative linked education, research and practice together in ways that reinforced the fundamental importance of place, Hawaiian culture and philosophy. Her academic areas of interest span indigenous planning, Hawaiian land use and education. She has professional planning experience and holds a PhD in urban and regional planning.

Richard Friend is a lecturer in human geography in the Environment Department at the University of York (UK). He has a PhD in development studies based on extensive anthropological fieldwork in southern Thailand. He has been working in the Mekong Region for 20 years in areas related to human development, governance and natural resource management, leading regional implementation and capacity building programmes as well as research networks and working for international NGOs, donors, intergovernmental organisations and independent think tanks. He has published widely, most recently writing on governance and poverty dimensions of urban climate resilience.

Matthias Garschagen is the head of Vulnerability Assessment, Risk Management and Adaptive Planning (VARMAP) at United Nations University Institute for Environment and Human Security (UNU-EHS). His research focuses on social vulnerability and risk governance in the context of natural hazards and climate change impacts, particularly in Asia. Amongst other functions, Dr Garschagen has served as an invited author in the IPCC's Fifth Assessment Report and in the writing team of the UN's New Urban Agenda. He also coleads the Expert Working Group on Environmental Risks within the German Association of Geographers and the TWIN-SEA expert network on coastal adaptation in Indonesia and Southeast Asia. He is a member of several editorial boards, including the *Journal of Extreme Events* and the journal *Urban Climate*. Dr Garschagen leads a number of international research projects, and his research findings have been published in high-ranking international journals, e.g. on the need to strengthen resilience in small and mid-sized cities (in *Nature*), on the relationship between global urbanisation trends and climate change vulnerability (in *Climatic Change*), on the effectiveness of decentralising risk reduction in Vietnam (in *Habitat International*) and on the applicability of urban resilience concepts across different cultural contexts (in *Natural Hazards*). Matthias Garschagen holds a PhD in geography from the University of Cologne, Germany, and has an additional background in economics and cultural anthropology. He received scholarships from the German National Academic Foundation and the German Academic Exchange Service.

Alain Guilloux holds a PhD from the University of Hong Kong and a master in management from École des Hautes Etudes Commerciales–Paris. He is an independent scholar with over 20 years of experience as a volunteer, programme officer, project evaluator, policy analyst, CEO, board member and treasurer in humanitarian aid, development, philanthropy and NGO management. He was formerly a visiting assistant professor at the Department of Asian and International Studies in the City University of Hong Kong, where he taught programme management, policy and advocacy, disaster management and research methods in the master of social sciences/development studies programme. His research focuses on humanitarian action, disaster management, foreign aid, humanitarian interventions and global and regional governance, primarily in the context of East and Southeast Asia.

Andrea Haefner has a PhD in political science from Griffith University in Australia. She is currently an adjunct research member at the Griffith Asia Institute and works with the Faculty of Water Resources at the National University of Laos. Overall, her research focuses on international relations in the Asia Pacific region, with a specific interest in the Mekong Subregion, especially in regard to nontraditional security issues focusing on environmental security and water governance. Andrea Haefner previously worked for the German local government and in international organisations in Southeast Asia. She recently published a book with Routledge titled *Negotiating for Water Resources: Bridging Transboundary River Basins* focusing on the Mekong, Danube and La Plata River Basins.

John Hannigan is professor of sociology and associate chair of the Graduate Studies (sociology) at the University of Toronto, where he teaches courses in cultural policy, urban political economy and environmental sociology. He has published four books: *Environmental Sociology* (1995, 2006, 2014), *Fantasy City: Pleasure and Profit in the Postmodern City* (1998), *Disasters Without Borders: The International Politics of Natural Disasters* (2012) and *The Geopolitics of Deep Oceans* (2015). *Fantasy City* was nominated for the 1999–2000 John Porter Award of the Canadian Sociology and Anthropology Association. *Environmental Sociology* has been translated into Chinese, Japanese, Korean and Portuguese. In his most recent book, Dr. Hannigan argues that our understanding of the deep ocean depends on whether we see it primarily as a resource cornucopia, a global political chessboard, a shared commons or a unique and threatened ecology. He is currently coediting the *SAGE Handbook of New Urban Studies* (with Greg Richards).

Karl Kim is chair of the US Federal Emergency Management Agency, National Domestic Preparedness Consortium, and the executive director of the National Disaster Preparedness Training Center. At the University of Hawaii, he is the director of the Disaster Management and Humanitarian Assistance Program. He is professor and former chair of the Department of Urban and Regional Planning at the University of Hawaii, where he also served as vice-chancellor for academic affairs. He leads an annual training academy for senior executives on the science of disasters at the Emergency Management Institute and has conducted training programmes funded by USAID and several foundations in Indonesia, Philippines and Vietnam as well as many Pacific Island nations.

Michelle Ann Miller is a senior research fellow at the Asia Research Institute, National University of Singapore. Trained in political science, her research focuses on intersections between urban and regional governance in the context of human conflict and environmental change. She leads the Disaster Governance theme of the Asian Urbanisms Cluster at ARI. Her interdisciplinary publications speak to contemporary theoretical debates and key policy issues in environmental disaster governance, decentralisation, urban change and citizenship and belonging. A reoccurring concern throughout her work is with the policy potential and lived experience of decentralisation in generating more inclusive and effective forms of governance, especially in Indonesia but across Asia more broadly. Before joining ARI, she taught at Deakin University and Charles Darwin University in Australia, and she has held visiting research fellowships in Indonesia at both the Centre for Strategic and International Studies (Jakarta) and Ar-Raniry Institut Agama Islam Negeri (Banda Aceh). She is international advisor of the Varieties of Peace research programme, a global initiative of Umeå University, supported by the Swedish Foundation for Humanities and Social Sciences.

Yenny Rahmayati is a postdoctoral research fellow at the Centre for Design Innovation (CDI), Swinburne University of Technology, Melbourne. She is a lead researcher for Humanitarian Habitat and Design Project, one of CDI's core programmes. Yenny received her PhD in architecture from the School of Design and Environment, National University of Singapore, and she has a master's in sustainable heritage development and management from the Australian National University. She holds a bachelor's degree in architecture from the University of North Sumatra, Indonesia. She previously worked with recognised international aid agencies and non-profit organisations, primarily in the areas of disaster and reconstruction, architecture, cultural heritage, community development, housing and urban studies. She also established and managed a local non-profit community-based organisation concerned with architectural and cultural-heritage issues in the post-disaster context of Aceh, Indonesia, for more than 10 years.

Anthony Reid is a Southeast Asian historian and emeritus professor at the Australian National University. Professor Reid served as founding director of the Center for Southeast Asian Studies at UCLA (1999–2002) and of the Asia Research Institute at NUS, Singapore (2002–2007). Since 2004, he has been increasingly interested in the impact of natural disasters on Southeast Asian history. His books include *The Contest for North Sumatra: Aceh, the Netherlands and Britain, 1858–1998* (1969); *The Indonesian National Revolution* (1974); *The Blood of the People: Revolution and the End of Traditional Rule in Northern Sumatra* (1979); *Southeast Asia in the Age of Commerce, c.1450–1680* (2 vols. 1988–1993); *An Indonesian Frontier: Acehnese and Other Histories of Sumatra* (2004); *Imperial Alchemy: Nationalism and Political Identity in Southeast Asia* (2010); *To Nation by Revolution: Indonesia in the 20th Century* (2011); and *A History of Southeast Asia: Critical Crossroads* (2015).

Matthew A. Shapiro is associate professor of political science at the Illinois Institute of Technology and an East Asia Institute fellow. He is also a research affiliate at the Argonne National Laboratory's Joint Center for Energy Storage Research. He was trained in political science, economics and public policy at the University of Southern California, and his published and ongoing research focuses on how national innovation systems are formed and contribute to sustainable development, how climate change is addressed and impacted by relevant policies and political forces and how communications from politicians, scientists and the media impact both of these areas. Dr. Shapiro's work has been published in *The Pacific Review*, *Environmental Communication*, *American Politics Research*, *Environment and Planning A*, *International Journal of Public Policy* and *Scientometrics*, amongst others.

Pakamas Thinphanga is a research director at the Thailand Environment Institute Foundation (TEI). She leads the Urban Climate Resilience Programme, with extensive involvement in urban climate resilience projects in the region. She currently acts as codirector of the Urban Climate Resilience in Southeast Asia Partnership (UCRSEA), a five-year project supported by the IDRC and SSHRC of Canada, focusing on building research capacity in regionalisation, urbanisation, climate change and vulnerability assessment. The project is being implemented in Cambodia, Myanmar, Thailand and Vietnam, linking with universities in Canada. Previously, she was responsible for the implementation of the Asian Cities Climate Change Resilience Network (ACCCRN) and Mekong-Building Climate Resilient Asian Cities (M-BRACE) projects in Thai secondary cities.

Robert James Wasson is a senior research fellow in the Institute of Water Policy at the Lee Kuan Yew School of Public Policy, National University of Singapore. Professor Wasson previously served as director of the Centre for Resource and Environmental Studies, dean of Science and head of the Department of Geography and Human Ecology at the Australian National University, and Deputy Vice-Chancellor (Research and International) at Charles Darwin University, Australia. He was trained in geomorphology, and his research interests are causes of change in river catchments, environmental history, extreme hydrologic events in the tropics, cross-disciplinary methods and the integration of science into both public and private sector policies. His current research is on disaster governance, particularly of floods in India and Thailand, and is extending his interests to interconnected water-related cascading disasters in Asia.

Fiona Williamson is a research fellow at the National University of Singapore and project coordinator for the Southeast Asian arm of the Atmospheric Circulation Reconstructions over the Earth (ACRE) initiative. Her current projects involve working with Southeast Asian meteorological services in order to locate, recover and digitise historic weather observations alongside a series of initiatives towards exploring extreme weather histories. Her research interests range from the history of natural disasters to climate change and urban development in the Straits Settlements, urban history and the global history of British meteorology.

Chapter 1
Crossing Borders: Governing the Globalising Urban Matrix of Compound Disasters in Asia and the Pacific

Michelle Ann Miller and Michael Douglass

1.1 Introduction

The Asia-Pacific[1] is more prone to cross-border disasters than any other region on earth. Home to four of the world's most active fault lines, which traverse multiple countries across thousands of kilometres (UNESCAP 2016), the region is highly vulnerable to cross-border earthquakes, tropical cyclones, transboundary floods, volcanic eruptions, erratic monsoons and droughts. With its major riparian regions originating from the receding glaciers of the Tibetan Plateau, it is also experiencing increasing risks of disasters in food production, livelihoods, water distribution and natural habitats that are affecting billions of people.

In pace and magnitude, the Asia-Pacific is also one of the most rapidly urbanising parts of the world. Its spectacular urban transition,[2] industrialisation and rural to urban migration flows have generated additional vulnerabilities through the production of extended urban agglomerations, many of which are located at or below sea level along coastlines or in river basins. In this context of unabated urban growth, these expanding spatial concentrations are exposing the inhabitants of dense settlements to environmental threats linked to the effects of climate change, as oceans rise and as unusual weather events become more frequent. This urban transition has been accompanied by environmentally unsound practices in cities that have produced new forms of vulnerability, exacerbated by anthropogenic transformations of coastal zones, marshlands and floodplains, creating unstable hillsides and other geographical conditions unsafe for human habitation (Miller and Douglass 2016a).

---

[1] We use the term Asia-Pacific in this volume to denote our geographical focus on East, South and Southeast Asia and Pacific island nation-states.

[2] Sixty-four percent of Asia's population is projected to be urban by 2050, meaning that the existing urban population of 1.6 billion is envisaged to almost double to 3.1 billion people (United Nations 2014, p. 1; Asia Development Bank 2011, p. 6).

M.A. Miller (✉) • M. Douglass
Asia Research Institute, National University of Singapore, Singapore, Singapore
e-mail: arimam@nus.edu.sg

M.A. Miller et al. (eds.), *Crossing Borders*, DOI 10.1007/978-981-10-6126-4_1

As urbanisation generates extensive flows of resources to and from rural and remote areas to cities, the Asia-Pacific region is becoming enmeshed in a highly interdependent urban matrix that further increases vulnerabilities through the compounding impacts of disasters, raising new questions about governance across international borders. For example, the great riparian regions of Asia map onto multiple nation-state borders that were arbitrarily drawn along post-colonial lines. With greater interdependencies along these river systems, there is increased potential for long-standing territorial disputes to erupt in times of crisis and displacement (Lebel et al. 2005). Likewise, along the mainland border corridors of the Asia-Pacific, anthropogenic interventions in nature—in the form of environmentally damaging megaprojects that service the growing demands of rapidly urbanising populations—are incubating new forms of cross-border environmental harm.

This book explores how, and to what effect, environmental disasters with compounding outcomes are being governed as they traverse national borders in the urbanising societies of Asia and the Pacific. Cross-border disasters are becoming ever more frequent and costly in our global urban age, just as scholars and policy makers are becoming ever more aware of the cross-border governance dimensions of these long-standing phenomena. Yet the division of the world into sovereign nation-states has meant that environmental disruptions continue to be treated primarily as domestic concerns in which the balance of power rests with national governments. Despite vocal calls from non-governmental organisations (NGOs), social movements, activists and private businesses, among others, for the creation of an enforceable global environmental governance system to address "accountability deficits" in responding to what Michael Mason (2008) calls the flow of "transnational environmental harm", attention to the inter-scalar challenges of cross-border disaster governance remains a nascent field of enquiry (Reed and Bruyneel 2010, p. 649; see also Malets 2013). In practice, nation-state borders present intractable barriers to cooperation and collaboration, even in such basic actions as information-sharing between neighbouring administrations within the same country. Sub-national authorities are rarely included in cooperative international cross-border disaster efforts, which tend to overlook or ignore local coping strategies and inter-generational knowledge of mechanisms for generating resilience to future or reoccurring disaster risks and impacts.

In preparing for large-scale disastrous events across borders, extensive collaborative planning is required to address the diverse needs of the populations at risk and to ensure that human and material resources can become readily available when official governing capacities become overwhelmed (Edwards 2009; Claringbull 2007; Wang 2013). As noted above, non-governmental organisations and international donors are cognisant of this need and many are engaged in efforts to traverse major jurisdictions to provide disaster relief, although the national legal frameworks that permit their activities on condition of political neutrality impose strict operational constraints (Hannigan 2012). National governments with relief agencies operating beyond disaster zones, international organisations such as the United Nations and non-profit organisations such as the Red Cross are taking on first responder roles in many instances. For major disasters, the militarisation of disaster governance has become a significant form of international intervention. These, too, signal a new era in which the political spaces of disaster governance extend well beyond borders and engage in issues of diplomacy as well as disaster relief.

City and municipal governments are also reaching beyond their own administrations to provide assistance when disasters occur or when environmental migrants appear from other localities. The shifting urban dimension of cross-border disasters in the Asia-Pacific and the multiple scales at which catastrophic events are occurring calls for a fundamental realignment in our thinking about how the complex dynamics of disasters should be spatially conceived and governed across scales.

Sometimes, the political ruptures resulting from a disastrous event usher in progressive transborder networks, relationships and agreements around shared problems, technologies and converging knowledge systems. These networks of cooperation regarding issues such as environmental conservation, social justice and the politics of land ownership are bringing people into conversation across administrative divides and are encouraging new voices in disaster governance activities and programs. As the decentralisation of governance to sub-national administrations continues to make headway in the Asia-Pacific, cities and towns rather than nation-states are emerging as engines of policy innovation in dealing with disasters that are difficult to map onto administrative jurisdictions (Miller and Douglass 2016b). City governments, for instance, have been strident advocates of urban governance for climate protection via purposeful interventions in urban socio-technical systems that traverse, and occasionally overcome, traditional distinctions between local, national and global environmental politics within and between countries (Bulkeley and Betsill 2005; Bulkeley and Broto 2013). Such mobilities act as spaces of hope through which to facilitate more inclusive and effective regimes of cross-border governance.

The contributors to this book came together at the *Conference on Crossing Borders: Governing Environmental Disasters in a Global Urban Age in Asia and the Pacific*, held at the Asia Research Institute, National University of Singapore in November 2015. They set out to address policy-relevant questions about the governance of compound disasters across national borders in the urbanising societies of Asia and the Pacific. Key questions guided the discussions. How can the kinds of environmental disasters that are traversing sovereign territories in the Asia-Pacific be conceptualised as problems of inclusive and collaborative governance rather than as technical, expert-driven managerial tasks? What policy innovations are providing redress for the multi-sector impacts of compound disasters reaching across national borders, and how well are they working? To what extent are governing institutions able to securely resettle environmentally displaced people who are forced to migrate across international borders? In what ways do shared experiences of cross-border disasters highlight or transform power relations within and between localities? And, can we use the lens of compound disasters to think about urban networks as part of a changing planetary ecology in the service of building more meaningful collaborations across nation-state borders?

Each of the chapters in this book considers a sub-set of these questions, interrogating the theoretical and empirical dimensions of scale and spatiality in governing disasters across urbanising countries in the Asia-Pacific. Collectively, they address a number of important areas that have been both understudied in the literature on cross-border disaster governance and which remain largely absent from policy programs. A reoccurring theme throughout this volume is the urban imperative of cross-border disaster governance that requires mainstreaming in all dimensions of preparedness, response, adaptation and resilience. Most scholarship on cross-border disasters locates analysis of the politics of scale at the level of the nation-state, while overlooking the paradox of urban settlements as both perpetrators and victims of cross-border disasters (Hodson and Marvin 2010; Miller and Douglass 2015). Yet energy-demanding cities are the primary producers of greenhouse gas emissions that substantially contribute to global climate change, just as dense populations in vulnerable locations are disproportionately impacted by climate change-related severe weather events and rising sea levels. In our global urban age of human-generated changes to the earth's life support systems, it is therefore cities that require greater attention in studies of the anthropogenic risks and compounding impacts of disasters once thought to have been natural and confined to specific sites and moments.

1.2 Rethinking Disaster Geographies

Our focus on the urban dimension of cross-border disasters locates urban settlements within an inter-connected urban-rural matrix of delineated spatial scales of governance ranging from the neighbourhood to the city, province, nation-state, transborder riparian region and even the global scale. This matrix allows us to consider how the expanding ecological reach and demands of urban agglomerations into rural and remote areas have contributed to an escalation in environmental disasters with compounding and far-reaching effects (Douglass 2016). The growing ecological footprint of urban energy demands has also contributed to a spike in what Alexander Betts (2013) coined "survival migrants", denoting those vulnerable irregular migrants who are forced to leave their home country because of an existential threat such as climate change, food insecurity or livelihood collapse, against which no domestic remedy is readily available. Urban energy demands are reshaping the countryside through mega-projects such as hydropower dams, reservoirs and oil and gas pipelines as well as through environmentally degrading modes of commercial production. These in turn are increasing the prevalence of floods, landslides and other environmental disasters, with cascading impacts that flow across national borders to create threats, vulnerabilities and hazards at multiple scales. All of these dynamics attending the urban transition in the Asia-Pacific require new ways of thinking about how the emerging geographies of compound environmental disasters can translate into more inclusive and innovative modes of cross-border governance.

A related conceptual contribution of this volume is our approach to compound disasters, a term used with growing frequency since the mid-1990s to describe the adverse consequences stemming from different but related disaster agents (ICLA 1996;cited in Wachira 1997, p. 109). Compound disasters may occur either in series or in parallel, such as an extreme weather event that floods a city and contaminates water supplies to produce a health pandemic with feedbacks to regional economies, with implications for building social resilience to future disasters. In the disaster-prone Asia-Pacific region, cross-border events with compounding impacts across multiple countries are all too common. In July 2015, for instance, Cyclone Kommen swept across India, Bangladesh and Myanmar, triggering floods and landslides that killed almost two million people. An earthquake in Nepal in April the same year generated multiple landslides (many of which were linked to environmentally damaging development projects) that blocked rivers and heightened flood risks in neighbouring Bangladesh, China (including the Autonomous Region of Tibet) and India (UNESCAP 2015, p. 6).

Through our focus on the urban dimension of compound cross-border disasters, we extend our conceptual lens to include a phase of incubation that begins well before the initial catastrophic event. By this, we mean that compound disasters are incubated in the socially and spatially uneven geographies of urbanising populations, which impact societies disproportionately and render certain groups especially vulnerable, namely the poor, ethnic minorities, informal settlers, the elderly, handicapped, women and children. Compound disasters are equally incubated in the anthropogenic interventions into nature that generate multiple causalities of environmental harm that become magnified through their socioeconomic impacts. Activities that serve the growth of cities such as deforestation, burning croplands for palm oil plantations, hydraulic fracking for oil, mining, mega-dams generating hydropower, and the production of nuclear energy assault local ecologies. They boomerang back to cities in the form of floods, air pollution and radioactive fallout from nuclear power plant failures. These spatially extensive compound effects increasingly traverse national political boundaries, with the extreme example being global climate change that is largely being generated by cities.

Posing the problem of cross-border compound disasters as a question of governance rather than disaster management is another contribution of this volume to the disaster studies literature. The task of the following chapters is to move beyond prevailing managerial approaches that privilege sector-driven expertise with its disproportionate emphasis on physical infrastructure, and to treat cross-border disasters as political phenomena embedded in unequal human geographies. Such an approach necessitates comprehensive multi-sector, multi-scalar analysis and responses. To the degree that governance has featured in scholarship on cross-border disasters, its main focus has been on disaster diplomacy underpinned by international agreements. Such diplomacy is typically aimed at responding to immediate threats and crises, especially in relation to traditional and non-traditional security issues and international humanitarian missions, usually at the expense of grounded empirical research on the underlying complexities of cross-border disaster causalities (Webersik 2010; Brauch et al. 2011; Elliot and Caballero-Anthony 2013). Locating cross-border disasters within the realm of governance also allows us to move away from silos of technical expertise and operational procedures to illuminate the politics of governance that encompass the full range of formal and informal actors, institutions and processes across transborder spaces at every scale. Through this wide approach to governance, the overall goal of the volume is to enhance our understanding of the complexities of cross-border disasters, with a view to developing more inclusive and effective policy choices that can link knowledge to practice in the dynamics of messy, real-world situations fraught with socio-spatial disparities.

1.3 Governing Disasters Across National Borders

The role of the national border in times of disaster has changed since the emergence of nation-states. These changes manifest in shifts in the degree to which borders are open or closed to such flows as resources, capital, people, information, and, more generally, the power to govern over a territorially defined space. Nation-states formed in seventeenth century Europe around ideas of religious homogeneity were designed to confer territorial divisions of power aimed at ending internecine wars among imperial domains by instilling respect for the boundaries of sovereign states and encouraging non-interference in domestic affairs. In Asia, at the height of Western imperialism, these same principles were used to organise and protect trade routes along colonial borders, as is described by Fiona Williamson in Chap. 3 of this volume.

In the modern nation-state, border control has fluctuated in the governance of flows of transnational environmental harm. Just as national boundaries loosely map the jurisdictional realm of governmental responsibility for environmental protection, they also obscure the terrain of responsibility for ecological destruction (Mason 2008; Elinoff and Vaughan forthcoming). The border continues to be regarded as a fixed space of containment because disasters are typically treated as a function of sovereignty in which national governments retain responsibility for dealing with crises that originate within their sovereign territories, only accepting offers of outside assistance when domestic coping capacities become overwhelmed.

Disasters are challenging these contemporary uses of the border to control external relations in three interrelated ways. First, disasters increasingly have sources and consequences that transcend national boundaries. Chemically poisoned rivers, riverbed erosion and destabilisation caused by water released from upstream megadams in nearby countries, and severe air pollution create cross-border political confrontations. The second challenge to border control, as examined by Anthony Reid in Chap. 2 of this book, manifests through the intersections between international humanitarian assistance and the spread of information through digital media that open disaster-prone countries to political leverage tied to financial and moral claims for political reform. The third challenge is disaster-induced migration, which does not tend to result in official recognition for disaster refugees, but are increasingly visible as collateral migration from collapsing economies, political upheaval and failed or failing production systems.

Taken together, these overlapping factors bring into view the symbolism of the border as a political line of negotiation over how disasters in the twenty-first century need to be governed across rather than simply within nation-states. As this collection highlights, the prevailing pattern is one of slow, piecemeal accommodation of external political pressures for greater transparency and the standardisation of processes of disaster mitigation with very little change to national migration systems. However, the cases brought together here also show how the disruptions of disasters can, and occasionally do, create unexpected openings for political reform and cooperative forms of cross-border governance.

The Asia-Pacific region has no supra-national body with powers to enforce agreements among national governments in managing the environmental commons or in adjudicating claims arising from cross-border disasters. Cross-border disaster governance thus becomes a process of negotiation that can remain indefinitely unresolved. With international standards and accountability frameworks lacking, donor organisations are pushing for sweeping changes in the governance of assistance within recipient developing countries in particular. For example, the Organisation for Economic Cooperation and Development (OECD) calls for standardising aid processes through reforms that include popular participation in implementing and giving feedback via coherent domestic policies and through compliance with international humanitarian law. Similarly, the United Nations, which organised the first World Humanitarian Summit in 2016, promotes the idea of a "Global Compact" to take advantage of all actors' complementary roles in strengthening supra-national regional capacities to prevent, manage and respond to disasters. These moves represent a substantial leap in attempts to govern disasters above the level of the nation-state through assistance programs. They also signal the emergence of a new era of cross-border disaster governance. As Anthony Reid and John Hannigan highlight in Chaps. 2 and 5 of this book, respectively, this is an era in which international organisations now see an unending, long-term need for providing humanitarian assistance via the establishment of global rules. It is also an era in which disasters have spawned a global aid industry, which in 2015 employed at least a quarter of a million people worldwide.

International disaster migration efforts in the Asia-Pacific have been complicated by the demarcation of national borders along colonial lines. One result of this colonial inheritance has been the artificial incorporation of dispersed ethnic minorities into national spheres without being assimilated into a shared national identity or related acceptance of state sovereignty (Miller 2012, pp. 2–3).[3] These sovereignty disputes have sometimes been brought to the political fore by, and settled through, catalytic ruptures of a disaster event. For example, the borders of Indonesia, Sri Lanka and Thailand were reinforced by the 2004 Indian Ocean tsunami and undersea earthquake. A protracted armed nationalist struggle in Aceh was democratically

[3] Even in Thailand, where no formal colonisation took place, the borders of Siam, as Thailand was called until 1939, and from 1945 to 1949, were to shaped by the independence agreements between British and French colonial powers and the newly independent nation-states of neighbouring Malaysia, Myanmar, Laos, Vietnam and Cambodia.

resolved in Indonesia's favour in the window of political opportunity that was generated by the disaster (Miller 2009). By contrast, the governments of Sri Lanka and Thailand used the opportunity created by the tsunami to militarily repress their separatist insurgencies (Åkebo 2016).

More commonly, however, environmental disasters in the Asia-Pacific have not produced high levels of armed conflict or the redrawing of national borders. Usually, disasters expose the everyday differential porosities of state boundaries, including the limits of territorially demarcated understandings of responsibility and containment. Such porosities vary not only by country and governance regime, but also by the types of flows of environmental harm and the narratives that surround them at different political moments (Mitchell 1997, p. 105; see also Cunningham 2012, pp. 373–374). For instance, how a national government responds to the perpetrators of air and water pollution within its territories can determine the extent of the flow of ecological harm into surrounding countries, as well as the likelihood of recurrent episodes of chronic cross-border pollution. When states are unable to deal domestically with the impacts of cross-border disasters, national political systems may be destabilised, igniting wider unrest. This can happen, for example, when people displaced by a drought or a famine in rural areas converge upon an urban centre to demand compensation, or when "survival migrants" seek refuge in neighbouring countries, sparking regional instability (Kelman 2003, p. 119; Betts 2013; Global Humanitarian Assistance 2015).

Projections along several fronts suggest that the movement of disaster victims across national borders will continue to escalate in the coming years. The 25–30 million environmental refugees estimated to exist globally in 2007 are projected to increase to a total of between 200 and 300 million by 2050 (Asia Development Bank 2011). This includes migrants impacted by global climate change, especially sea rise. Pacific islands are particularly susceptible to sea rise, and since many are small island states, any migration is necessarily international. While the number of environmental disaster refugees going across national borders is reportedly not large, it would likely increase manifold times if the compound effects of disasters were included. Land degradation, mega-dam impacts, water pollution and many other factors related to environmental change can result in migration not registered as disaster refugee movement. In addition, the millions of people who work abroad in remittance economies across the Asia-Pacific is expected to rise from negative impacts on traditional livelihoods and ways of life. As evidenced in Nepal following the 2015 earthquake, both emigration to foreign countries and remittances back to Nepal surged following the disaster. With 30 % of its GDP already coming from money sent back to Nepal by its workers abroad, the earthquake heightened Nepal's slide into a remittance economy.

National governments representing nation-states play an active role in rendering the border either obdurate or flexible in the face of environmental harm. When investment opportunities and national strategic interests are at stake, state actors and institutions are more likely to subvert the territorial logic of the state, or to deploy what Mountz and Hiemstra call a strategy of "flexible sovereignty'" (2012, p. 468). They do this by pursuing destructive development projects in borderland zones where national environmental regulations are less enforceable, and in ways that obscure state responsibility for environmental harm through, for example, collusion with local and national companies, intra-regional firms and multinational corporations.

This rendering of the idea of the border as a space of exception has been the subject of growing scrutiny in the aftermath of the 9/11 terror attacks on the United States, when the offshore detainment of suspected terrorists cast a critical spotlight on the paradox of the de-territorialised virtual border that excludes local interests set against the thickening heavily securitised border (Muller 2009; Eilenberg 2014). The borderland zones that service the expanding energy demands of urbanising populations across mainland Southeast Asia have become such spaces of exception. These spaces of exception exist for the exploitation of human and environmental resources by small-scale subcontractors who manage unregulated multinational megaprojects such as hydropower dams, reservoirs and gas pipelines. As borderland subcontractors are not regarded as agents of corporate social responsibility for environmental protection, their flexible labour standards do not attract the sorts of political protests for environmental and social justice that could be reasonably expected in towns and cities (Pangsapa and Smith 2008). Moreover, the bureaucratic mechanisms that govern decision-making between the countries and communities that traverse these fluid borderland zones are as deliberately open-ended as the borders are selectively porous. The higher the spatial scale of decision-making, the more opaque planning guidelines are likely to be, concealing cross-scalar issues of sustainable development and environmental justice to smooth the way for modernisation projects. In the Mekong Delta, for example, China and Myanmar have elected to be excluded from, to avoid being bound by, the regional institutional framework for water resource development, sharing and use. The boundaries of what constitutes the "Mekong Basin" are also continually being redefined to serve the interests of mega-development projects, including the selective exclusion of tributaries from the basin's governing bureaucracy and operational procedures to enable tributary-based projects to proceed, even when such projects have cross-border implications (Lebel et al. 2005).

Finally, the border must be understood as a vehicle for the exchange, transformation or blockage of knowledge, ideas and technologies in the governance of cross-border disasters. In other words, if all disasters occur in political spaces then the political contingencies of the international border are a potent determinant of the receptiveness of impacted countries and communities to outside offers of assistance in times of crisis. Border politics also influence the extent to which receiving countries feel the need to adapt or mutate foreign recovery resources for domestic consumption. There are often good reasons for this cautionary approach to idea of the national border as a conduit for potential political harm in addition to welcome humanitarian interventions. As Karl Kim and Konia Freitas warn in Chap. 9 of this book, introducing disaster governance programs into the isolated island communities of the Asia-Pacific that lack knowledge of local conditions or ignore cultural traditions can fuel social conflict or further destabilise impacted communities, with clear implications for their capacities to build resilience to future disasters.

As the towns and cities of Asia and the Pacific become ever more interconnected through their imprints onto people and places extending beyond urban nodes, cross-border urban networks for sustainable environmental governance have never been more possible, or important. The rapidly urbanising Asia-Pacific warrants special attention as a site where dense concentrations of people and resources can be mobilised in the service of grappling with the growing complexities of cross-border disastrous events that have multiple causalities and far reaching impacts. To this end, we consider in the following pages how the border is being navigated in the governance of cross-border disasters across the Asia-Pacific in the theoretical and empirical contributions that comprise this volume.

1.4 About This Volume

Taken together, the chapters in this book address key issues in cross-border disaster governance in the urbanising societies of Asia and the Pacific. They consider the constantly shifting permeability of the national border within the complicating contexts of modernising development projects and postcolonial nation-building agendas, and how assumptions about territorial sovereignty challenge cross-border cooperation in times of crisis, rupture and displacement. Above all, they are concerned with the many ways in which the transition from rural to urban settlements across the Asia-Pacific is fundamentally realigning the possibilities for cross-border disaster governance, while at the same time raising new problems for socioeconomic resilience and stability at multiple scales. This includes evaluating how the expanding ecological footprint of cities into increasingly remote and rural localities is creating new chains of ecological harm with cascading and unpredictable impacts that cannot be contained within neatly drawn jurisdictional boundaries. It also includes exploring the possibilities for innovative cross-border disaster governance in urbanising societies that will increase our understanding of how to mobilise technologies, ideas and knowledge in more effective and inclusive ways.

The role of the border in disaster governance in historical and contemporary perspective is the focus of the first section of this book. Considerable research has been devoted to the changing nature of the border through the experiences of disaster and displacement. In Chap. 2, Anthony Reid speaks to this scholarship through his provocation to realign the traditional function of militaries in defending national security interests to respond to the twenty-first century priority of protecting vulnerable urbanising populations across state borders in the face of ever more frequent and large-scale environmental disasters. According to Reid, this shift was precipitated by the historical transition away from the polarising sovereignty wars that disrupted much of the twentieth century through the consecutive periods of colonisation and decolonisation (and, in some cases such as East Timor, re-colonisation by Indonesia following the end of Portuguese colonial rule). In the twenty-first century, by contrast, Reid contends that relative peacetime conditions in the Asia-Pacific, coupled with an unprecedented global awareness of, and experience of responding to, some of the biggest environmental disasters in memory have ushered in a planetary imperative to forge coordinated regional networks in dealing with the growing threat of environmental disasters in increasingly risk-prone urbanising societies.

Zooming in on a slice of the colonial history of Southeast Asia, Fiona Williamson interrogates in Chap. 3 the role of the border in the circulation of scientific knowledge about tropical climates, deforestation, climate change and urban resilience across the Straits Settlement colonies of the British Empire. Through her examination of how British colonial administrative thinking about governing recurrent flooding episodes in the cities of Kuala Lumpur and Singapore shifted during the nineteenth century, Williamson highlights discrepancies between official records detailing British technical efficiency and the real-world limitations of cross-border governance that fuelled social discontent and spatial inequalities. In doing so, she makes a powerful case for why wider historical perspectives of disasters that take into account colonial urban planning regimes and the flows of knowledge across complex historical geographies are vital to our contextual understanding of contemporary events, with a view to improving the efficacy of cross-border disaster governance systems in the future.

In different but overlapping ways, Chaps. 4 and 5 present theoretical and policy insights into contemporary thinking about cross-border disaster governance in our global urban age. In Chap. 4, Matthias Garschagen identifies knowledge gaps and research needs in extant scholarship on cross-border disaster governance in East, South and Southeast Asia to highlight the limitations of established regional platforms such as ASEAN (Association of Southeast Asian Nations) in developing and implementing transnational policy frameworks. Emphasising the role of urbanisation in shaping cross-border disaster risks and impacts, Garschagen shows how uneven analysis of this emerging policy terrain can offer a heuristic device for formulating more comprehensive recommendations in cross-border disaster governance regimes. For John Hannigan (Chap. 5), the modern nation-state border is itself the principal object of interrogation and a potential space of hope in the search for greater efficacy and inclusiveness in cross-border disaster governance. Describing different ways of conceptualising the "border" and "bordering", Hannigan explains how a fluid, transitional approach to the border allows us to open up spaces of hope through which to establish collective socioecological identities that can transcend conventional territorial constraints. This involves, for example, examining how crossing borders can create policy options for human agency via the mobilisation of humanitarian corridors to deal with disaster displacement, and by rethinking the spatial potential of borderlands as zones of shared ecological units tied to sustainable community livelihoods.

Part Two of this book considers the challenges and opportunities of governing cross-border disasters in Asia's transboundary riparian regions. In these megaregions, or regions of regions, that include multiple countries, the governance of increasingly scarce water resources is becoming a source of political conflict and regional crisis. This crisis is being exacerbated by the proliferation of hydropower dams and large-scale irrigation projects that are disrupting dependent ecologies and displacing large populations (for example, through forced evictions and loss of land associated with decreased sedimentation or coastal erosion), while rendering other settlements more vulnerable to floods, droughts, landslides, environmental pollutants and diseases. The Mekong Delta Region is emblematic of this growing crisis in cross-border governance within a matrix of interlinked urban systems that are dependent upon shared ecosystems undergoing intensive anthropogenic transformations. Chapters 6 and 7 examine different dimensions of the challenges and opportunities involved in cross-border governance in the face of increasing environmental disruptions and diminishing resources in the Mekong Delta. In Chap. 6, Richard Friend and Pakamas Thinphanga set out the implications for climate change adaptation in the Mekong by showing how accelerated urbanisation and capital investment are creating new patterns of risk and vulnerability that extend well beyond localities and events to connect with regional and global processes. By analysing the complexities of multi-scalar, interlinked and interlocked urban systems, Friend and Thinphanga explain how any meaningful approach to governing disasters and climate change adaptation in the Mekong Delta must involve cross-border policy interventions at all scales. Moreover, they argue that such interventions need to consider the cascading impacts of shocks and crises across countries and the uneven risks and socioeconomic vulnerabilities they produce in the interests of building more resilient urban futures that can withstand the effects of global climate change.

Like Friend and Thinphanga, Andrea Haefner (Chap. 7) emphasises the importance of adopting a regional perspective in dealing with the interrelated ways in which rural to urban migration and multinational investment in environmentally degrading megaprojects are contributing to cross-border compound disasters in the Mekong Delta. Focusing on the case of the Xayabouri hydropower dam construction project in the lower Mekong basin of northern Laos, Haefner examines how the dam's construction is raising regional tensions by threatening food and water security and through the displacement of millions of people and their livelihoods. More broadly, Haefner highlights a critical opportunity for inclusive transboundary governance around the vexed issue of hydropower energy, not only by sharing responsibility for the downstream vulnerabilities that hydropower dams produce and the uneven distribution of their environmental impacts, but also through the introduction of cross-border incentives to promote alternative patterns of sustainable energy consumption.

The Pearl River Delta, as described by Alain Guilloux in Chap. 8, is another example of how accelerated urbanisation coupled with heavy industrial development in a region fraught with historical sensitivities over contested national borders is complicating efforts to establish a coordinated approach to cross-border disaster governance. The low-lying Pearl River Basin, which extends from Hong Kong to Guangzhou and includes northeast Vietnam, is the second most densely populated region on earth and one of China's primary engines of economic growth. The disaster-prone region is also critically unprepared to deal with persistent recurrent flooding, typhoons, storm surges, rising sea levels and increasingly severe weather events linked to climate change. Guilloux points out how mounting public pressure experienced by delta authorities, coupled with residual tensions over border issues with Hong Kong and the absence of shared legislative arrangements, are creating a crisis in coordinating disaster programs across multiple bureaucracies. To address these problems, Guilloux makes a case for facilitating a greater role for civil society and the private sector in building the collaborative cross-border capacities in this extremely vulnerable urban megaregion.

Part Three of this book examines how collaborative networks across national borders can either assist or impede the transfer of knowledge, ideas and technologies aimed at building more resilient urbanising societies across the Asia-Pacific. In this, the border not only demarcates the scope of environmental risk, harm and responsibility, but it also acts as a conduit for the transfer or blockage of disaster knowledge and resources. In Chap. 9, Karl Kim and Konia Freitas examine the role of the border as a vehicle for negotiating outside offers of disaster assistance in small island communities that have indigenous traditions of intergenerational knowledge, cultural systems of community resource management and a lived awareness of the physicality of the border born from their experience of relative isolation. Through examples from Hawaii, Samoa, Tonga and Indonesia (Simeulue Island), Kim and Freitas highlight the ways in which globalisation, climate change and the loss of traditional knowledge through localised urbanisation processes are reshaping the geographies of risk and vulnerability in small island communities. They consider how the border serves as a mechanism for perpetuating these processes as well as for navigating offers of external help in times of crisis. Kim and Freitas offer insights into the ways in which military assistance and international humanitarian aid could become more attentive to the possibilities for integrating aspects of indigenous knowledge into imported resilience programs. They also raise questions about the extent to which aspects of indigenous knowledge could be transferred as policy interventions for strengthening resilience in disaster-prone communities elsewhere.

City actors and institutions are playing a growing leadership role in establishing collaborative cross-border networks around the governance of disaster risk, response, recovery and resilience. The urban orientation and ecological stewardship of cross-border disaster governance networks is the focus of Chap. 10 by Kristoffer Berse. Through his study of CITYNET, a regional association of city-level authorities in the Asia-Pacific region, Berse examines the obstacles and opportunities presented by city-to-city relationships in post-disaster recovery and rehabilitation programs. More broadly, Berse is concerned with the potential of decentralised urban networks to function as agents of globalised care via the horizontal mobilisation of disaster aid and services across international borders that could parallel and complement existing country-to-country disaster programs at the subnational scale.

Scaling down further to the level of civil society organisations, Yenny Rahmayati describes in Chap. 11 how the flood of international humanitarian organisations into Indonesia's westernmost city of Banda Aceh following the 2004 Indian Ocean tsunami and undersea earthquake created unprecedented opportunities for urban-based civil society actors to forge empowering cross-border networks of collaboration. These cooperative networks formed around issues such as cultural heritage conservation, ending child labour in post-disaster economies and the recruitment of women and young people into disaster governance programs. Rahmayati shows how the urban orientation of these international partnerships and regional networks largely overlooked rural organisations while privileging Banda Aceh-based NGOs and community groups, which subsequently became more capable of meeting their organisational goals and establishing sustainable programs. To provide redress for this rural-urban imbalance and to strengthen the overall role of civil society in future disaster governance programs, Rahmayati argues for the establishment of closer collaborative relationships between community groups and government at all levels, especially in the phase of building resilience after the recovery phase has ended and when international organisations have departed.

The final part of this volume assesses the growing potential for transnational flows of environmental harm and cross-border conflict over resource scarcity. The "water wars" between India and Bangladesh are emblematic of this new front in cross-border conflict that has become associated with twenty-first century border dynamics. As Sarfaraz Alam describes in Chap. 12, India's diversion of the dry season flow of the River Ganges away from Bangladesh has produced catastrophic long-term consequences for millions of Bangladeshis whose livelihoods have relied for generations on the river and its downstream ecologies. Squabbles between India and Bangladesh over escalating irregular migration linked to environmental destruction and dwindling shared ecological resources are becoming intractable as both countries retreat into defensive nationalist rhetoric about where the responsibilities of one state ends and the other begins. For Alam, the resolution of this hyper-politicisation of the India-Bangladesh border must begin by educating governing authorities in both countries about their interconnected resource dependencies and causalities related to the transnational flow of environmental harm. Without such a bilateral approach to disaster governance, Alam warns that it will be impossible to manage the growing stresses on the River Ganges unilaterally, and that the existing ecological hazards confronting urbanising populations on both sides of the border will continue to generate ever more frequent and destructive environmental crises.

For Matthew Shapiro, in Chap. 13, the regional politics of transboundary air pollution and the yellow sand/dust storms that emanate from mainland China and blow across East Asia are posing an equally intractable problem in identifying and prosecuting the perpetrators of environmental harm. Focusing on the role of EANET (the East Asian Acid Deposition Monitoring Network), Shapiro examines how regional networks of scientific researchers are being constrained in their efforts to collect data and disseminate information about environmental pollution, both domestically and within the context of the sensitive regional politics that produce multilateral agreements such as those that created EANET. Arguing that epistemic communities are vital to identifying and addressing regional environmental challenges, Shapiro points out that the very existence of organisations such as EANET and the willingness of Chinese urban planners to adopt best practices from abroad show considerable promise for East Asia's ability to deal with transboundary air pollution. On the other hand, however, the geographical shift of worsening air pollution away from China's coastal cities to rural hinterlands creates an imperative for China's urban centres to prioritise sustainable modes of resource consumption and to play a more progressive role in national and regional disaster governance networks to diffuse mounting regional tensions.

Going against the grain of the other contributions in this volume, Robert Wasson (Chap. 14) sees less, rather than more, regional connectedness as being potentially desirable to minimise the cross-border threats posed by geomagnetic solar storms. This is because solar storms produce geo-electric currents that can destroy cross-border power grids and disrupt water management systems, with cascading impacts on all aspects of human settlement ranging from health to food production, waste disposal, livelihoods and social stability. For this reason, Wasson argues for a more decentralised, localised and modulated approach to electricity and water production and distribution across Asia and the Pacific than is currently in effect.

What all of our authors emphasise through their contributions to this book is that the tremendous range of contextual variables within and between the urbanising societies of the Asia-Pacific necessitates a multi-sector, multi-disciplinary and multi-stakeholder approach to cross-border disaster governance. Equally, they point to the need for far greater attention to the range of causalities that incubate cross-border disasters before the actual moment of crisis. These interconnected causalities in turn contribute to the compounding impacts of catastrophic events that have far-reaching effects, often across multiple countries, and with long-term legacies that shape the capacities of future generations to build strategies of resilience to disasters. These factors, combined with the urbanising and industrialising processes that are transforming the Asia-Pacific region and creating new geographies of risk and vulnerability, call for a flexible and adaptable approach to forging collaborative networks in disaster governance regimes across countries.

Through this collection we hope to raise awareness of the need to recalibrate our spatial and scalar understandings of environmental disasters in the service of developing more effective and inclusive forms of cross-border disaster governance. By invoking the idea of the national border as an entity that is at once fixed and fluid, we aim to infuse a sense of the complex political dynamics that must be negotiated in the transfer of knowledge, technologies and disaster resources across sovereign territories in a region awash with contested colonial histories and shifting alliances around development projects and strategic national interests. Within these complicated and diverse transnational contexts, the border can either function as a vehicle for the transfer of flows of environmental harm or as a conduit for progressive approaches to disaster governance and ecological conservation. It is the goal of this book to provide a platform for researchers and policy makers to develop more innovative participatory approaches to cross-border disaster governance that can nurture resilient urbanising societies within and beyond Asia and the Pacific.

References

Åkebo, M. (2016). Disaster governance in war-torn societies: Tsunami recovery in urbanising Aceh and Sri Lanka. In M. A. Miller & M. Douglass (Eds.), *Disaster Governance in Urbanising Asia* (pp. 85–108). Singapore: Springer.

Asia Development Bank. (2011). *Asia 2050: Realizing the Asia Century*. Singapore: Asia Development Bank.

Betts, A. (2013). *Survival migration. Failed governance and the crisis of displacement*. Ithaca: Cornell University Press.

Brauch, H. G., Spring, Ú. O., Mesjasz, C., Grin, J., Kameri-Mbote, P., Chourou, B., Dunay, P., & Birkmann, J. (Eds.). (2011). *Coping with global environmental change, disasters and security. Threats, challenges, vulnerabilities and risks*. Berlin: Verlag. and Heidelberg: Springer.

Bulkeley, H., & Betsill, M. (2005). Rethinking sustainable cities: Multilevel governance and the 'urban' politics of climate change. *Environmental Politics, 14*(1), 42–63.

Bulkeley, H., & Broto, V. C. (2013). Government by experiment? Global cities and the governing of climate change. *Transactions of the Institute of British Geographers, 38*(3), 361–375.

Claringbull, N. (2007). The case for regional post-natural disaster preparation. *Journal of Business Continuity and Emergency Planning, 2*(2), 152–160.

Cunningham, H. (2012). Permeabilities, ecology and geopolitical boundaries. In T. M. Wilson & H. Donnan (Eds.), *A companion to border studies* (pp. 371–386). Malden: Wiley Blackwell.

Douglass, M. (2016). The urban transition of disaster governance in Asia. In M. A. Miller & M. Douglass (Eds.), *Disaster governance in urbanising Asia* (pp. 13–44). Singapore: Springer.

Edwards, F. L. (2009). Effective disaster response in cross border events. *Journal of Contingencies and Crisis Management, 17*(4), 255–265.

Eilenberg, M. (2014). Frontier constellations: agrarian expansion and sovereignty on the Indonesian-Malaysian border. *The Journal of Peasant Studies, 41*(2), 157–182.

Elinoff, Eli, Tyson Vaughan (forthcoming). Introduction. In Eli Elinoff & Tyson Vaughan, (Eds.), *The quotidian anthropocene: Reconfiguring environments in urbanizing Asia*. Philadelphia: University of Pennsylvania Press.

Elliot, L. M., & Caballero-Anthony, M. (Eds.). (2013). *Human security and climate change in Southeast Asia. Managing risk and resilience*. London: Routledge.

Global Humanitarian Assistance. (2015). *Global humanitarian assistance report*. Bristol: Global Humanitarian Assistance.

Hannigan, J. (2012). *Disasters without borders: The international politics of natural disasters*. Cambridge, UK: Polity Press.

Hodson, M., & Marvin, S. (2010). Urbanism in the anthropocene. Ecological urbanism or premium ecological enclaves? *City, 14*(3), 299–313.

ICLA [International Conference on Local Authorities]. (1996). International Conference on Local Authorities Confronting Disasters and Emergencies, Background Documents, Amsterdam.

Kelman, I. (2003). Beyond disaster, beyond diplomacy. In M. Pelling (Ed.), *Natural disasters and development in a globalizing world* (pp. 110–123). London: Routledge.

Lebel, L., Garden, P., & Imamura, M. (2005). The politics of scale, position and place in the governance of water resources in the Mekong Region. *Ecology and Society, 10*(2), 18. [online], last accessed 13 Apr 2016.

Malets, O. (2013). Governing our environment: Standardising across borders. In L. Dobusch, P. Mader, & S. Quack (Eds.), *Governance across borders: Transnational fields and transversal themes* (pp. 91–114). Berlin: epubli GmbH Publisher.

Mason, M. (2008). The governance of transnational environmental harm: Addressing new modes of accountability/ responsibility. *Global Environmental Politics, 8*(3), 8–24.

Miller, M. A. (2009). *Rebellion and reform in Indonesia. Jakarta's security and autonomy policies in Aceh*. London: Routledge.

Miller, M. A. (2012). The problem of armed separatism: Is autonomy the answer? In M. A. Miller (Ed.), *Autonomy and armed separatism in South and Southeast Asia* (pp. 1–15). Singapore: ISEAS.

Miller, M. A., & Douglass, M. (2015). Governing flooding in Asia's urban transition. *Pacific Affairs, 88*(3), 499–515.

Miller, M. A., & Douglass, M. (Eds.). (2016a). *Disaster governance in urbanising Asia*. Singapore: Springer.

Miller, M. A., & Douglass, M. (2016b). Decentralising disaster governance in urbanising Asia. *Habitat International, 52*, 1–4.

Mitchell, K. (1997). Transnational discourse: Bringing geography back in. *Antipode, 29*(2), 101–114.

Mountz, A., & Hiemstra, N. (2012). Spatial strategies for rebordering human migration at sea. In T. Wilson & H. Donnan (Eds.), *A companion to border studies* (pp. 455–472). Oxford, UK: Blackwell.

Muller, B. J. (2009). Borders, risks, exclusions. *Studies in Social Justice, 3*(1), 67–78.

Pangsapa, P., & Smith, M. J. (2008). Political economy of Southeast Asian borderlands: Migration, environment and developing country firms. *Journal of Contemporary Asia, 38*(4), 485–514.

Reed, M. G., & Bruyneel, S. (2010). Rescaling environmental governance, rethinking the state: A three-dimensional review. *Progress in Human Geography, 34*(5), 646–653.

UNESCAP (2015) *Disasters in Asia and the Pacific: 2015 year in review*. http://www.unescap.org/sites/default/files/2015_Year%20in%20Review_final_PDF_1.pdf, 21pp. Last accessed 18 May 2016.

UNESCAP. (2016). *Disasters without borders. Regional resilience for sustainable development*. Bangkok: UNESCAP (United Nations Economic and Social Commission for Asia and the Pacific).

United Nations (2014). *World urbanisation prospects. 2014 revision*. United Nations Department of Social and Economic Affairs. http://esa.un.org/unpd/wup/Publications/Files/WUP2014-Highlights.pdf

Wachira, G. (1997). Conflicts in Africa as compound disasters: Complex crises requiring comprehensive responses. *Journal of Contingencies and Crisis Management, 5*(2), 109–117.

Wang, J.-J. (2013). Post-disaster cross-nation mutual aid in natural hazards: Case analysis from sociology of disaster and disaster politics perspectives. *Natural Hazards, 66*(2), 413–438.

Webersik, C. (2010). *Climate change and security: A gathering storm of global challenges*. Santa Barbara: Greenwood Publishing Group.

Part I
Cross-Border Disasters in Historical and Contemporary Perspective

Chapter 2
Recognising Global Interdependence Through Disasters

Anthony Reid

2.1 Introduction. From Isolation to Global Responsiveness

In 2015 we commemorated the bicentenary of the eruption of Gunung Tambora, in Sumbawa Island of south-eastern Indonesia, on 10 April 1815. It was the most devastating eruption anywhere in the last 500 years, and brought disasters to the whole planet. It is interesting to speculate how different the world's responses would be if a comparable eruption occurred tomorrow. The closest recent analogy to help us must be the eruption of Mount Pinatubo in 1991, the biggest of the last 50 years, although with less than a tenth the explosive capacity of Tambora.

Tambora was the last great pre-telegraph disaster. Although the sound of the explosion was heard in Batavia (modern Jakarta) and Padang, information about the eruption that caused the noise didn't reach these places for weeks. Europeans and North Americans experienced their 'year without summer' in 1816, when crops failed and thousands died, but had no idea why. Only 150 years later did it begin to be understood that these global disasters were caused by an eruption on the other side of the world. The people of Indonesia's southeastern island chain—Bali, Lombok, Sumbawa—died without warning, without help, in isolation. Hot gases and pyroclastic flows killed the 8000 people of the Tambora Peninsula of Sumbawa almost immediately. The disaster wiped out the Tambora language, which had up to then been the most westerly survival of a Papuan-type (non-Austronesian) language (Donahue 2007). The three islands were covered with ash that destroyed cultivation. In Lombok 'the depth of ashes which fell…varied…from one to two feet in depth. This not only destroyed the growing crops, but for some years prevented the sowing of corn, and the result was famine, disease and the cutting off of much of the popula-

The original version of this chapter was revised.
An erratum to this chapter can be found at https://doi.org/10.1007/978-981-10-6126-4_15

A. Reid (✉)
The Australian National University, Canberra, ACT, Australia
e-mail: anthony.reid@anu.edu.au

M.A. Miller et al. (eds.), *Crossing Borders*, DOI 10.1007/978-981-10-6126-4_2

tion' (Crawfurd 1856 p. 220). The population of Sumbawa in the first post-disaster count, in 1847, had still only recovered to 74,500, about 40% of the pre-eruption population, even though more than 10,000 immigrants had arrived to take advantage of the vacant land newly fertilised by the phosphates (Crawfurd 1856, pp. 420–1). Bali, further west again, suffered terrible famines for 5 years before crops could again flourish. It normally exported chiefly rice, but for 15 years after the eruption exported only slaves, desperate to escape starvation by selling themselves to whoever would feed them. These three islands must have lost well over 100,000 people, probably about a quarter of their total population, without any effective relief from outside except for the buying of slaves (Reid 2014).

The eruption of Mount Pinatubo in 1991 was a remarkable demonstration of how a modern state supported by the best international science and assistance can moderate such effects. Fortunately Pinatubo also behaved well, from the viewpoint of geologists from the Philippines, the US Geological Survey and elsewhere, by escalating its outbursts to confirm the increasingly urgent warnings and evacuation orders. About 60,000 people were evacuated in stages during the month before the eruption, including 14,000 from the US Air Force base at Tarlac, 15 km from the volcano, which was never re-established. It was estimated that 20,000 of these would have died immediately from the major eruption if not evacuated. Only 700 did so, mostly due to roofs collapsing from the weight of the pyroclastic material deposited. Globally there was an estimated one degree drop in temperatures, and the sulphur dioxide in the atmosphere boosted the ozone hole over the Antarctic to the largest so far recorded. The longer-term effects on agriculture were also moderated by national and international aid, with the Filipino diaspora proving once again its great value. Two years after the event, it was estimated that 2.1 million people had been affected, 8000 houses and 81,900 hectares of rice land destroyed, and 779,000 head of livestock and poultry killed. The Department of Agriculture in Central Luzon calculated total losses from lahars and ash deposits as 1.5 billion pesos (US$55 million) (Mercado et al. 1993; Pappas 2011). Over a million people were cared for in evacuation centres until they could be resettled in new farms or different occupations. Aid flowed in from 27 countries, and from international organisations like WHO, UNDP, UNICEF, UNDRO and WFP (Guzman n.d.).

There were at least 20 encounters of aircraft with the volcanic ash, two of which caused engine failure (although a fatal crash was averted). These near-disasters, compounding the near-loss of two jets over Java as a result of the Galunggung eruption in 1982, caused a tightening of restrictions on flights through volcanic ash. A consequence was the chaos of thousands of flight cancellations in Europe in 2010 (Webster et al. 2013), as well as closure of five airports and cancellation of hundreds of flights to Bali and elsewhere in July and November 2015, as a result of three moderate eruptions at Iceland's Eyjafjallajökull, East Java's Raung and Lombok's Rinjani respectively.

In short, global awareness of and response to such disasters have been totally transformed in the two centuries since Tambora. The immediate casualties of another Tambora-scale eruption in our day would be much reduced by advance warning, evacuation and relief measures. The impact of the agricultural destruction would be evened out by national and international food distribution and resettlement efforts. Today's effects would be immediately understood as global in scale in our interdependent world. In ways unimagined two centuries ago, global air traffic would be devastated by the first Tambora-scale eruption of the age of aviation, while the world's scientists would at once begin to calculate the effects on weather, global temperatures, sulphuric acid dispersion and ozone depletion. Environmental disasters have become spectacular reminders of our interdependence.

2.2 Explaining Disasters, East and West

The global sense of responsibility for disaster victims is relatively new, though one can trace a few ancestors. I will oversimplify greatly by suggesting three broad historic patterns of response to natural disaster in the Eurasian civilisational pool: viewing it as a punishment for sin (scriptural), a cosmic judgement on the ruler (classic empires), or a cosmic omen (especially southern Asia). Blaming and dehumanising victims for their sins, rendering them undeserving of empathy like enemies in war, was by no means the only response of the ancient world, but it is the best remembered because it is firmly embedded in scripture.

Otherwise inexplicable catastrophes were seen as divine punishment in the three great Abrahamic religions that shared the basic mythology of the early chapters of the Torah or Old Testament. The first archetype was the great flood in the mythological remote past that destroyed all humans except the family of Noah in the Old Testament/Torah (Genesis 6 and 7) or Nuh in the Qur'an (Surah 11 and 71). The explanation was their general godlessness and immorality. More specific sins were held to cause the destruction of Sodom and Gomorrah in the fertile plain to the south of the present Dead Sea, in the proto-historic time of Abraham (Genesis 13 and 19; Deuteronomy 29), known as Ibrahim in the Qur'an (Sura 11). Christian and Islamic scripture has insisted that this divine punishment was because of sexual transgressions, especially homosexuality, hence the English term sodomy (and most other European languages, from the Latin Vulgate's *sodomiticum*). But there are other Biblical traditions (reflected in Isaiah 1 and 3) that have a much broader list. Collectively, these traditions tend to suggest a major volcanic eruption:

> The Lord rained down sulphurous fire upon Sodom and Gomorrah. He overthrew those cities and the whole Plain ….[Abraham looked back on the Plain and] saw dense smoke over the land rising like fumes from a furnace. (Genesis 19: 24–5)

In the Qur'an's more extensive treatment, 'We turned (the cities) upside down, and rained down on them brimstones hard as baked clay, spread, layer on layer' (Qur'an 11: 82, in Yusuf Ali translation).

In the Sinic world, blame was more likely to be cast on the ruler. The classical Chinese authors had taught that the occurrence of natural disasters was evidence that the emperor had lost the mandate of heaven. His 'virtue' (*de*), the quality that demanded grateful submission in subjects, was no longer sufficient to guarantee popular welfare (Sanft 2014, pp. 47–48). Disasters were interpreted as sent by heaven (*tian*) to mark the decline of a dynasty, or at best to test the mettle of a ruler to overcome the disaster (Janku 2009). Chroniclers therefore paid particular attention to such disasters, giving us a better record of earthquakes, floods and droughts in Northeast Asia than in other pre-modern societies. The cosmos was basically knowable and benign, and Emperors were responsible for maintaining that beneficence. China, it has been argued, pioneered not just the bureaucratic and meritocratic state, but the welfare state (Mair and Kelley 2015, p. 342, citing Bin Wong).

Because of this sense of imperial responsibility, substantial resources were devoted to preventing and responding to disasters. Records of imperial relief to victims of disaster go back at least as far as the Han Dynasty, and an established 'natural disaster policy' (*huang zheng*) was often appealed to thereafter. Pierre-Etienne Will (1990) has shown that when the Qing Dynasty was at its expansive eighteenth century peak, famine relief was at a level of efficiency never reached (or indeed needed) by Europe, in marked contrast with the mishandled disasters of nineteenth century China and India (see below). During the El Niño drought and famine that struck Zhili (now Hebei) province in 1743–1744, some two million otherwise starving peasants in North China were kept alive for 8 months by massive government shipments of grain from the South. The food security of the people remained a major preoccupation of the imperial system for the remainder of the century.

The same imperial ideology held Vietnamese rulers to account for climatic disasters like droughts, floods and typhoons. Only in the nineteenth century, however, did the reach of Vietnamese emperors and their literati advisors extend to the more vulnerable coastline of what is today central and southern Viet Nam. For this last dynasty, the Nguyen (1802–1883), there seemed no respite from disasters somewhere. The chronicles tell us much about them, and about the government's response. The emperors devoted considerable resources to organising the correct 'calling for wind and rain' (*cầu đảo*) rituals, even for epidemics and other disasters not related to drought. They diverged from their Chinese counterparts, however, in providing little by way of practical relief throughout their extensive domains. Even the costs of the ritual requirements to satisfy the spirits were imposed on local authorities wherever possible (Dyt 2015).

State concern for disaster victims may have been more a feature of imperial world rulers than of Confucian ideas *per se*. Roman emperors from the time of Augustus (27BCE–14CE) also established their legitimacy and benevolence by assisting the victims of natural disasters. Right at the beginning of his reign, Augustus responded generously to a petition carried to him from the island of Chios where a city had been destroyed in an earthquake. It was rebuilt from imperial funds, and renamed Caesarea in a show of imperial propaganda (Higgins 2009, pp. 64–5). Subsequent emperors appear to have followed this example, providing tax relief and monetary grants to the survivors of major earthquakes in both Italy and the eastern Mediterranean (Dillon and Garland 2015, p. 678; Higgins 2009, pp. 64–74). Some pre-Christian writers saw natural disasters as signs of divine intervention in human affairs, and were not above blaming the victims to make moral points about who was selected for destruction. The emperors, however, appear to have seen these events as an opportunity to demonstrate their power and benevolence—even including the notoriously less benevolent Caligula and Vitellius.

Once the Roman emperors became Christian, they merged this tradition with the scriptural idea of natural disaster as divine punishment for sin. Some Christian emperors appear to have taken these signs of divine disapproval personally, publicly demonstrating their repentance and humility. Justinian responded dramatically to an earthquake in Antioch in 526:

> He threw aside his crown and imperial robes, and, dressed in dirty rags, he wept for many days, and even on feast days he entered the temple in wretched garments, for he could not bear to wear any symbols of power. And all those who were in the city gathered in their rags… and for seven days they fasted and prayed. (Cedrenus, cited in Higgins 2009, p. 75)

Outside the sphere of Sinic influence, the Asian response was largely of the third type: natural disaster was seen as a signal to the living from gods and spirits, but not necessarily of a negative or judgemental kind. The Indic and Southeast Asian literatures record little by way of benevolence by the wealthy and powerful towards victims of disaster. The principal means of assisting those dying of hunger was to accept them as slaves or bondsmen. The chronicles and inscriptions do not, indeed, see natural disaster as tragic so much as portentous. An indication of this is the way Buddhist and Hindu temples were built in stone even in earthquake-prone Burma, Sumatra and Java, presumably with the knowledge that these would be destroyed sooner or later by earthquakes. These were the abodes of gods, not men, and gods had a right to destroy them, giving humans the opportunity to build them again and gain further merit. The places where humans congregated, including hermitages, monasteries, religious schools and mosques, were always built of wood and bamboo and had little to fear from earthquakes.

The frequent volcanic eruptions of Java and the earthquakes of Sumatra were reported in the chronicles and inscriptions as omens, frequently of the birth, death or great achievement of some divinely-inspired figure (Reid 2015, pp. 66–68). The Javanese regularly built temples on the slopes of active volcanoes, and developed ritual responses to great events. The major eruption of Mount Merapi in 1822 was widely interpreted to presage the coming of the messianic 'just king' (*ratu adil*), a figure the anti-Dutch rebel Diponegoro then sought to embody (Christie 2015). A probable mega-tsunami in 1618 may have been used by the rising dynasty of Mataram as a sign that Sultan Agung and his successors alone could harness the supernatural powers of the Queen of the South Seas (Ratu Kidul) by mating with her (Reid 2016a, pp. 99–107). Going further back, the Javanese court poet Mpu Prapanca, who wrote his *Desawarnena* (or *Nagarakertagama*) in 1365, described a 1334 eruption thus:

> The earth quaked and rumbled, there was a rain of ash, thunder and lightning zigzagged through the sky,
>
> Mount Kampud [Kelud] erupted and the wretched evildoers were annihilated and died without a gasp. (Robson 1995, p. 26)

Prapanca saw this as supernatural confirmation of the birth of a great king, his patron King Hayam Wuruk, whom he exonerated from blame by insisting that only evil people were killed.

Burmese traditions popularised the Buddha's explanation of earthquakes to his favourite disciple, Ananda. The first of the eight causes he gave was about the fragile structure of the cosmos, with the earth resting on water, which in turn rested on unstable air. The other seven all had to do with marking the progress of some gifted individual toward Buddhahood (Shway Yoe 1896, p. 575).

Throughout Eurasia, the great religions encouraged compassion towards the suffering victims of life, on the grounds that all men are essentially brothers with similar destinies. Everywhere there are exemplary stories of compassion. But it is difficult to trace the rise of empathy towards distant victims, beyond the specific responsibility of rulers towards their subjects, until the transformation of European communications and sensibility in the eighteenth and nineteenth centuries. Hence we must return to that western extremity of Eurasia, itself relatively free of natural disasters, to take the story forward.

2.3 1755 and the Beginnings of European Transnational Compassion

Europe developed its economic leadership in the past millennium in part because it was unusually favoured climatically and geologically. Its disasters were chiefly man-made in the form of endless wars. After reviewing the ancient writers who attributed the fall of the Roman Empire to divine punishment of its excesses through natural disasters, Edward Gibbon sagely pointed out that in reality:

> Man has much more to fear from the passions of his fellow-creatures than from the convulsions of the elements. The mischievous effects of an earthquake or deluge, a hurricane, or the eruption of a volcano, bear a very inconsiderable proportion to the ordinary calamities of war. (Gibbon 1789/2008, p. 70)

This certainly was the case in war-ravaged Europe, if not perhaps in the 'ring of fire' unknown to Gibbon. The 1755 Lisbon earthquake therefore came as a considerable shock. Most of the buildings of Europe's fourth-largest city were destroyed, with probably 4000–8000 of its population killed (though this number was greatly inflated by posterity) and much of the remaining 150,000 made homeless (Aguirre 2012). The disaster appeared to have no remembered antecedents in Western Europe. Though European colonists in Latin America, the Philippines and Java had reported drastic events in these places, a delay of months in the news weakened the impact of these faraway events. The destruction by earthquake and tsunami of Port-Royal (Jamaica), the Caribbean's busiest port, in 1692, and most of Lima, one of South America's richest cities, in 1746, may have been comparable disasters, but had little impact in Europe. Catania, in Sicily, destroyed by Mount Etna's eruption in 1669, was poor and peripheral. As Goethe rhapsodised, 'Raging volcanoes rise up in the distance, seeming to threaten the world with destruction. Yet the bedrock of my refuge remains unshaken, while those who live on distant shores and islands are buried beneath the faithless land' (Goethe 1998, p. 133).

But Lisbon was close, and the growing newspaper culture reported the disaster around Europe within days. 'For the first time in the western world, the press helped to create the illusion of proximity and unity among the peoples of different European nations' (Araújo 2006). The philosophers all took it up. For Voltaire, famously, it was proof that God was either not benign, or not omnipotent. Goethe too thought it destroyed the benign fatherly figure of God. Rousseau took it as proof that crowded cities were not part of the divine plan, and one should return to nature. John Wesley, the founder of Methodism, was more representative of popular ideas in his continued insistence that divine intervention was the best explanation for this and the other more distant natural disasters. He did not exactly blame the victims, but saw the earthquake as a timely reminder that no amount of wealth or science could protect man against such disasters, but only prayer (Wesley 1755).

The earthquake marked a shift in European worldview, whereby the catastrophic spasms of the planet became subjects for intense interest, an essential part of Enlightenment enquiry. The buried city of Pompeii, near Naples, had been rediscovered in 1748, and excavation proceeded throughout the remainder of the century. It and the active volcano Vesuvius, erupting frequently between 1744 and 1761, became crucial agenda items on the grand tours which Enlightenment intellectuals made to Italy. The volcanists championed by Alexander von Humboldt argued that earthquakes and eruptions revealed a necessary but decentred process of ongoing creation. The rival Neptunists, inspired by Abraham Gottlieb Werner, preferred a more coherent design for creation, in which water played the central role. Volcanoes became such a fashionable topic that artificial ones were built, such as that in the famous garden of Prince Franz of Anhalt-Dassau (Brodey 2008, p. 30).

For many, however, 1755 represented a humanitarian disaster for the innocent people of a small country, deserving of international support. The Spanish king arranged to send as much money to Lisbon as a courier could manage every day. The German cities of Hamburg and Danzig sent off several shiploads of building supplies. King George II requested the British Parliament to vote £100,000 for his faithful ally in Lisbon, to arrange 'such speedy and effectual relief, as may be suitable for so afflicting and pressing an exigency' (cited in Murteira 2004). International responsibility for innocent victims of natural disasters had begun.

2.4 A Genealogy of Globalising Relief

For Western Europe, 1755 was an exception. Warfare remained the great killer in that sub-continent, and it was war that provoked the first international relief organisation. The Swiss businessman Jean-Henri Dunant witnessed the Battle of Solferino in Italy in 1859, where 40,000 men were left dead or wounded on the battlefield in a single day. He did his best in the following days to help bury the dead and tend the wounded, but it was his 1862 book about the event, *A Memory of Solferino* (1986) that had the greater effect. Its plea for international treaties to provide and protect health personnel bore fruit in a conference around these ideas in October 1863. Thirteen European governments and various non-government groups there agreed on the establishment of national relief societies in each country for care of the war wounded, whose personnel should be protected from attack. The following Geneva Convention in 1864 included representatives from outside Europe, from the US, Brazil and Mexico. In 1876 the international character of the movement was given structure and a name: the International Committee of the Red Cross.

While wars remained the primary scourge of geologically safe Europe, the non-European members of the Red Cross pressed to add natural disasters to their mandate. The American branch begun by Clara Barton was always more concerned with natural disasters. It aided victims of a forest fire in Michigan in 1881, a Mississippi River flood in 1884, another flood from a dam break in Pennsylvania in 1889, and a hurricane in a South Carolina island in 1893. At the Third International Red Cross Conference in Geneva in 1884, Barton pressed for the extension of the official Red Cross mandate to natural disaster relief. Her 'American amendment' eventually prevailed over much European resistance. The American branch led in the internationalisation of relief to subsequent victims of volcanic eruptions and earthquakes in Chile, Colombia, Ecuador, Costa Rica, Iran, Japan and Turkey (Red Cross 2015). The spread of telegraphic communications around the world in the 1860s and 1870s helped give a sense of immediacy to these disasters, so that newspaper readers in Europe and North America could begin to empathise and mobilise support within days of a disaster.

The Save the Children Fund was also founded to ameliorate the horrors of Europe's wars, working on the effective strategy of exempting 'innocent' children from the dehumanisation that had been accorded to enemies in war. The well-connected Jebb sisters founded the organisation at a surprisingly successful meeting at the Albert Hall in 1919. They were aghast at the destitution of Germany and Austria, whose peoples had been demonised during the war. Younger sister Eglantyne had the organisational skill to turn this tide of sympathy into an international organisation centred in Geneva the following year, and its activities moved on from crisis to crisis around the world.

The internationalising of responses to epidemics took place in roughly the same globalising period of improved communications. It was however more of a top-down process, led by governments concerned to inform and protect their own populations against border-crossing epidemics. Intergovernmental 'sanitation' conferences began in Europe in 1851 to coordinate the reporting of and response to the outbreak of epidemics. The eleventh such conference, in 1903, recommended the formation of what became the Office International d'Hygiène Publique in 1907. This was overshadowed after the war by the Health Organisation of the League of Nations, set up in Geneva in 1921 to monitor and exchange epidemic information (Charles 1968). The WHO is its much more ambitious descendent, established as a UN agency in 1948.

Most of the other international organisations that we now expect to respond to the latest disaster owe their origins only to the post-colonial moment since 1945. Responsibility for poor societies shifted from colonial powers to international organisations at a time when the gulf between rich West and poor rest was at an all-time peak. Even that most ancient of international organisations, the Catholic Church, had no international structure for its relief activities until the 1950s. The Acting Vatican Secretary of State, Monsignor Montini (the future Pope Paul VI), convened a meeting in Rome of thirteen national Caritas organisations in 1951. That of Germany was the oldest and strongest, founded in 1897, and it took a lead in implementing Montini's call for an 'international organism' to coordinate relief. In 1954 Caritas International was formed. It began relief immediately in response to floods in Italy, Holland and Belgium, and in following years to China, Viet Nam and Ethiopia. By 1962 it had 74 national members (Caritas 2015).

Oxfam had been founded in Oxford for famine relief in 1942, but became an international organisation only in 1995. Médecins sans Frontières was established in Paris in 1971 as a reaction to the perceived failure of the Red Cross to speak out against atrocities during the Nigerian war against Biafra. Its first mission the following year, like many that followed, was to relieve a natural disaster, when an earthquake destroyed much of the Nicaraguan capital, Managua. Though less rigorously non-political than the Red Cross, it did learn the benefits of strict neutrality during the Lebanese conflicts. It developed an international network in the 1980s (Newell 2005).

2.5 Asian Disasters and Western Empathy

This process of developing a global infrastructure of relief took place during the late imperial period, and more intensively as that system died after 1945. In the century 1860–1960 the gap had grown ever wider between wealthy and powerful western countries and populous but poor Asian ones. The latter were prone to horrendous disasters that dwarfed anything in European and American experience. Better communications, the telegraph, the press, a growing public sphere, and graphic photography gradually drew sympathy and relief efforts from the rich West to the poor East.

A major turning point was the exceptionally severe El Niño of 1876–1879, which appears to have been responsible for the worst series of droughts and crop failures around the tropical world known to history. Well over 30 million people must have died in the resulting famines—probably 'the worst ever to afflict the human species' (Davis 2001, p. 1, quoting John Hidore). They devastated agricultural regions of South India, Java, the western Visayas (Philippines), North China, Brazil, New South Wales and eastern Africa. Research has focussed on India and China, which appear to have lost 6–10 million and 9–15 million people respectively through famine deaths in this period (Davis 2001, p. 7; Janku 2009).

The two responsible imperial governments, British and Manchu, were for different reasons unable or unwilling to provide effective relief. The Indian Viceroy Lord Lytton and his key advisors were perversely committed to Adam Smith's doctrine that interference with the market only made shortage more acute, and rewarded unviable inefficiency. The Government had responded effectively to the earlier Bihar famine in 1873–1874 by importing rice from Burma, which had kept mortality commendably low through an exceptional drought. This policy had been rejected by Lytton for disturbing the rice market, creating dependency and endangering the budget needed for expensive wars in the Northwest. He imposed a much harsher line this time, allowing Indian grain to be exported for better prices in Britain while millions starved (Hall-Matthews 1996; Davis 2001, pp. 25–54).

In China the precociously advanced Qing system of managing state granaries to provide for famine assistance had broken down before a century of environmental degradation, destructive war and rebellion, and what Davis (2001, p. 344) calls 'the commercialization of subsistence'. By 1820 the grain shipped north to the capital on the Grand Canal had already ceased to provide the stocks needed as a buffer against recurrent famines in the north. In mid-century, foreign-owned coastal steamers became the principal means of shifting grain around, guided by profit margins rather than need. The Qing government could respond to the prolonged drought only by urging local authorities to do their best, and providing them pitiful assistance in cash.

As British and American military successes unlocked China, Japan and India to the universality of trade and investment in the middle third of the nineteenth century, global interdependence became manifested through the new norm of rich west and poor rest. For the first time the telegraph communicated the scale of disaster and coordinated responses to it, while photography displayed the stark horror of events. Horror, guilt and pity were mobilised among the global rich (however unevenly) towards the poor. Debate began almost at once about the causes of the famines. While the British establishment viewed Asian despair as a product of Malthusian overpopulation crisis and 'natural' disaster, its many critics focussed on the opium trade, cynical government misallocation of resources, and indifference. The case against the Indian government's criminal irresponsibility was made by radical journalists William Digby and Robert Knight, and later by economic historian R.C. Dutt, contributing to the inter-communal nationalism of the first Indian National Congress in 1885 (Davis 2001, pp. 55–9).

Despite the stern hostility of the Viceroy to publicity or fund-raising for the victims, the Madras Governor chaired a public meeting in his capital in August 1877. It formed a famine relief committee which telegraphed the Lord Mayor of London, whose Mansion House had already became the first address for disasters closer to home. From there the appeal went out to other cities around Great Britain, and after some offended protests from them, also to distant but wealthy British cities such as Melbourne. Indian Famine Relief Committees were formed in many cities and gained the support of radical and Christian reformers like Florence Nightingale. These committees raised £689,000, of which £52,000 was from Australia (Twomey and May 2012, pp. 233–37).

Imperial solidarity appeared to be the central motive for the generous British and Australian response, in which civic networks played a larger role than religious ones. Other appeals for disasters in the 1870s rested largely on the solidarity of ethnic diasporas towards their countries of origin. Australian donations reached only £3600 for the great Chinese famine (1878) and £1087 for a Persian one (1872), but £95,000 for an Irish one (1880), reflecting the numerical and financial strength in Australia of the Chinese, Jewish and Irish diasporas respectively on which these three campaigns rested (Twomey and May 2012, pp. 247–50). The Indian fund was different, with Anglo-Australian establishment support on grounds of imperial solidarity. 'The vastness of empire and identity of interests,' the Archbishop of Sydney declared, tended 'towards softening asperities of creed, and towards breaking down the isolation of antagonistic race' (cited in *ibid.*, p. 239).

In the case of China, by contrast, it was largely western missionaries who publicised the despair of the famine victims around the world, and sought to mobilise empathy beyond 'antagonistic race' through description, photos and sketches. British missionary Timothy Richard drew foreign attention to the famine with his heart-rending account of starving people lying helpless, wives and children being sold, and children being eaten. The Shandong Famine Relief Committee was formed by the foreign community in Shanghai in March 1877. Missionary accounts were circulated to comparable committees in London (where the great opium trading firms provided support), Europe, America and Australia. Donations were tiny in Britain and Australia in comparison with the India campaign. The warmest response was in the United States, where missionary pressure produced a bill in Congress to return to China some of the indemnity that had been extracted from its government in 1859. The bill failed to gain support, however, overcome by the wave of hysteria in California against Chinese immigration. In total the campaign is estimated to have raised some US$400,000. The Chinese state's once-formidable capacity for famine relief was shown to be in ruins, and private fund-raising both among affluent Chinese (in the south and in diaspora) and around the world began to take its place (Janku 2009, pp. 236–7; Davis 2001, pp. 64–79).

Despite the unifying effects of telegraph and of international organizations such as the Red Cross, in other words, international responses continued to be extremely uneven, dependent on the presence or absence of particular sources of empathy. Two fateful stereotypes were formed in this crisis. In the West the problem of 'Asia's starving millions' became entrenched, breeding concern, anger, or Malthusian fatalism in different quarters. In China the inadequate efforts of missionaries to rescue some from the disaster fostered resentment at 'rice Christians' and the 'buying' of babies.

Tectonic traumas—earthquakes, tsunamis and eruptions—were less destructive of lives than these terrible famines, but more spectacular in their visual effects and in the way they affected wealthy cities rather than poor peasants. Two deadly Northeast Asian earthquakes in the 1920s again demonstrated a marked difference in international response. Essentially the 1920 Gansu earthquake in northern China and the 1923 Kanto earthquake in the Tokyo area of Japan each killed at least 140,000 people and rendered millions homeless, but the latter evoked the first truly global relief effort while the former was scarcely known outside China, and received only local relief.

Tokyo was already a significant global city, East Asia's largest and most modern, well connected to the world by telegraph and press networks. The destruction of urban buildings was spectacularly obvious, and well reported both verbally and graphically. Japan was relatively well-organised in its response, with a domestic Red Cross Society connected to sister organisations throughout the world. About US$100 million was channelled to the Japanese societies by its sister organisations. Sympathy was highest in the United States, where the Red Cross was already specialised in disaster relief, and where the Japanese diaspora provided a visible and organised source of organisation and empathy. President Coolidge gave a strong and public lead in calling for a sympathetic response as soon as the news reached the US, and American ships from the China station were sent promptly to Tokyo Bay to render assistance. In proportion to GDP at the time, this may have been the largest international relief effort to date. Even though domestically the disaster could still be interpreted as divine punishment, provoking a nasty pogrom against 'polluting' Koreans, it was the first major example of international structures working effectively for aid and reconstruction (Schencking 2013, pp. 116–152).

The Gansu earthquake, centred in remote Haiyuan, affected mostly rural Hui Muslim people, many of them living in cave-like homes dug out of the fine yellow loess. Landslides and collapses of such dwellings may have accounted for about 100,000 deaths, while a comparable number died of starvation, thirst or exposure as they wandered about looking for help. This occurred amidst another drought-induced famine affecting some 20 million people on the North China plain, which absorbed what relief supplies the struggling Chinese Republic then had to offer. Military units in the immediate vicinity offered a little assistance in the form of tents and food in some areas, while gentry committees also mobilised some rice and shelter. The Gansu diaspora in the eastern cities was able to organise a little aid, while foreign missionaries in the area also helped with reconstruction in the year following. But like many disasters in China before and since, the broader international community knew almost nothing about this disaster, and its victims suffered much as they always had, in relative isolation (Fuller 2015).

2.6 Cold War and Nationalism Interrupt the Globalising Trend

Interdependence was not a linear progression. The profound political divisions of the twentieth century interrupted the expansion of global empathy and coordination that better communication would otherwise have made possible. After each of the World Wars there were great steps forward towards global coordination and understanding, quickly undermined by polarisation and belligerence. The Cold War (1948–1989) poisoned empathy and ruptured cooperation across the iron and bamboo curtains, while the self-righteous nationalism of some fragile new nations made them reluctant to accept outside aid. Sukarno's Indonesia was busy launching its confrontation of Malaysia at the time of the eruption of Bali's Gunung Agung in 1963, the most severe in Indonesia's relatively moderate twentieth century. The pyroclastic flows killed 1580 people immediately, though over 50,000 more were 'missing' from the 1971 census in the four affected eastern districts, their crops destroyed by ash and acid (Reid and Rangkuty 2014). Bali and neighbouring eastern Java were already in near-famine conditions as a result of Sukarno's mismanagement. The Bali Governor reported that 100,000 hectares of rice land would be unproductive for many years to come. 'We have to feed 85,000 refugees and we simply do not have the food to do it' (Robinson 1995, pp. 239–40).

The 1976 Tangshan earthquake in Northeast China, although probably the deadliest of the twentieth century anywhere (at least 250,000 died), was poorly reported both inside and outside China in a year of political crisis in which both Zhou Enlai and Mao Zedung died. Mao's successor Hua Guofeng did gain stature in China by visiting the disaster site and showing sympathy for the victims, but his radical rival, Mao's widow Jiang Qing, prioritised her campaign against the comeback of Deng Xiaoping. She was widely quoted in China as having said, 'There were merely several hundred thousand deaths. So what? Denouncing Deng Xiaoping concerns 800 million people' (Palmer 2011, p. 189). The international assistance that was offered, even that of the United Nations, was rejected by the Chinese government.

2.7 The Globalised Present

The period since the end of the Cold War in 1989 marks the most remarkable extension of global information exchange, preparedness and response to natural disasters. The 2004 Indian Ocean tsunami was the most dramatic example, with amateur videos keeping graphic images on TV screens around the world as never before. Aid for the relief effort was pledged from at least 54 countries, most of them having lost their own citizens in the exceptionally transnational disaster. International organisations also moved quickly to provide assistance, to a record total in excess of US$13.5 billion. Nearly $6 billion of this was provided by private donations through NGOs of all kinds. It was the most generous international disaster response in history, and indeed raised many questions about whether such funds should have been distributed more equitably for development purposes (Telford 2012; Brauman 2009).

Even a purely China-based phenomenon like the Sichuan earthquake of 2008 (over 80,000 killed) was reported widely and responded to internationally. Although there was protest in China (well publicised by Ai Weiwei) at the government's withholding information about the carnage resulting from badly constructed schools, the contrast with the 1976 event was enormous. The Chinese government response was effective, centralised and military-led on this occasion, but billions of dollars in aid were also delivered by other governments, Red Cross societies and international agencies (IFRC 2012; UNICEF 2009). Regional rivals Japan and Russia were particularly quick to offer aid in money and kind.

Both sides of the old iron curtain now cooperate in the sharing of data, while satellite imaging and other forms of global monitoring provides better information than most governments can generate. Bodies such as the US Geological Survey have established a global reach, and web sites proliferate to assemble data on a global basis. The 24-h news cycle, the ease of electronic communication and the universality of mobile phones with a video capability combine to ensure that graphic, emotive images of disasters anywhere in the world are on our screens within hours. International journalists are on the scene of every disaster within days, and recovery teams and aid workers swiftly follow. There is increasing coordination between the various UN agencies concerned with disaster relief (WHO, UNHCR, WFP, UNICEF, UNDP) and increasingly globalised NGOs such as Red Cross/Red Crescent, OXFAM, Caritas, Médecins sans Frontières and World Vision. The UN's Office for the Coordination of Humanitarian Affairs (OCHA) seeks to coordinate the UN's response to these disasters, while an Inter-Agency Standing Committee (IASC) attempts the even more difficult task of coordinating the work of all agencies, government, UN and private.

Subjectively, we are all aware of what appears to be an increasing frequency of mega-disasters in the Asia-Pacific. The devastating tsunamis of 2004 (Indian Ocean) and 2011 (Honshu) in March 2011 were brought to our living rooms with unprecedented vividness. Cyclone Nargis in the Irrawaddy delta in 2008 (138,000 dead) appeared to have no Burmese precedent. Typhoon Haiyan in the Central Philippines in 2013 created a record for the destruction of homes, and we need to go back at least to 1897 to find a comparable event, with many fewer casualties (Switzer 2015). Nor had the Philippines experienced for centuries a volcanic eruption on the scale of Mount Pinatubo in 1991. The floods that affected Bangkok and all the deltaic areas of Indochina in 2011 were, in dollar terms, as destructive as any fresh-water flood in human history (Reid 2016b, pp. 46–50).

How much of this impression reflects reality, as opposed to the effect of intensified coverage and journalistic fashion? Firstly the reality. The numbers for property damage do tend to increase with higher coverage, inflation and rising living standards. Increased population, especially in vulnerable cities, also increases the numbers rendered homeless or otherwise affected by a disaster. The numbers killed in disasters (Fig. 2.1), however, are dropping, at least relative to population. The reasons are better preparedness both nationally and internationally, and more effective systems of relief.

In the past 30 years Asia has provided the overwhelming majority (almost 90%) of people affected by disasters, due to its very high population densities, and the rich countries relatively few. The incidence of disasters in Asia was rather low in relation to Asia's share of world population (Table 2.1). If we go further back in time, it is clear that the most destructive volcanic eruptions, earthquakes and floods of the last thousand years have been in Asia, and can be expected to recur there in the twenty-first century to a greater extent than they did in the twentieth. Indonesia's Tambora (Sumbawa) and Samalas/Rinjani (Lombok) eruptions in 1815 and 1257, respectively,

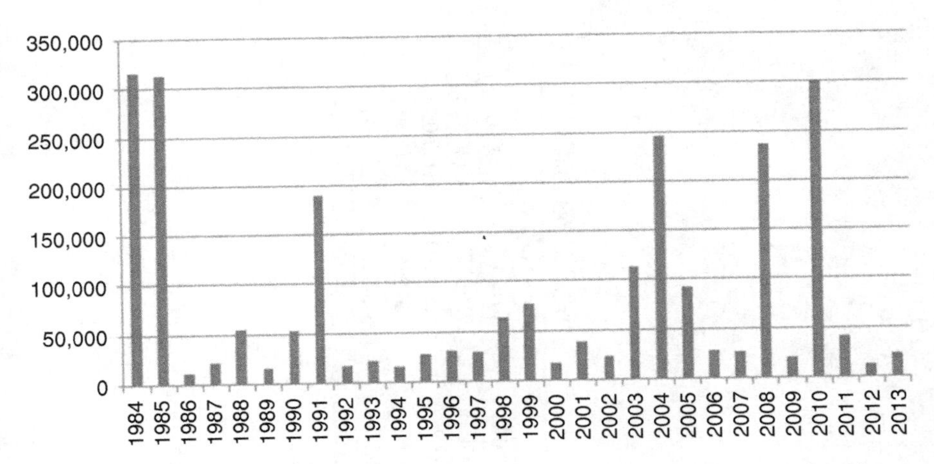

Fig. 2.1 Numbers killed by natural disasters worldwide, 1984–2013 (Source: Asian Disaster Reduction Center 2014)

were of a magnitude more than ten times greater than anything that has occurred since 1900. Of the many earthquakes in China over the past century, none has come close to the Shaanxi earthquake of 1556 in the mortality it caused (over 800,000), despite the far greater population of modern China's cities. It may be fair to say, therefore, that although deaths from natural causes are not tending to increase in absolute terms during the time-scale when we have credible data, Asia is the likeliest place for a future mega-disaster of a very different order of magnitude.

The media coverage of natural disasters appears to be increasing both in quantity and universality. A Red Cross report on the media pointed out that:

> News must be new. Editors sort stories by death tolls. Disasters that are unusual yet explicable, and that cause considerable death or destruction in accessible places which the audience is believed to care about, get covered. Baffling stories get less attention.
>
> The commercial imperative has sharpened journalists' quest for ratings. Today, TV news is part news and part entertainment. So it's understandable that sudden, dramatic disasters like volcanoes or tsunamis are intensely newsworthy, whereas long-drawn-out crises (difficult to describe, let alone film) are not. (ICRC 2005)

Popular consciousness of the depredations of natural disasters has undoubtedly increased in the last 40 years, out of proportion to any identifiable increase in the victims of natural disaster. Part of this is explained by the visual images now available from spectacular natural events, and the capacity of modern media to deliver them to our living rooms. Perhaps partly for this reason, there appears to have been a gradual growth of empathy for victims perceived as innocent fellow-humans in great need.

The other sea change that has been occurring in my lifetime is the reduction of wars and armed conflicts, now well documented by Steven Pinker (2011). Observable globally since 1945, this trend was slower to manifest itself in the Asia-Pacific region, where the Cold War was at its hottest. There too the period of peace since

Table 2.1 Impacts of natural disasters by region, 1984–2013

Region	*Impact*							
	Occurrence		Killed		Affected		Damage (US$ million)	
	(Share in %)		(Share in %)		(Share in %)		(Share in %)	
Africa	2099	(20.2%)	726,996	(29.3%)	423,394,194	(7.0%)	19,064	(0.3%)
Americas	2495	(24.0%)	339,148	(15.7%)	204,311,734	(3.4%)	916,177	(36.3%)
Asia	3952	(33.1%)	1,136,437	(47.7%)	5,396,306,705	(88.7%)	1,169,341	(47.0%)
Europe	1398	(13.5%)	176,505	(7.1%)	35,344,415	(0.6%)	320,256	(12.9%)
Oceania	432	(4.2%)	5753	(0.2%)	20,431,165	(0.3%)	64,662	(2.6%)
Total	10,376	(100.0%)	2,434,339	(109.0%)	6,030,733,213	(100.0%)	2,439,499	(100.0%)

Source: Asian Disaster Reduction Center (2014)

1980 has been unprecedented. Wars and conflicts dominated the news media for much of the twentieth century, and still do even when the violence-hungry media have to make do with dozens killed by terrorists rather than millions in earlier conflicts in Indochina, China, Korea, India-Pakistan, Nigeria, Congo or Indonesia. Fortunately we appear to be a little less desensitised to mass killing than our grandparents, a little more outraged that violence continues. But natural disasters have come into their own as sources of shock and horror. It has been noted that Hollywood's current output of disaster movies began as US coverage of armed conflict in Indochina ceased in the 1970s (Mauch 2009, p. 2).

This shift of attention is fraught with imbalance, inconsistency and chaotic competition between agencies. But it has fortunately nudged forward the desperate need of *Homo sapiens* to think as global citizens whose destinies are interdependent, before we destroy our species and our planetary environment. Environmental disasters and the threat of their future recurrence are powerful incentives to establish regional and global regimes regarding nuclear energy and weapons, water rights, pollution and carbon emissions, epidemics and solar storms, to minimise their severity and destructive impact on humans. The monitoring of such regimes is highly technical, shifting some supra-national authority to the international community of scientists and technicians. Though such experts can provoke populist backlash, as in the European Community, they are less politically unpalatable than the clash of global contestants for power.

2.8 The Way Ahead

We remain grossly unbalanced in the resources every country devotes to preparing for war to destroy each other, as against preparing for a common effort to save our species and our planet as it faces future threats. The more we can encourage a trend in this direction, the better it will be for all. For all the chaos, competitive infighting and waste that marked the unprecedented response to what seemed an unprecedented disaster in the 2004 tsunami, this had the effect of ending a secessionist war in Aceh. The global spotlight was briefly turned to a nasty but forgotten war, and the contestants were shamed into ending it. Although unfortunately the tsunami had no such effect on equally conflicted South Thailand or Sri Lanka, the potential of disaster response to promote interdependence was clear.

Singapore and Australia, the two rich countries little threatened by natural disasters in an otherwise poor and vulnerable region, have an opportunity to play a constructive role in regional preparedness. They are by far the biggest military spenders in the region, with a total expenditure of US$9.8 billion and $25 billion respectively in 2014. Such expenditure dwarfs that of their giant, but poor but vulnerable, neighbours Indonesia ($7 billion) and the Philippines ($3.2 billion), let alone that of equally vulnerable Papua New Guinea, Vanuatu and the Solomons with their negligible military forces. In per capita terms these two rich regionals are among the biggest military spenders in the world, at more than tenfold the rate of their neighbours (SIPRI 2015).

Both countries justify their large expenditures, big arms purchases, and in Singapore's case compulsory military service, in terms of insecurities deriving from the threats of the 1940s (Japan in Australia's case) and the 1960s (Malaysia and Indonesia in Singapore's). These imagined dangers have little to do with present realities, though military and industrial interests have persuasive reasons to keep funds flowing their way. Regional disaster governance provides a powerful twenty-first century reason for these two militaries to retain their critical position while tilting their objectives towards the certainty of environmental disasters in the region rather than the memory of obsolete threats. The bloated mega-cities of maritime Asia are particularly at risk. Military cooperation with poorer neighbours in modelling and planning for transnational disaster governance is a positive path towards trust and a sense of common interests. It provides a more urgent regional justification for the new US base in Darwin and docking facilities in Singapore than insurance against hypothetical military threats.[1]

References

Aguirre, B. E. (2012). Better disaster statistics: The Lisbon earthquake. *Journal of Interdisciplinary History, 43*(1), 27–42.

Araújo, A. C. (2006). The Lisbon earthquake of 1755: Public distress and political propaganda. *E-journal of Portuguese History, 4*(1), 1–11. viewed 9 May 2016, https://www.brown.edu/Departments/Portuguese_Brazilian_Studies/ejph/html/issue7/html/aaraujo_main.html.

Brauman, R. (2009). Global media and the myths of humanitarian relief: The case of the 2004 tsunami. In R. A. Wilson & R. D. Brown (Eds.), *Humanitarianism and suffering: The mobilization of empathy* (pp. 108–117). Cambridge: Cambridge University Press.

Brodey, I. S. (2008). *Ruined by design: Shaping novels and gardens in the culture of sensibility*. New York: Routledge.

Caritas 2015. History: Who we are, viewed 9 May 2016, http://www.caritas.org/who-we-are/history/

Asian Disaster Reduction Center. (2014). *Natural disaster data book 2013*. Kobe: Asian Disaster Reduction Center (ADRC).

Charles, J. (1968). Origins, history and achievements of the World Health Organization. *British Medical Journal, 168*(2), 293–296.

Christie, J. W. (2015). Under the volcano: Stabilizing the early Javanese state in an unstable environment. In D. Henley & H. Schulte-Nordholt (Eds.), *Environment, trade and society in Southeast Asia: A longue durée perspective* (pp. 46–61). Leiden: Brill/KITLV.

Crawfurd, J. (1856). *A descriptive dictionary of the Indian islands and adjacent countries*. London: Bradbury & Evans.

Davis, M. (2001). *Late Victorian Holocausts: El niño famines and the making of the third world*. London: Verso.

[1] It was welcome news that the new flagship of the Australian Navy, *HMAS Canberra*, commissioned in November 2015, was designed in part for 'conducting large-scale humanitarian and disaster relief missions' (Royal Australian Navy n.d.). The same unfortunately cannot be said for vast projected expenditure on twelve submarines.

Dillon, M., & Garland, L. (2015). *Ancient Rome: Social and historical documents from the early Republic to the death of Augustus* (2nd ed.). Abingdon: Routledge.
Donahue, M. (2007). The Papuan language of Tambora. *Oceanic Linguistics, 46*(2), 520–537.
Dunant, H. (1986). *A Memory of Solferino [1862]*. Geneva: International Committee of the Red Cross.
Dyt, K. (2015). Calling for wind and rain: Rituals, environment, emotion, and governance in Nguyễn Vietnam, 1802–1883. *Journal of Vietnamese Studies, 10*(2), 1–42.
Fuller, P. (2015). Writing disaster: A Chinese earthquake and the pitfalls of historical investigation. *History Workshop Journal, 80*(1), 201–217.
Gibbon, E [1789] 2008. The history of the decline and fall of the Roman Empire, 7, Cosimo, New York.
Goethe, J. W. (1988). In D. Miller (Ed.), *The collected works, scientific studies*. New York: Princeton University Press.
Guzman, (n.d.). Eruption of Mount Pinatubo in the Philippines in June 1991 publicized', Asian Disaster Reduction Center, retrieved October 2015, www.adrc.asia/publications/recovery_reports/pdf/Pinatubo.pdf.
Hall-Matthews, D. (1996). Historical roots of famine relief paradigms: Ideas on dependency and free trade in India in the 1870s. *Disasters, 20*(3), 216–230.
Higgins, C 2009. Popular and imperial response to earthquakes in the Roman Empire, Unpublished MA Thesis, Ohio University.
ICRC 2005. International Committee of the Red Cross, World Disasters Report 2005, Chapter 6: Humanitarian media coverage in the digital age. http://www.ifrc.org/en/publications-and-reports/world-disasters-report/wdr2005/wdr-2005.
IFRC 2012. Emergency appeal final report: China: Sichuan earthquake, emergency appeal no. MDRCN003, GLIDE no. EQ-2008-000062-CHN, International Federation of Red Cross and Red Crescent Societies (IFRC), Geneva, retrieved 9 May 2016, https://www.ifrc.org/docs/Appeals/08/MDRCN003fr.pdf.
Janku, A. (2009). Heaven-sent disasters in Late Imperial China: The scope of the state and beyond. In C. Mauch & C. Pfiste (Eds.), *Natural disasters, cultural responses: Case studies toward a global environmental history* (pp. 233–264). New York: Roman & Littlefield.
Mair, V. H., & Kelley, L. (2015). *Imperial China and its southern neighbours*. Singapore: ISEAS.
Mauch, C. (2009). Introduction. In C. Mauch & C. Pfiste (Eds.), *Natural disasters, cultural responses: Case studies toward a global environmental history* (pp. 1–16). New York: Roman & Littlefield.
Mercado, RA, Lacsamana, JBT, Pineda, GL 1993. Socioeconomic impacts of the Mount Pinatubo eruption, viewed 2013, http://pubs.usgs.gov/pinatubo/mercado/
Murteira, H. (2004). The Lisbon earthquake of 1755: The catastrophe and its European repercussions. *Economia Global e Gestão (Global Economics and Management Review), 10*, 79–99.
Newell, C 2005. 'A working history of Médecins sans Frontières: The changing face of humanitarian aid', University of Ottawa, viewed 10 May 2016, http://www.med.uottawa.ca/historyofmedicine/hetenyi/newell.html.
Palmer, J. (2011). *Heaven cracks, earth shakes: The Tangshan earthquake and the death of Mao's China*. New York: Basic Books.
Pappas, S 2011. 'Pinatubo: Why the biggest volcanic eruption wasn't the deadliest', *Live Science*, 15 June, viewed 2013, http://www.livescience.com/14603-pinatubo-eruption-20-anniversary.html.
Pinker, S. (2011). *The better angels of our nature: Why violence has declined*. New York: Penguin Books.
Red Cross 2015. American Red Cross History: Founder, Clara Barton, American Red Cross, viewed 9 May 2016, http://www.redcross.org/about-us/history/clara-barton.
Reid, A. (2014). Population history in a dangerous environment: How important may natural disasters have been? *Masyarakat Indonesia, 39*(2), 505–526.

Reid, A. (2015). History and seismology in the Ring of Fire: Punctuating the Indonesian past. In D. Henley & H. Schulte-Nordholt (Eds.), *Environment, trade and society in Southeast Asia: A longue durée perspective* (pp. 62–77). Leiden: Brill/KITLV.

Reid, A. (2016b). Building cities in a subduction zone: Some Indonesian dangers. In M. Douglass & M. Miller (Eds.), *Disaster governance in urbanizing Asia* (pp. 45–6059). Singapore: Springer.

Reid, A. (2016a). Two hitherto unknown Indonesian tsunamis of the seventeenth century: Probabilities and context. *Journal of Southeast Asian Studies, 47*(1), 88–108.

Reid, A, Rangkuty, H 2014. 'Testing in Bali the demographic effects of tectonic and political disaster', unpublished paper presented to the Indonesia Study Group, Australian National University.

Robinson, G. (1995). *The dark side of paradise: Political violence in Bali*. Ithaca: Cornell University Press.

Robson, S. (Ed.). (1995). *Desawarnana (Nagarakrtagama), by Mpu Prapañca*. Leiden: KITLV Press.

Royal Australian Navy 2016. Ships, boats & craft: HMAS Canberra (III), viewed 9 May 2016, http://www.navy.gov.au/hmas-canberra-iii

Sanft, C. (2014). *Communication and cooperation in early Imperial China: Publicizing the Qin Dynasty*. Albany: SUNY Press.

Schencking, J. C. (2013). *The great Kanto earthquake and the chimera of national reconstruction in Japan*. New York: Columbia University Press.

Shway Yoe [Sir J.G. Scott]. (1896). The Burman: His life and notions, Macmillan, London.

SIPRI 2015. Military expenditure data base: World military expenditure by Country per Capita, Stockholm International Peace Research Institute, SIPRI, Sweden.

Switzer, A 2015. 'The significance of historical typhoon records: Notes from a comparative study of Super-Typhoon Haiyan and its 1897 predecessor in the Philippines', Paper presented to Euroseas Conference, European Association for Southeast Asian Studies, Vienna.

Telford, J. (2012). Disaster recovery: an international humanitarian challenge? In P. Daly, M. Feener, & A. Reid (Eds.), *From the ground up: Perspectives on post-tsunami and post-conflict Aceh* (pp. 1–22). Singapore: ISEAS.

Twomey, C., & May, A. J. (2012). Australian responses to the Indian Famine, 1876–78: Sympathy, photography and the British Empire. *Australian Historical Studies, 43*(2), 233–252.

UNICEF 2009. Sichuan earthquake: One year report, retrieved 9 May 2016, http://www.unicef.org/eapro/UNICEF-China_Sichuan_Earthquake_One_Year_Report.pdf.

Webster, H. N., Witham, C. S., Hort, M. C., Jones, A. R., & Thomson, D. J. (2013). *NAME modelling of aircraft encounters with volcanic ash plumes from historic eruptions, forecasting research technical report no. 552*. Exeter: Met Office.

Wesley, J 1755. Some serious thoughts occasioned by the late earthquake at Lisbon, Furman University, viewed 9 May 2016, http://history.furman.edu/benson/hst11/docs/wesley.htm.

Will, PE 1990. Bureaucracy and famine in eighteenth century China, Forster, E (trans.), Stanford, Stanford University Press.

Chapter 3
Crossing Colonial Borders: Governing Environmental Disasters in Historic Context

Fiona Williamson

3.1 Introduction

Historic urban floods provide a window into how knowledge was shaped, shared and developed across borders. Borders here are conceptualised as those created by nineteenth century imperial governance regimes which cut across geographic boundaries. Knowledge relates to contemporaneous hydraulic management, meteorological science, and urban development. It draws from a recent strand of British imperial history has been concerned with understanding how scientific knowledge was transmitted across geographic, legislative and cultural borders. Networks comprising individual scholars and scientific institutions linked the disparate countries of the British Empire through a shared interest in progress and development (MacLeod 1987). Knowledge was transmitted country-to-country by books, letters, newspapers, journals and later, the telegraph, as well as the direct transfer of knowledge between individuals engaged in research. The spatial framing of these networks was bounded by shipping routes, telegraph cables, and job postings. Although access to knowledge was often uneven, favouring ruler rather than ruled (Prakash 1999, pp. 84–5), scientific and technological developments in the British colonies owed much to ideas developed across the globe, as well as at home. Few studies, however, have considered the development of knowledge concerning hydrometeorological hazards and urban development (Ng et al. 2014). The exception would be studies of Indian history, where regional scholars have led the field (D'Souza 2006).

Thinking about flood management in cities and riparian zones underwent a shift during the nineteenth century. In British India and the Straits Settlements, the British saw water as a commodity, one that needed to be strictly managed for economic purposes. That a controlled environment was a productive environment had been

---

F. Williamson (✉)
Asia Research Institute, National University of Singapore,
AS8, #07-22, 10 Kent Ridge Crescent, 119260 Singapore, Singapore
e-mail: F.Williamson@uea.ac.uk

M.A. Miller et al. (eds.), *Crossing Borders*, DOI 10.1007/978-981-10-6126-4_3

core to British philosophy on imperial expansion since at least the seventeenth century (Irving 2008). Floods were sources of immediate disruption, but their cascading impacts also had a profound effect on contemporary society and economy. It was increasingly considered imperative to devise ways of mitigating the socio-economic impact of hydro-meteorological hazards.

During the nineteenth century, British flood mitigation schemes built on the technical developments then transforming riparian environments—canal building and irrigation, for instance—whilst denigrating forms of 'traditional' knowledge. British methods of flood control largely ignored indigenous methods of avoidance and early warning (Weil 2006, p. 6). Embankments were the first formal flood measure implemented under British rule; whilst certainly not original to Britain or unprecedented in Asia, their use stemmed directly from specific British experiences at home. Flood mitigation was, in large part, more concerned with economic efficiency rather than disaster reduction *per se*. Coterminous irrigation schemes, canal building, and river management often ran counter to disaster prevention, in fact: alterations to water courses, siltation, and poor urban water management actually created flash floods and contributed to burst river banks after heavy rains. As Weil (2006, p. 15) argues, the 'British arrived at an approach to flood control that required ever more massive and complex technological systems designed to exclude natural processes as much as possible'. These systems focused on physical barriers including embankments (and much later trees), diversions and drainage from natural watercourses, and even the shifting of people and settlements. For example, there were proposals to shift sections of Kuala Lumpur in the 1910s and 1920s but the plan never came to fruition. As British experience in Asia expanded, it became increasingly apparent that embankments had a negative impact on drainage systems, actually increased flood waters in many cases, and were costly to build and maintain. Sir Proby Cautley, the English-born engineer who made his name building the Ganges Canal in the 1840s, was later to write of the absolute necessity of avoiding 'intercepting the natural drainage of the country' when attempting major water works (Cautley 1864, p. 46).

It was their colonial experiences that introduced the British to drastically different sets of environmental challenges, whereby familiar strategies of environmental management known in Britain were proved woefully inadequate in the face of drastically different climates and diverse landscapes. This forced change, although its pace was often slow. Progressive development was observable only from the late 1840s 'as specialists trained in narrow reductionist science replaced an older generation of generalists … and engineering became the dominant mode for managing the environment…' (Weil 2006, p. 4).

This chapter considers the nineteenth-century history of serious floods in colonial Singapore and, to a lesser extent, Kuala Lumpur. It pays close attention to how the British authorities understood and reacted to serious inundations and adapted policy in accordance (or not). Cross-border governance shaped the development of both towns, ruled as they were through a combination of local municipal councils, the British government in India, and that in London. This was an arrangement that both hindered and advanced flood management, as will be seen later. It will consider how contemporaries thought about floods, from cause, to impact, to future mitigation. Moreover, it will contend that long-term context and precedent allows fresh perspectives on urban disaster today. We can learn from patterns of flood frequency, intensity, and geographic impact—and assess their correspondence to rainfall and/or El Niño/Southern Oscillation (ENSO), or human urban environmental impact. We can also learn from failures and successes in prevention methods used historically and determine not to repeat the same errors. This is important to advancing the recently identified 'de-centred, multi-scalar approach that is multidisciplinary and inclusive of many voices and forms of knowledge' in addressing disasters in todays complex anthropogenic world (Miller and Douglass 2016, p. 2). Based on primary archival sources relating to governance and urban development in the British Straits Settlements, including colonial records and contemporary newspapers, this research builds on several interrelated fields. Inspired by recent studies of entangled histories, global scientific networks, and Victorian meteorology (Anderson 2005; Laidlaw 2005; Lester 2001; McAleer 2011; Roberts 2009), it will remain firmly grounded within an historiographical framework, yet draw from the interdisciplinary influence of historical climatology and hydrology (Brázdil et al. 1999; Rohr 2005; Wetter et al. 2011), and the modern-day field of disaster analysis and prevention. The chapter will begin with an introductory discussion of contemporary weather and flood understanding, before moving on to look more closely at inundations in Singapore and Kuala Lumpur.

3.2 British Meteorology and Cross-Border Knowledge in Colonial Asia

Despite unsystematic governmental support (Bennett 2011), between 1800 and 1840 a globally diffuse network of people became engaged in meteorological studies (MacKeown 2011; Moore 2015). This achievement was the result of several factors: the importance of improving information for seafarers upon whom global imperial trade relied, a period of extreme drought and monsoon failures affecting late eighteenth century India, the connection made between health and disease in tropical environments, and a culture of scientific colonialism and improvement (Macleod 1987). The British were strongly influenced by the Baconian methodology of collating large quantities of international observational weather data, which can be seen in the magnetic experiments of the 1830s and 1840s, whereby vast quantities of simultaneous global datasets were made. Observatories at St Helena, Toronto, Tasmania, Singapore, and Madras established between the 1820s and the 1840s made this possible, alongside the pioneering efforts of individuals including Henry Piddington, William Reid, Francis Beaufort and Robert Fitzroy, amongst many others who pushed for consistency and expediency in data gathering (Adamson 2015; Moore 2015).

In the Asian colonies, where monsoon, tropical storms, and typhoons were unpredictable, dangerous, and potentially costly, developing nascent meteorology and a system of effective country-to-country communication of data were imperative. The new colonial ports, towns and plantations created a unique microcosm and rare opportunity for the study of meteorology. Rapid development had resulted in deforestation, changes in natural rainwater runoff and watercourses, and soil erosion, especially in and around urban areas. Contemporary observers connected this type of environmental impact with changes in localised climates and rainfall, and the frequency and severity of floods and drought. This came to be known as 'desiccation' and the theory gained momentum in France, Germany and Britain during the 1830s and 1840s (Grove 1995). The 'desiccationist school' became firmly established in scientific thought in the Straits Settlements. Key work in the field was undertaken by J. D. Hooker in St Helena and Ascension and by T. J. Newbold of the Madras Light Infantry in India (Skott 2012). Newbold was later posted to the Straits Settlements where he continued his research and publications. James Richardson Logan, publisher of the *Journal of the Indian Archipelago and Eastern Asia* and prominent Straits Settlements reformer, argued based on his own observations that forest denudation was having a serious negative impact on localized rainfall. He advocated protecting forests, especially those situated on higher slopes and mountains because of their important role 'in attracting and condensing clouds, diminishing local temperature, and increasing humidity' and praised Singapore's forest conservancy practices (Logan 1848, pp. 535–6). Interestingly, these practices were imported from colonial experience in India. In the 1880s, the Settlements' chief medical officer T. Irving Rowell likewise observed how wanton jungle desiccation had affected localized rainfall in Selangor (Rowell 1889).

The 1850s and 1860s engendered political and technological developments critical to advancing meteorological studies. Changes to colonial governance structures from 1854 enabled a subtle shift toward greater top-down investment in scientific research. The new Government of India Office (1857) created 'their own interlinked systems of scientific institutions, bureaucracies, experts, and cultures' building on extant scientific networks (Bennett 2011, p. 31). An emerging interest in forecasting (under the auspices of the first dedicated British government meteorological office in London, established in 1854), combined with the standardization of observations globally and faster communication of the same via the new underwater telegraph cable, transformed the scale of regional and global activities (Anderson 2005, pp. 1–2).

In Asia, cross-border communication in meteorological studies, especially in the field of prediction and forecasting, became the norm. The large, permanent observatories in China, Hong Kong, the Philippines and India acted as information hubs, storing and transmitting observations and issuing storm warnings. Newspapers and journals, including the *Canton Register, Journal of the Royal Asiatic Society*, and the *Hong Kong Government Gazette*, published weather reports and commentary (MacKeown 2011, pp. 22–3). Communications sped up after Singapore and Hong Kong were linked by telegraph in 1868 and the British Indian Submarine Telegraph Company and Extension Company laid underwater telegraph cables between Britain and India in 1870, extending to Penang and Singapore in 1871 (Huurdeman 2003, p. 136). These developments created a cross-border knowledge community linking British Asian colonies. Between the founding of Singapore in 1819 and the establishment of Kuala Lumpur as a British base in 1880, the study of the weather had become an integral part of colonial life.

3.3 Floods, Governance and Urban Development in Nineteenth-Century Singapore and Kuala Lumpur

The subject of borders in nineteenth-century Singapore is complex. Between 1819 and 1824, only the right to establish a factory for trade was ceded to the British under the authority of the Sultan of Johore and the Temenggong of Johore (a local. In 1824, a new treaty was signed granting the British the right to the land itself, with governance established under a town committee (and then municipal council) locally—which was, however, ultimately accountable to the British government in Bengal (Turnbull 2009). In 1826, Singapore became the administrative centre of the Straits Settlements and, in 1867, power was handed formally from India to the British government in London. Physically, Singapore is an island state, but the early settlement was small, with limited regulation over immigration or expansion. Its population was composed of migrants from across Southeast Asia, India and China, as well as a smaller number of Europeans. The notion of borders then is perhaps incongruous; it would be more accurate to consider early Singapore as a hub in a shifting landscape of people, ideas, and forms of governance.

Singapore's first dedicated Town Committee was appointed in 1822, and it was to them that responsibility for drainage, sanitation and flood mitigation fell (Dobbs 2003, pp. 21–4; Yeoh 2003, p. 71). Escaping the worst ravages of the monsoon and largely free of typhoon, nevertheless 'few days elapse[d] without the occurrence of rain' in Singapore (Penny Magazine 1842, p. 140). The heavy rainfall created many public health and urban development challenges. The colonial settlement was situated on the north bank of the Singapore River, amid swampland and mangrove forest. The south bank appeared to be largely covered at high tide and, at other times, to be muddy, boggy and full of creeks (Dobbs 2003, p. 20). This immediately created a social divide: wealthier inhabitants developed the higher land to the north where flooding was less likely, and poorer communities of migrants sprang up haphazardly in the south.

It was essential to the survival of the colony that swamplands were drained and made habitable, and that the regular tidal flooding was allayed. To achieve this, the local government mooted proposals for building an embankment to defend the southern riverside section of the town from the sea. Despite reluctance on the part of the British government in India who were asked to provide the money, the plan went ahead (Dobbs 2003, p. 27). Although the embankment went some way to providing protection from the sea, it did not prevent the regular flooding of the low-lying swampy land between the Rochor and Singapore Rivers. Despite the efforts of the town's engineers, the Bukit Timah Road and Rochor Canal area flooded time and again. Interestingly, this is still the most flood-prone area today despite extensive works by the modern Singaporean government (Singapore Public Utilities Board 2015).

The Town Committee had limited powers and inadequate resources for public works at its inception. Civic funds were raised by donation until the introduction of rates assessed on housing in 1826, and extraordinary expenditure on public works had to be approved by the British Government in India, an arduous and ponderous process (Makepeace et al. 1921). By the late 1830s, the British government was facing increasing pressure from the wealthier inhabitants of Singapore for autonomy in managing their own affairs. Singapore's inability to finance major public works was a particular bugbear. It is perhaps no coincidence that the inhabitants' crusade toward decentralisation of power peaked around 1839, the year of the heaviest rainfall then on record for Singapore. In 'twenty-six rainy days … twenty-eight inches of rain fell' on each and 'in one heavy thunderstorm, the rain gauge, which held only 2 ½ inches, overflowed in the course of one hour' (Buckley 1965, pp. 338, 738). The extraordinarily wet weather starkly revealed the inadequacies of the contemporary road, bridge and drainage system. Complaints that the British authorities were not sufficiently attending to water management were widespread across the Empire. Colonel George Chesney argued that 'until almost the termination of their existence, [they] did not recognize the prosecution of public works as a necessary part of their policy' (cited in Baber 1996, pp. 164–5). Certainly, the provision of water management improvements in Singapore was largely reactive, usually dealt with by the creation of temporary ad hoc committees. In 1855, a severe flood unambiguously highlighted inefficiencies in governance and the urgent need for public works improvements.

The last week of November 1855 saw 'an unusual amount of rain, the country in many parts being flooded and the roads out of town being almost impassable'. Then, from 7 a.m. on 30 November to 4 a.m. on 2 December 'it rained without intermission' (Straits Times 1855, p. 4). Main roads were awash with flood waters of up to 2 ft in depth, drains failed and damages to bridges and roads necessitated extensive recovery work. As one anonymous inhabitant anticipated in an open letter to the *Straits Times*, 'our Scavengers and Road-menders will find employment for at least a month in repairing the roads in and near town' adding somewhat sarcastically that, 'we trust, however, the Municipal Committee will avail of any labour offering to effect the repairs as promptly as possible' (*ibid.*). Repair works were as sluggish as the author predicted, impeded by the grinding wheels of administration. It was 14 December before the Resident Councillor wrote to the Secretary to the Governor of the Straits Settlements to request that convicts be made available to hasten repair works. 'I regret to say,' wrote Thomas Church, that

> 'the recent unprecedented heavy rains have caused most extensive damage to the whole of the public roads, as to render some of them impassable for carriages [and as] the means of the Municipal Council are limited, and there is some difficulty in engaging a large number of daily coolies, I therefore trust, where the comfort of the community generally is so

intimately concerned, His Honor [sic] the Governor will be pleased, as [a] special case, to allow the aid of Convicts for a limited period' (Straits Times 1856, p. 5).

The request was granted but offered only a temporary solution. The wider issue concerned 'the fact that some of the chief roads in the country having been made through swampy ground and filled in with fascines, the Committee are satisfied, at no remote period, it will be requisite to incur a large outlay in reconstructing some of the roads in question' (Straits Times 1856, p. 5). These circumstances added to inhabitants' list of complaints and the following year the British government finally acquiesced and permitted a Municipal Board to be established (Singapore Free Press and Mercantile Advertiser 1935, p. 16). This move was intended to make the job of directing and administering funds for public works more efficient. One of the Board's first orders was to raise the streets that regularly suffered flooding (Yeoh 2003, pp. 31–3; Buckley 1965, p. 626).

The years 1864–1891 have been considered an intense period of climatic change, characterised by rainier than normal conditions and more frequent flooding (Quinn and Neal 1995, p. 638). This was certainly the case in the *Straits Settlements*. Despite many improvements under the guidance of renowned engineer J. F. A. McNair, including raising Bras Basah, South Bridge Road and 'Middle Road from North Bridge Road to Beach Road', serious floods exposed the deficiencies of the urban infrastructure during extraordinary weather conditions (Straits Times 1885a, p. 3). The flood of 8 December 1884, for instance, was caused by a number of internal and external factors coalescing. Heavy rainfall, inadequate drainage, poorly maintained waterways, and the inability of the Municipal Commission to muster a quick response combined to create one of the worst disasters in inhabitants' memory (Municipal Fund of Singapore 1885). According to the *Straits* Times,

> Orchard Road from the Police Station was under water in some places to a depth of upwards of two feet … The road … between Dhobie Green and Mr Lambert's [and] the lower floors of the houses at the roadside … [were] under water. All the land bordering Orchard Road on the south side from Killiney Road to the entrance of Abbotsford had about 2 feet of water on it … and on the north side … from Scott's Road to and including part of Dhobie Green were flooded. (Straits Times 1885b, p. 3)

The list of damages was long and costly. Bridges, especially those still made from 'jungle wood' instead of brick, were damaged or washed away. Recently installed gas street lamps in Orchard, Killiney and River Valley Roads failed after the mains flooded. Paradoxically, the Municipal Commission were quick to point out the failings of the private lighting company—perhaps to detract from their own shortcomings—claiming 'that had the gas mains been in an efficient and proper state they could not have been affected by the rain'. They also demanded 'that the necessary reduction must therefore be made in the gas bill for the month of December' (Municipal Fund of Singapore 1885).

Interestingly, the Municipal Engineer's correspondence with the Municipal Commission reveals how, in his opinion, continued inattention to essential public works were more of a contributing factor to the flood than heavy rainfall. This is borne out by the meteorological records, which show that Singapore did not in fact have record rainfall that year (Jarman 1988, p. 212). Ongoing improvement works had been protracted and, on occasion, suspended entirely due to the lack of construction materials (Straits Times 1884, p. 3). The Municipal Engineer, James MacRitchie, noted how his previous reports highlighting key issues had been ignored:

> I had opportunities of observing the cause of this flooding, and my observations confirm me in the opinion I formerly gave, that the principal in fact the sole cause was the insufficient openings in the various Bridges over the Stamford Canal to allow flood waters to escape with ease and rapidity. Institution Bridge … was some years ago widened by 12 feet. The addition consisted of walls in which were built two frames to carry sluice gates … the sluice gates were found to be an utter failure and were removed some time ago, but the obstructions caused by the stone work which carried them is so great, that during the last rain storm the water was backed to a depth of 4 feet 6 inches … The Bridge at the junction of Bencoolen Street and the built in channel beyond it, are also much too narrow and cause serious backing up … during the last flood the sectional area of water opposite Dhobie Green was 360 sq feet, and that passing through the bridge only 70 sq feet (Straits Times 1885b, p. 3).

He also remarked that the various culverts and bridges leading to private properties off Orchard Road, Abbotsford and the adjacent Chinese Cemetery, Paterson Road and Scott's Road were not fit for purpose. In his opinion, the culverts should be widened and reinforced and works undertaken to implement grading (shaping and sloping to direct surface drainage) in all flood-prone areas, and notice must be served on proprietors and homeowners who had culverts or bridges obstructing the free flow of rainwater. He also mooted the need for a comprehensive set of rules for managing severe weather. Where sluice walls were used and in good repair, for instance, there were no guidelines to enforce that they be opened during heavy rains. The worst culprit in 1884 was, ironically, the Public Works Department, whose conduit sluice gates had remained closed throughout the inundation (Straits Times 1885b, p. 3).

The records also reveal the cascading impacts of the inundation. Interruptions to water supplies caused by the 'the sinking of the pipes from the ground becoming saturated', blocked storm water drains, the pollution of domestic water supplies and the time the municipal authorities took to make good after the inundation created a serious set of public health issues. Exactly the impact of the flood on the health of Singapore's inhabitants is hard to ascertain. The official monthly statement as to the town's health considered it 'satisfactory' with no epidemics to report (Municipal Fund of Singapore 1885, f. 941). However, Singapore was already notorious for the general state of its water management and drainage, where the river, canals and drains were continually blocked by weeds and refuse. During heavy rains, dirty water would seep into houses and pool on the streets (Manderson 1996, p. 101). Indeed, the poor state of the Singapore River likely contributed to the severity of the 1884 flood. In 1877, a special Commission had been appointed to assist McNair in investigating the herculean task of clearing and deepening the river. No less a man than Sir John Coode had been tasked to report on the state of the river, to which he responded that the 'refuse and filth' dumped from the boats and the quay were the chief reason behind the river's shoaling (Municipal Fund of Singapore 1884, f. 933).

On 8 December 1884—somewhat ironically the day of the flood—the matter had once again been raised before the Municipal Commission with questions raised as to why progress was so slow, despite the purchase of a steam-dredger earlier that year. Post-inundation, the clean-up was delayed by confusion over who was to be appointed the 'successful contractor for the next financial year' (Dobbs 2003, p. 55). The consequence was that 'piles of liquid mud that have recently been scraped off the surface of the roads … [were] then left to spread about in front of houses and godowns'. Inhabitants, exasperated as to why the municipal authorities 'do not come along and take the stuff away instead of leaving it for days', complained vociferously to the local papers (Straits Times 1885a, p. 2). Even if the causes of disease were not fully understood at this time (miasmatic theory was still prevalent), it was known that poor drainage and flood waters led to insanitary conditions (Cautley 1864, pp. 83, 97–8). Urban planning and public health were dealt with in unison in most nineteenth-century British towns and colonies (Corburn 2009; Frumkin et al. 2004), and Singapore was no exception. The Town Committee and the Municipal Board were responsible for public health until the Municipal Ordinance of 1887 established a separate Municipal Health Department (Makepeace et al. 1921, p. 321).

The year 1891 also saw serious flooding in Singapore. Heavy rains across the afternoon and night of 27 April 'brought down quite a flood of water from the Bukit Timah and Thompson Road districts, so as to completely inundate some of the low lying land in the vicinity'. The problem was exacerbated by the inadequate size of 'the open drains in which the water flows along to the sea … and as a consequence the roads … were under water'. The worst-affected area was Bukit Timah Road to Kampong Java Road and 'all the land in the vicinity … the greater portion of the racecourse and the adjoining thoroughfares' (Straits Times Weekly Issue 1891, p. 5). The racecourse was reclaimed paddy and much of the land affected abutted the Rochor River (now the Rochor Canal). Central Singapore from Thomson Road to Bukit Timah and Orchard Road—the flood plain created by the Rochor and Singapore River—was to suffer two more major inundations that decade, in 1892 and 1899.

The 1892 inundation occurred on Sunday 29 May after 9.25 in. of rain fell between 7 a.m. and 11 a.m. On Orchard Road, the flood water was measured at 2 ft 6 in. deep at the gates of Government House and, whilst one 'gentleman had swum down Orchard Road with a 3 foot rule to gauge the depths', another 'canoed from Tanglin to the sea' just for the sport (Buckley 1965, p. 738). Fun aside, the speed of the rising flood waters 'caused a good deal of damage at the Impounding Reservoir Works at Thompson Road', broke bridges and burst through embankments (Singapore Free Press and Mercantile Advertiser 1892b, p. 2.). Such a 'phenomenally heavy rainstorm' had never been known within the memory of the oldest residents. Rainfall of almost 8.5 in. was noted at Kandang Kerbau Hospital registering station, nearly 1.5 in. more than the record-breaking 7.0 in. measured at the Sepoy Lines on 8 December 1884. The 1884 inundation had 'previously [been] the heaviest known rainfall since 1869, in fact since observations were first taken at Singapore' (Singapore Free Press and Mercantile Advertiser 1892a, p. 2).

The cascading impact was significant. Many local businesses suffered huge losses (Singapore Free Press and Mercantile Advertiser 1892c, p. 3). An interesting account survives of Fillis' Circus: not only was the business forced to close for 4 days, but the company had to spend $1,000 to 'keep the place from becoming a swamp'. The latter point was also a public health issue, a fact that this anonymous letter writer was keen to point out to the press (Daily Advertiser 1892, p. 3). The close connection between public health and public works in the 1890s was made plain by the author, who asked why the Municipal Health Officer and Engineer were not made available to assist circus staff. To further argue this point, it was the Municipal Engineer who, in June 1892, requested an inquiry into the high death rate (602 deaths) and the location of the same for the month immediately following the flood (Municipal Fund of Singapore 1892, f. 1884).

In Kuala Lumpur, urban problems including social inequality, lack of political or economic agency for non-European races, and inadequate housing regulation were starkly highlighted during floods. The town had developed some decades later than Singapore, emerging out of an 1850s mining frontier. Yap Ah Loy, a powerful Chinese clan leader and astute businessman, had de facto ruled the prospecting town from 1868, with power devolving to a British Resident in 1874. The town became the state capital of Selangor in 1880 (Gullick 1994, pp. v–vii). Yap Ah Loy had invested time and money into building projects, but a civil war followed by a fire and flood had limited urban development. A majority of houses had sprung up without regulation, and the cheap local materials used proved both inefficient and dangerous.

On 21–22 December 1881, D. W. Daly, then superintendent for Public Works in Selangor, was eyewitness to the rains that lashed the city for 30 h without respite. He described in great detail how the 'flood started to rise … around noon on the 21st and the bridge over the Klang River was … destroyed … The waters then rose into the shops and, mingling with the earthen walls, undermined them and the houses fell in every direction' (Public Works Department 1881). The flood was the worst recorded that century. Strong currents coursed across the cricket ground, which became submerged under 3–4 m of water, and seagoing vessels took up new moorings 'in the streets around the market place' (*ibid.*; Straits Times Overland Journal 1881, p. 7). In the aftermath, it was discovered that the flood waters had destroyed 92 mud houses and many more made of bamboo, planks and attap (Straits Times Overland Journal 1881, p. 7). The suddenness of the inundation—the water level rising by 3 ft in 1 h—resulted in the deaths of six people, of whom only three bodies were recovered by the police. The flood revealed the serious infrastructural problems of the least developed and most impoverished areas. As is so often the case in Southeast Asian cities, then as now, the poorest inhabitants tended towards the riverside in overcrowded, poor quality housing. The cascading impact of floods on the poor was addressed only by providing emergency assistance for selected individuals, as communities were expected to provide for their own. The British Resident did call for urgent improvement works, however. It was proposed that the government 'buy the site of the market place and farms from Captain China', for houses to be rebuilt with bricks and tiles and, where appropriate, to be raised on piles

(Superintendent of Works 1881a). Nevertheless, these methods did little to alleviate the real source of the problem as the homes of the poor were not included in public works budgets. It could also be argued that there was self-interest in the land purchase, as it coincided with a decision to consolidate British power in the town. The purchase of central, important land was a direct physical symbol of British aspirations. Indeed, even into the 1910s and 1920s, disagreements over how best to develop the city in combination with a lack of money meant that urban inequalities and flood mitigation schemes had not yet been sufficiently tackled. Indeed, the issue was considered so severe by this point in time that plans including moving parts of the town and altering the course of the river were being debated (State Engineer 1915).

3.4 Local and Regional: Governance and Desiccationist Theory

The building of roads and industrial works and the clunky processes of colonial administration had hindered, rather than advanced, progress in flood mitigation. There was a lack of immediate central governmental responsiveness or involvement in post-inundation operations, stemming from the sudden nature of inundations and the amount of time necessary for communications between London, India, and Singapore, especially before the telegraph. News sent by ship to and from India or London might take months to arrive, and months again for a response, thus leaving large gaps in communication (Bell et al. 1995; Godlewska and Smith 1994; Lester 2001). The repeated calls by the Straits Settlements elite over the course of the nineteenth century for more autonomy starkly reveal that the Legislative and Municipal Councils felt the distant overseas government had little inclination to micro-manage in the Southeast Asian colonies. At the same time, although irrigation and river management were considered important economically, flood control was given lower priority (Chaturvedi 2012, p. 49). However, a number of acts designed for flood management were sanctioned prior to the twentieth century, many of which stemmed from British experiences overseas. In India, successive water and drainage acts of 1873 and 1876 had helped clear up some problems of flood mitigation resulting from the overlapping and complex jurisdiction of state versus private water control (Peissker 2013, p. 37). It was also India that provided the British with their most extensive experience of water management. The mammoth task of irrigation and the redistribution of water, combined with mitigating inundations during monsoon, preoccupied the colonial engineers in that vast country. The process drew on a global discourse of water engineering based on the challenges of imperial (yet also indigenous) authorities around the world. In India, the effort to transform arid wastelands into productive areas of cultivation was a driver in the expansion of the 'hydraulic environment' but the lessons learned—including the associated issues of siltation, waterlogging, increased water tables and the spread of malarial mosquito breeding grounds—were to be applied across the Empire (Gilmartin 1995).

Cross-border knowledge tended to move with the colonial engineers. British India was a world leader in water management and river engineering. People like Sir Richard Strachey and Sir Arthur Cotton led the field in river, irrigation and flood management, taking responsibility for major canal and dam projects in British India, writing treatises and offering advice even after retirement. River engineers were sought after across Britain's empire in the east after a stint of working in India. Hugh McKinney, for example, was an irrigation engineer who had spent 10 years in India before being hired in New South Wales as Director of the Water Conservation Commission, and the retired British Indian Colonel Frederick Home was consulted on plans for irrigation and river navigation for the same region (O'Gorman 2012, pp. 120, 124). Sir John Coode, the renowned harbour engineer, worked on schemes in England, Table Bay, Colombo, Melbourne and New South Wales. He was also part of the Suez Canal Commission. What is less well known is that he also worked and advised on the coastal land reclamation schemes undertaken in 1880s Penang and Singapore, the latter at Telok Ayer, as well as the Singapore River problems noted earlier. He also made recommendations for the new pier constructed at Singapore in 1888 which, as Cecil Clementi Smith noted, was an essential public work (Crawford 1969).

Reports written directly after the 1881 Kuala Lumpur floods, and in the two decades beyond, demonstrate thinking inspired by the desiccationist school of thought. Urban development, jungle clearances, and river siltation caused by mining and deforestation operations, as well as human neglect, were considered serious failings. Although exceptional rainfall was noted, it was not given as the chief reason behind the extraordinary late nineteenth-century inundations. After the 1881 Kuala Lumpur flood, for example, D. W. Daly had noted that there was little to be done to combat the threat of future flooding unless deforestation was addressed (Superintendent of Works 1881b) and, in 1893, floods in Serendah, Malaysia, were ascribed to 'the enormous traffic, chiefly railway materials, daily passing' (Public Works Department 1893). A few years later, H. F. Bellamy, the retired Executive Engineer for the Kuala Lumpur Public Works Department, wrote a report on the floods that he had experienced during his working life. These were, in his opinion, caused by a combination of 'abnormal rainfall; the clearing of the jungle in the watershed of the Klang River and its tributaries above Kuala Lumpur; and the prevention of the free flow of the river below Kuala Lumpur, by silting up and fallen trees'. He concluded that he could see no evidence that the rainfall in the watershed had increased; rather, he believed that annual rainfall had actually decreased as a 'natural result of the removal of trees' in the area (Bellamy 1912, pp. 1–2). The Selangor State Engineer corroborated, stating that since the 'commencement of mining on a large scale in the valleys above [Kuala Kubu]' there had been 'a rapid tendency towards silting and, after heavy rains and landslips … the silt deposited from upstream became more rapid and pronounced [until] eventually the original channel of the river disappeared completely' (State Engineer 1915).

3.5 Conclusion: Learning from History?

In all the flood disasters to have affected Singapore and Kuala Lumpur during the nineteenth century, it can clearly be seen that human, rather than natural, exigencies exacerbated their severity. The municipal engineers of Kuala Lumpur and Singapore noted the urgent need to improve drainage and employ flood prevention schemes. Knowledge circulated across borders from imperial experiences that preventative work—including the use of good quality building materials, the widening and improvement of drainage systems, grading and raising roads, understanding the location of flood-prone areas, and governmental intervention in private property—were all methods of flood mitigation. The reality was different. The system of cross-border governance prevented significant changes from taking place because of disinterest caused by distance, limited funding, and a cultural reluctance for governmental interference in urban development at home, or abroad. This latter point was especially crucial. Despite late nineteenth-century Britain's reputation for great public building schemes, the practice of urban investment in the same was somewhat different. Most urban improvement schemes were funded from local rates, the spending of which on public works was deeply unpopular (Haggard 2001, p. 162). Even in London, despite several high-profile pilot schemes overseen by the government, it was 1855 before a Metropolitan Board of Works was established to coordinate sanitation and infrastructure in the capital (this would become the London County Council in 1888). Policy in the colonies frequently derived from precedent at home. The Straits Settlements thus relied on local rates or private donation to fund improvement schemes, and the result was limited investment in flood mitigation. This situation was only to change into the twentieth century after previously colonial regions transformed into nation-states.

Disasters always happen in political spaces. In Southeast Asian cities, political, social and economic inequalities abound. The complexity of such societies and political realities necessitates a complex disaster governance regime (Miller and Douglass 2016, p. 2). The description of a modern city, characterised by rapid urbanisation, a rural to urban population shift, high migrant population, severe infrastructural inadequacies and substantial social inequality, is also an apt description of a colonial city. In which case, can we not learn from the experiences of the past? In colonial Singapore or Kuala Lumpur, floods affected the urban infrastructure for days or weeks at a time. Road and rail closures disrupted travel, commerce, and communications. This had a direct impact on trade across Asia and beyond. In urban areas, inadequate drainage facilitated the spread of disease. Social inequality was also starkly highlighted. In colonial cities the poor were the most affected by disaster. The poor and unprotected migrant communities clustered in the low-lying riverside areas most prone to flooding. Their housing was the least durable and their livings the most precarious. A flood could reduce a family to ruin overnight. The same issues are true of many modern Asian cities. Today's urban disasters have the most impact on those people living in riparian zones, who often encompass some of the poorest inhabitants. It is these people who are at the least resilient and have inadequate access to public services to mitigate risk. There are also parallels in governance structures in colonial cities and cities in Asia's developing countries, in particular a lack of decentralised government and 'bottom-up' agency. Although we should be wary of drawing too strong a comparison, enough similarities exist to warrant an investigation into how past governments managed disasters, even if only to understand what tactics to avoid. Narrative accounts of disaster provide information relevant to coping strategies from which we can deduce effective or ineffective mechanisms for recovery. This is essential in improving the social learning process today. Certainly, more contemporary forms of coping strategy, such as population relocation, had been mooted over a hundred years previously, and abandoned. Today, such strategies are also being found inadequate.

Historic studies likewise add to our knowledge of long-term flood frequency and intensity. Historical archival records of flood data and rainfall are increasingly used in probabilistic prediction models used by meteorologists, climatologists, and reinsurance industries (Williamson et al. 2015; Compo et al. 2011). Documentary evidence can be usefully employed alongside paleohydrological analyses, providing additional context. Indeed, there is increasing acknowledgement of their contemporary relevance and efficacy (Heymann 2010). The point is succinctly made by J. R. McNeill, who argues that 'the more one unpacks the concept of climate change into its components, the more the record of the past becomes relevant to imagining the future' (2008, p. 45). Indeed, as Mark Carey writes, the 'record of the past is particularly important for understanding the broader historical processes that have led to anthropogenic climate change and created the unequal geographies of vulnerability that exist in the world today' (2012, p. 233).

This chapter does not seek to offer more than an introduction to the history of floods in these cities, and a background to the knowledge that underpinned colonial flood management. It serves as a preliminary introduction to research in the field that has the potential to move in many new directions. Using historical records advances a multi-disciplinary approach to disasters that incorporates historical analysis as one element, rather than placing knowledge into silos. There remains a great deal that we can learn from history.

References

Adamson, G. (2015). Colonial private diaries and their potential for reconstructing historical climate in Bombay 1799–1828. In V. Damodaran, A. Winterbottom, & A. Lester (Eds.), *The East India Company and the natural world* (pp. 102–127). Basingstoke: Palgrave Macmillan.

Anderson, K. (2005). *Predicting the weather: Victorians and the science of meteorology*. Chicago: University of Chicago Press.

Baber, Z. (1996). *The science of empire: Scientific knowledge, civilization, and colonial rule in India*. Albany: State University of New York Press.

Bell, M., Butlin, R., & Heffernan, M. (1995). *Geography and imperialism 1820–1940*. Manchester: Manchester University Press.

Bellamy, H. F. (1912). *Remarks on the flooding of KL in December 1911, 23 January, 1358/1912.* Kuala Lumpur: National Archives of Malaysia.

Bennett, B. M. (2011). The consolidation and reconfiguration of 'British' networks of science, 1800–1970. In B. M. Bennett & J. M. Hodge (Eds.), *Science and empire: Knowledge and networks of science across the British Empire, 1800–1970* (pp. 29–43). Basingstoke: Palgrave Macmillan.

Brázdil, R., Glaser, R., Pfister, C., Dobrovolny, J. M. A., Barriendos, M., Camuffo, D., Deutsh, M., Enzi, S., Guidoboni, E., Kotyza, O., & Sanchez Rodrigo, F. (1999). Flood events of selected European rivers in the sixteenth century. *Climatic Change, 43*(1), 239–285.

Buckley, C. B. (1965). *An anecdotal history of old times in Singapore, 1819–1867.* Kuala Lumpur: University of Malaysia Press.

Carey, M. (2012). Climate and history: A critical review of historical climatology and climate change historiography. *WIREs Climate Change, 3*, 233–249.

Cautley, P. (1864). *Ganges Canal: A disquisition on the heads of the Ganges of Jumna Canals, North-Western Provinces.* London.

Chaturvedi, M. C. (2012). *India's waters: Advances in development and management.* Boca Raton: CRC Press.

Compo, G. P., Whitaker, J. S., Sardeshmukh, P. D., Matsui, N., Allan, R. J., Yin, X., Gleason, B. E., Vose, R. S., Rutledge, G., Bessemoulin, P., Brönnimann, S., Brunet, M., Crouthamel, R. I., Grant, A. N., Groisman, P. Y., Jones, P. D., Kruk, M. C., Kruger, A. C., Marshall, G. J., Maugeri, M., Mok, H. Y., Nordli, Ø., Ross, T. F., Trigo, R. M., Wang, X. L., Woodruff, S. D., & Worley, S. J. (2011). The 20th century reanalysis project. *Quarterly Journal of the Royal Meteorological Society, 137*, 1–28.

Corburn, J. (2009). *Towards the healthy city.* Cambridge, MA: MIT Press.

Crawford, D.R. (1969). Coode, Sir John (1816–1892). In *Australian dictionary of biography.* National Centre of Biography, Australian National University, viewed 10 April 2016, http://adb.anu.edu.au/biography/coode-sir-john-3250/text4915.

D'Souza, R. (2006). Water in British India: The making of a "colonial hydrology". *History Compass, 4*(4), 621–628.

Daily Advertiser (1892). 3 June, p. 3, NewspapersSG, http://eresources.nlb.gov.sg/newspapers/Default.aspx.

Dobbs, S. (2003). *The Singapore river: A social history, 1819–2002.* Singapore: National University of Singapore Press.

Frumkin, H., Frank, L., & Jackson, R. (2004). *Urban sprawl and public health.* Washington, DC: Island Press.

Gilmartin, D. (1995). Models of the hydraulic environment: Colonial irrigation, state power and community in the Indus Basin. In D. Arnold & R. Guha (Eds.), *Nature, culture, imperialism* (pp. 210–236). Delhi: Oxford University Press.

Godlewska, A., & Smith, N. (1994). *Geography and empire.* Oxford: Blackwell.

Grove, R. (1995). *Green imperialism: Colonial expansion, tropical island Edens and the origins of environmentalism, 1600–1860.* Cambridge: Cambridge University Press.

Gullick, J. M. (1994). *Old Kuala Lumpur.* Oxford: Oxford University Press.

Haggard, R. F. (2001). *The persistence of Victorian liberalism: The politics of social reform in Britain, 1870–1900.* Westport: Greenwood Press.

Heymann, M. (2010). The evolution of climate ideas and knowledge. *Wiley Interdisciplinary Review: Climate Change, 1*, 581–597.

Huurdeman, A. A. (2003). *The worldwide history of telecommunications.* Hoboken: Wiley.

Irving, S. (2008). *Natural science and the origins of the British Empire.* London: Pickering and Chatto.

Jarman, R. L. (1988). *Annual reports of the straits settlements 1884–1891* (Vol. 3). Bath: Archive Editions.

Laidlaw, Z. (2005). *Colonial connections, 1815–45: Patronage, the information revolution and colonial government.* Manchester: Manchester University Press.

Lester, A. (2001). *Imperial networks: Creating identities in nineteenth-century South Africa and Britain*. London: Routledge.

Logan, J. R. (1848). The probable effects on the climate of Pinang of the continued destruction of its hill jungles. *Journal of the Indian Archipelago and Eastern Asia, 2*, 534–536.

MacKeown, P. K. (2011). *Early China coast meteorology: The role of Hong Kong*. Hong Kong: Hong Kong University Press.

Macleod, R. M. (1987). On visiting the moving metropolis: Reflections on the architecture of imperial science. In N. Reingold & M. Rothenberg (Eds.), *Scientific colonialism: A cross-cultural comparison* (pp. 217–249). Washington, DC: Smithsonian Institution Press.

Makepeace, W. (1921). In G. E. Brooke & R. S. Braddell (Eds.), *One hundred years of Singapore* (Vol. I). London: John Murray. 1991 reprint.

Manderson, L. (1996). *Sickness and the state: Health and illness in colonial Malaya, 1870–1940*. Cambridge: Cambridge University Press.

McAleer, J. (2011). Stargazers at the world's end: Telescopes, observatories, and 'views' of Empire in the nineteenth century British Empire. *British Society for the History of Science, 46*(3), 389–413.

McNeill, J. R. (2008). Can history help us with global warming? In K. M. Campbell (Ed.), *Climatic cataclysm: The foreign policy and national security implications of climate change* (pp. 26–48). Washington, DC: Brookings Institution Press.

Miller, M. A., & Douglass, M. (Eds.). (2016). *Disaster governance in urbanising Asia*. Singapore: Springer.

Moore, P. (2015). *The weather experiment: The pioneers who sought to see the future*. London: Chatto and Windus.

Municipal Fund of Singapore. (1884). *Minute book of the Municipal Fund of Singapore (MFS), 8 and 23 December, NA425, f. 933*. Singapore: National Archives of Singapore.

Municipal Fund of Singapore. (1885). *Minute book of the Municipal Fund of Singapore (MFS), 9 and 18 January, NA425, f. 941*. Singapore: National Archives of Singapore.

Municipal Fund of Singapore. (1892). *Minute book of the Municipal Fund of Singapore (MFS), 24 June, NA426, f. 1884*. Singapore: National Archives of Singapore.

Ng, S., Wood, S. H., & Ziegler, A. D. (2014). Ancient floods, modern hazards: The Ping River, paleofloods and the 'lost city' of Wiang Kum Kam. *Natural Hazards*. doi:10.1007/s11069-014-1426-7.

O'Gorman, E. (2012). *Flood country: An environmental history of the Murray-Darling basin*. Collingwood: CSIRO Publishing.

Peissker, T. (2013). *The governance of climate change adaption in developing countries*. Hamburg: Anchor Academic Publishing.

Penny Magazine. (1842). *Singapore, D2009120149, 9 April* (p. 140). Singapore: National Archives of Singapore.

Prakash, G. (1999). *Another reason: Science and the imagination of modern India*. Princeton: Princeton University Press.

Public Works Department. (1881). *'A disastrous flood at KL on the 21 December 1881', PWD 566/81, 28 December*. Kuala Lumpur: National Archives of Malaysia.

Public Works Department. (1893). *Letter from the district Engineer Ulu Selangor relative to the late floods, PWD 4184/1893, 12 July*. Kuala Lumpur: National Archives of Malaysia.

Quinn, W. H., & Neal, V. T. (1995). The historical record of El Niño events. In R. S. Bradley & P. D. Jones (Eds.), *Climate since A.D. 1500* (2nd ed., pp. 623–648). London: Routledge.

Roberts, L. (2009). Situating science in global history: Local exchanges and networks of circulation. *Itinerario, 33*, 19–30.

Rohr, C. (2005). The Danube floods and their human response (14th to 17th c.) *History of Meteorology, 2*, 71–86.

Rowell, T. I. (1889). Meteorological report for the year 1885. *Journal of the Malaysian Branch of the Royal Asiatic Society, 16*, 385–412.

Singapore Free Press and Mercantile Advertiser (1892a). 30 May, p. 2, NewspapersSG, http://eresources.nlb.gov.sg/newspapers/Default.aspx.

Singapore Free Press and Mercantile Advertiser (1892b). 2 June, p. 2, NewspapersSG, http://eresources.nlb.gov.sg/newspapers/Default.aspx.

Singapore Free Press and Mercantile Advertiser (1892c). 11 June, p. 3, NewspapersSG, http://eresources.nlb.gov.sg/newspapers/Default.aspx.

Singapore Free Press and Mercantile Advertiser (1935). 8 October, p. 16, NewspapersSG, http://eresources.nlb.gov.sg/newspapers/Default.aspx.

Singapore Public Utilities Board (2015). *Managing flash floods: Frequently asked questions*, viewed 10 May 2016, http://www.pub.gov.sg/managingflashfloods/faqs/Pages/default.aspx.

Skott, C. (2012). Climate, ecology and cultivation in early Penang. *Proceedings of the PIO Conference.* Malaysia (pp. 99–110).

State Engineer. (1915). *Memorandum regarding action taken to combat the flooding of Kuala Kubu and to safe guard private properties since the flooding of December 1911, 12 October, 5074/1915.* Kuala Lumpur: National Archives of Malaysia.

Straits Times (1855). 4 December 1855, p. 4, NewspapersSG, http://eresources.nlb.gov.sg/newspapers/Default.aspx.

Straits Times (1856). 8 January, p. 5, NewspapersSG, http://eresources.nlb.gov.sg/newspapers/Default.aspx.

Straits Times (1884). 17 January, p. 3, NewspapersSG, http://eresources.nlb.gov.sg/newspapers/Default.aspx.

Straits Times (1885a). 7 January, p. 2, NewspapersSG, http://eresources.nlb.gov.sg/newspapers/Default.aspx.

Straits Times (1885b). 21 January, p. 3, NewspapersSG, http://eresources.nlb.gov.sg/newspapers/Default.aspx.

Straits Times Overland Journal (1881). 31 December, p. 7, NewspapersSG, http://eresources.nlb.gov.sg/newspapers/Default.aspx.

Straits Times Weekly Issue (1891) 28 April, p. 5, NewspapersSG, http://eresources.nlb.gov.sg/newspapers/Default.aspx.

Superintendent of Works. (1881a). *Report with plan from Supt. of Works regarding the recent 1881 flood at KL, 50/82, 19 December*. Kuala Lumpur: National Archives of Malaysia.

Superintendent of Works. (1881b). *Report with plan from the Superintendent of Works regarding the recent 1881 flood at KL, 50/82, 29 December*. Kuala Lumpur: National Archives of Malaysia.

Turnbull, C. M. (2009). *A history of modern Singapore, 1819–2005*. Singapore: National University of Singapore Press.

Weil, B. (2006). The rivers come: Colonial flood control and knowledge systems in the Indus Basin, 1840s–1930s. *Environment and History, 12*, 3–29.

Wetter, O., Pfister, C., Weingartner, R., Luterbacher, J., Reist, T., & Trösch, J. (2011). The largest floods in the High Rhine basin since 1268 assessed from documentary and instrumental evidence. *Hydrological Sciences Journal, 56*(5), 733–758.

Williamson, F., Allan, R., Switzer, A. D., Chan, J. C. L., Wasson, R. J., D'Arrigo, R., & Gartner, R. (2015). New directions in hydro-climatic histories: Observational data recovery, proxy records and the Atmospheric Circulation Reconstructions over the Earth (ACRE) Initiative in Southeast Asia. *Geoscience Letters, 2*(2), 1–12.

Yeoh, B. S. A. (2003). *Contesting space in colonial Singapore: Power relations and the urban built environment*. Singapore: National University of Singapore Press.

Chapter 4
Governing Cross-Border Effects of Disasters in Urbanising Asia: What Do We Know?

Matthias Garschagen

4.1 Introduction

Most people working in the field of disaster risk reduction would probably agree that cross-border effects of disasters are of great importance when wanting to understand disasters, their impacts and the responses to them. A number of recent examples such as the 2011 floods in Bangkok or the Tohoku earthquake and tsunami served as a strong reminder that disasters and their impacts are often not contained by national or other jurisdictional boundaries. Firstly, bio-physical hazards such as floods, heat waves, tsunamis or earthquakes show little respect for such boundaries and frequently transcend them. Secondly, even if floods, earthquakes or other hazards strike only within a particularly country, they in most cases lead to impacts that are felt far beyond the boundaries of that particular country. This is due to the strong regional and global integration of economies, which are today linked through a complex web of trade, production, migration, travel, information exchange and many other factors. Such linkages materialise particularly in the nodal points of these integrated systems, i.e. in cities which function as hubs of trade, production, mobility, information exchange and decision-making.

In fact, it seems likely that the potential for cross-border disaster and/or disaster which at least has strong cross-border effects will continue to rise over the next years and decades. It is particularly driven by the confluence of three megatrends. First, there is a continuing, probably even intensifying, regional integration and globalisation of economic production, trade, human mobility, infrastructure dependency and information flows. The Asia-Pacific is a particularly dynamic region in this respect, as indicated, for instance, by the current negotiations on a new Trans-Pacific Free Trade Agreement. Secondly, there is a strong ongoing urbanisation, particularly in Asia, leading to a further concentration of services and economic

M. Garschagen (✉)
United Nations University – Institute for Environment and Human Security (UNU-EHS),
Bonn, Nordrhein-Westfalen, Germany
e-mail: garschagen@ehs.unu.edu

M.A. Miller et al. (eds.), *Crossing Borders*, DOI 10.1007/978-981-10-6126-4_4

activity—in effect making the above-indicated nodal points even thicker and more important, yet also more prone to causing wider system disturbance. Thirdly, disaster risk is driven by environmental change and particularly climate change, which is expected to increase the frequency, intensity and reach of many of the existing natural hazards, particularly in the Asia-Pacific Region which is prone to hydro-meteorological hazards including typhoons, floods and droughts (Intergovernmental Panel on Climate Change [IPCC] 2012).

As a result of these convoluting risk factors and the increasingly connected disaster risk, there will be an equally increased need for improving transboundary disaster risk governance.[1] There will be a need not only for transboundary disaster response, but also for reducing risk in a preventive manner and reducing the vulnerabilities within integrated cross-border systems of social and economic activity.

Neither the acknowledgement of cross-border disaster impacts nor the recognition of transboundary risk governance is entirely new. However, the chain of recent disasters in the Asia-Pacific region as well as the outlook of increased cross-boundary disaster risk call for a thorough stock-taking of the engagement with cross-boundary disasters that can be observed hitherto. The paper therefore aims at reviewing and analysing the state of knowledge on cross-boundary effects of disasters and their governance. The paper will focus particularly on Asia and the Pacific and will pay special attention to the role of urbanisation.

Four fields of information will be covered. Firstly, peer-reviewed scientific publications on the topic will be reviewed, contributing the main pillar of the analysis. Secondly, grey literature is analysed, mainly covering reports by international organisations such as UN-ESCAP as well as NGOs and other civil society organisations. Thirdly, the main global providers of secondary statistical data on disasters are considered and analysed for their relevance and capacity to inform about cross-boundary disasters. Fourthly, key international agreements in the post-2015 agenda are examined for their provisions on cross-border disaster risk governance. The detailed steps but also limits of this methodology are discussed in the individual chapters below. Within the boundaries of this analysis, the paper concludes by synthesising the key achievements that have been made to date but also develops a list of pressing knowledge gaps and research needs.

---

[1] The terms 'risk management' and 'risk governance' are both used in the literature depending on the context and focus. While 'risk management' typically implies a notion of the technocratic manageability of risk—often closely related to state organisations and a top-down approach to administration—the term 'risk governance' puts more emphasis on the fact that risk is, and needs to be, negotiated and mediated between different political and social actors within a society, often with very different perspectives and priorities for risk reduction. This chapter uses both terms, depending on the context. Even though the author very much agrees with the emphasis and tenets within the risk governance debate, he uses the term 'risk management' here when referring to literature or policy concepts that use it.

4.2 Analysis

4.2.1 Peer-Reviewed Scientific Literature

In order to analyse scientific publications on cross-border disaster impacts and their governance, the two main databases for peer-reviewed literature—i.e. ISI Web of Science and Scopus—have been used as an entry point. Given that Scopus has a wider coverage than ISI Web of Science (also covering major book projects and conference proceedings), only the results of the Scopus searches are explored here. Strategic keyword searches were applied, targeting the title, abstract and keywords of the publications. Three main combinations have been explored: (1) 'disaster' and 'cross-border'; (2) 'disaster' and 'transboundary' and (3) 'disaster' and 'international'. These searches resulted 99, 89 and 9419 publications, respectively, from the earliest record in each of the categories to the end of 2015. Given the high number in the combination 'disaster' and 'international', the search for this pair has been limited to both words appearing in the title, still resulting in 394 publications which were considered for the analysis. This selection of search criteria necessarily has its limitations, as a wide range of engagements on cross-border disasters is not captured if the authors use different terms or discuss cross-border implications mostly indirectly. However, the selection is purposeful in that it captures the literature that deals with cross-border, transboundary or international dimensions of disasters in an explicit fashion. It is these explicit engagements that are of most relevance for this review. The analysis suggests however, that the three terms are not—and should not—be taken as synonyms; rather, they carry specific connotations. The term 'cross-border' implies that neighbouring countries are affected by the same hazard or disaster in their territory, calling for collaboration of states and their institutions. 'Transboundary', in contrast, can apply to state territories but is also applied to a wide range of boundaries, including sub-national administrative entities but also ecological or cultural boundaries. The term is, for instance, often used in the context of river and basin management. The term 'international', meanwhile, is applied to a wide range of contexts, including international policy frameworks (such as the International Decade for Disaster Risk Reduction in the 1990s) as well as international aid and relief amongst countries not sharing any territorial borders. Nevertheless, the literature also clearly shows that a good dose of fuzziness remains in the past and current use of the three terms. Throughout this chapter, all three terms are used, depending on the context.

Figure 4.1. shows the number of publications per year for the three searches. For all three categories, a clear increase in publications can be observed from the early 2000s onwards. Measured against the widespread acknowledgement of the importance of cross-border and transboundary disasters (see below) and the heavy increase in publications on disaster risk reduction more generally over the last decades (Garschagen 2014), the number of publications published *explicitly* in this field is remarkably low—especially when considering that some, even though not many, of

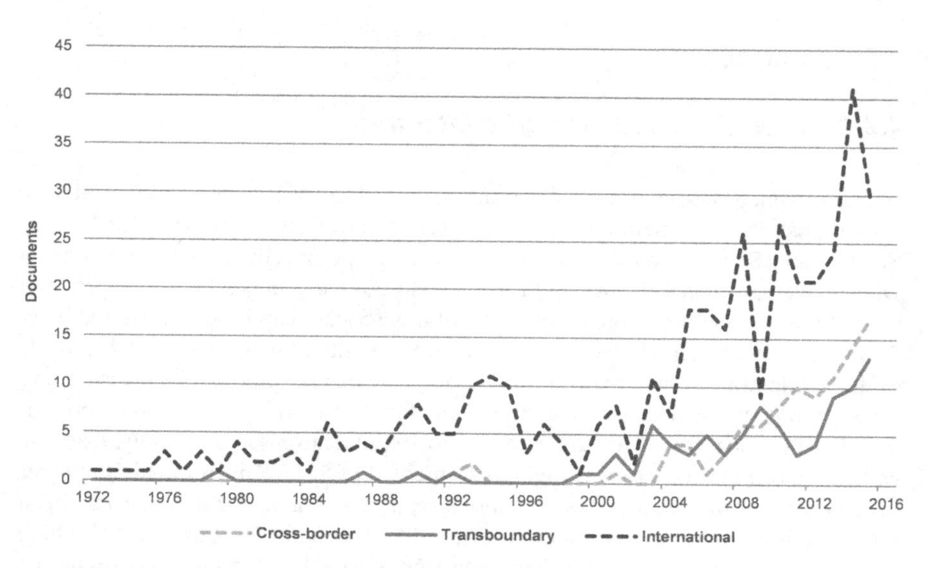

Fig. 4.1 Published items per year based on Scopus queries on 'disaster' and 'cross-border', 'disaster' and 'transboundary', and 'disaster' and 'international' (Own draft based on Scopus data)

the papers resulting from the three searches are featured in either two or all three of the groups.

Tracking the contributions of different disciplines has become increasingly difficult due to the growing number of multi- and transdisciplinary collaborations and mergers—especially in the field of disaster risk research, with its human-environment focus. Nevertheless, the existing data on the global record of peer-reviewed publications suggest a number of interesting points. In terms of publication numbers, the large contributions have been coming from the social sciences, environmental sciences and earth and planetary sciences. Interestingly, the field of medicine is the single largest contributor in the query using the 'cross-border' and 'international' terminology (almost one third of all publications in these groups) while only playing a minor role in the publications using the 'transboundary' terminology—see below for some interpretation. It is also worth noting that the number of contributions declared as belonging to the field of economics is comparatively low, contributing below 4% of the publications in all search queries. While a number of the papers in the other subject areas address economic dimensions of cross-border and transboundary disasters, of course, Fig. 4.2 still underscores that this field is heavily underemphasised.

While it would be impossible to review all of the articles resulting from this search in detail in this chapter, a number of thematic clusters can be identified from the review. One of the areas receiving the most attention is analysis of geophysical systems that are prone to producing cross-border disasters. Most prominently, these include transboundary river sheds with flood hazards (e.g. Bakker 2009), but also seismic zones (e.g. Parolai et al. 2010). This results in a strong emphasis on

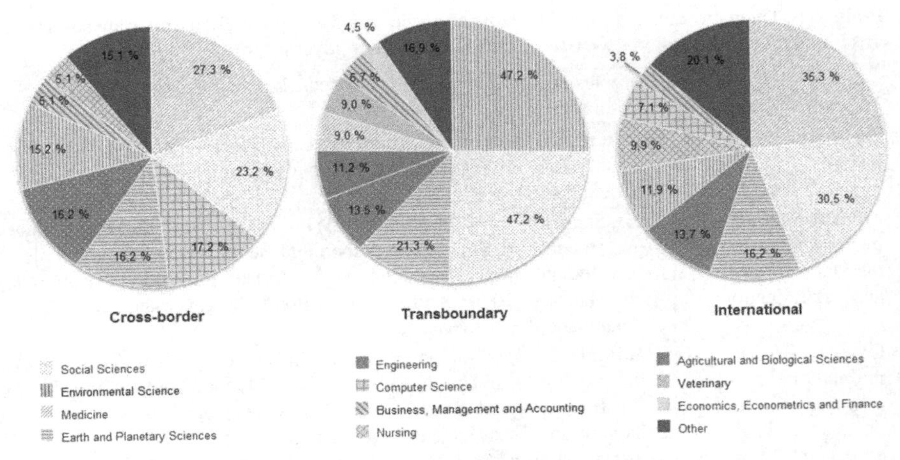

Fig. 4.2 Subject areas for results of Scopus query on 'disaster' and 'cross-border' (*left*) and 'disaster' and 'transboundary' (*center*), and 'disaster' and 'international'(*right*) (Own draft, based on Scopus data)

cross-border early warning systems and transboundary governance systems in order to mitigate disaster risk in a preventive way, particularly between upstream and downstream countries. Most prominently, this area of research includes the vast field of transboundary watershed governance, for instance, in central Europe (e.g. Janssen 2008) or the Himalayas (e.g. Katel et al. 2015). Many of these studies have close ties to concepts around Integrated Water Resources Management (IWRM).

Interestingly, however, a yet much larger body of literature concentrates not on disaster prevention but on transboundary, cross-border and international dimensions of post-disaster aid and relief. Using selected highlights, Table 4.1 illustrates the breadth and depth of this literature. It addresses organisational, legal, political and administrative layers and spans from arrangements between neighbouring countries all the way to regional or even global international agreements for disaster assistance. Lai et al. (2009), for example, explore current mechanisms for transboundary disaster risk management in ASEAN countries, concluding that the current provisions are not able to deliver the required level of effective risk management. Hence, they call for a new ASEAN disaster response, training and logistics centre and propose main criteria for its institutional design. Along similar lines, the majority of papers emphasise a need for improved institutional cooperation, synchronisation and preparation of cross-border international disaster response. A considerable body of literature even critiques the disaster relief actions of the past and examines their effectiveness as well as negative (unintended) side effects, especially in cases where aid was provided by the 'global north' to the 'global south' (e.g. Dudasik 1982; Habibzadeh et al. 2008). In principle, the literature on cross-border disaster response and relief includes the review of political frameworks and legal arrangements, e.g. the exchange of remote sensing information for disaster management or Geographic Information Systems for synchronising assistance by non-state humani-

Table 4.1 Thematic clusters and selected publications with explicit focus on transboundary, cross-border and international aspects of disasters and their governance

Thematic cluster	Examples of publications explicitly focusing on transboundary, cross-border and international aspects
Transboundary hazards	Parolai et al. (2010) on cross-border seismological monitoring networks in Central Asia; Bakker et al. (2009) on a global review of transboundary river floods
Transboundary institutions for disaster prevention and risk reduction	Katel, Schmidt-Vigt & Dendup (2015) on transboundary flood and water management in the Himalayas; Rubert and Beetlestone (2014) on tools for transboundary river management and disaster prevention in Southern Africa; Janssen (2008) on cross-border flood risk governance in Germany and the Netherlands
Cross-border and international disaster response and relief (state as well as non-state organisations; institutions and technology)	Boin et al. (2014) on the potential for transboundary crises management within the European Union; Caron et al. (2014) for a compendium on international law and disaster relief; Petiteville et al. (2014) on the collaboration of space agencies for strengthening national disaster risk reduction; Brattberg and Rhinard (2013) on the effectiveness in international disaster relief of the European Union and the United States; Rose and Kustra (2013) on economic considerations for designing transboundary emergency management institutions; Lai (2012) on cross-border disaster relief in Asia after the 2004 Indian Ocean Tsunami and the 2008 Wenchuan Earthquake; Stefanelli and Williams (2011) on regulatory barriers to effective international disaster assistance within the EU; Ansell et al. (2010) on the administrative capabilities required for managing transboundary crises; Tapia et al. (2010) on cross-border information systems used by NGOs to coordinate their humanitarian relief; Bas et al. (2010) on a Geographical Information System(GIS)-based management for cross-border evacuation in disaster situations in Europe; Edwards (2009) on the general principles of cross-border disaster response; Larsson (2009) on crisis management cooperation within the European Union; Becker et al. (2007) on challenges for transboundary flood management in the Rhine basin; Kapucu (2011) on an analysis of coordination and collaboration of international relief organisations; McCann and Cordi (2011) on international efforts to develop standards for disaster preparedness and response; Jia'nan et al. (2009) on the conceptual framework for a proposed international disaster compensation fund; James (2008), Habibzadeh et al. (2008) and Dudasik (1982) on critiques of international aid in disasters; Xu et al. (2008) on reviewing international cooperation in the case of the Wenchuan earthquake; Rokach et al. (2008) on standard-setting in collaborative search and rescue skills between Turkey, Greece and Israel; Abolghasemi et al. (2006) on lessons learned from the international response to the Bam earthquake in 2003; Lau et al. (2005) on Singapore's contribution to the victim identification effort in Thailand following the Indian Ocean Tsunami; Ito and Martinez (2005) on early lessons from the implementation of the International Charter for sharing remote sensing information for disaster management; Einhaus (1988) on a review of the operations of the Office of the United Nations Disaster Relief Coordinator; Kent (1987) on a review of international relief networks since 1945

(continued)

Table 4.1 (continued)

Thematic cluster	Examples of publications explicitly focusing on transboundary, cross-border and international aspects
Cross-border impacts of disasters	Oh and Reuveny (2010), Oh (2015) and Gassebner et al. (2010) on the impacts of disasters on international trade; Olivero et al. (2012) on a method for disaster risk assessment of cross-border infrastructure; Yang (2008) on a empirical analysis of the impacts of hurricanes on international financial flows over the last decades
International migration in relation to disasters and climate threats	Pourhashemi et al. (2012) on rights of climate refuguees; Kolmannskog and Trebbi (2010) on the protection of cross-border displaced victims of disasters; Cohen and Bradley (2010) on the protection gap in cross-border displacements following disasters
Pandemics and health emergencies	Grier et al. (2011) on information sharing for managing cross-border health emergencies; Fisher (2010) on legal and regulatory challenges in cross-border disaster medicine; Dopson (2009) on cross-border early warning for infectious diseases; Gao et al. (2008) on cross-border health mapping; Jones et al. (2008) on needs within cross-border preparation for health emergencies; Pohl-Meuthen et al. (2006) on obstacles for cross-border medical emergency services; Owens et al. (2005) on the US efforts for installing rapidly deployable assembly shelters and surgical hospital; Post (2004) on regulative barriers to cross-border medical assistance in Belgium, Germany and the Netherlands
Conceptual approaches	Sapountzaki and Daskalakis (2015) on a conceptual discourse on transboundary resilience to water scarcity and drought; Yang and Zhang (2014) on the evaluation of different systems for transboundary disaster management; Lagadec (2009) on new paradigms and concepts for capturing complex transboundary crises
Miscellaneous relevant topics	Engstrom (2013) on the military involvement in international disaster relief in East Asia and the projection of power; Raschky and Schwindt (2012) on a review of types and channels of international disaster assistance; Deebaj et al. (2011) on an analysis of whether airports in Sweden, Great Britain and Finland had been prepared to deal with injured and traumatised travellers returning after the Indian Ocean Tsunami in 2004; Cao et al. (2010) on the role of Chinese rescue teams in the Pakistan floods or 2010; Nelson (2010) on the international politics of aid refusal, providing a global data set of these cases between 1982 and 2006; Margesson (2010) on a review of the US legal framework and budget trends for international humanitarian assistance; Byard and Winskog (2010) on the challenges arising in victim identification during international disasters; Park and Reisinger (2010) on an empirical analysis on the relationship between perceived disaster risk and travel choices; Diaz (2008) on the integration of psychosocial support in the international assistance of the American Red Cross

tarian organisations (e.g. Ito and Martinez 2005), as well as case studies from specific disasters in the past. A considerable number of these examples are from Asia, most notably the Indian Ocean Tsunami (e.g. Lai 2012; Lau et al. 2005), but also Typhoon Nargis (e.g. Kapucu 2011), major earthquakes such as in Bam and Wenchuan or other major disasters (e.g. Abolghasemi et al. 2006; Xu et al. 2008).

An additional thematic cluster gravitates around the assessment of cross-border disaster impacts. These include changes in international trade (e.g. Oh 2015) or infrastructure disruption (e.g. Olivero et al. 2012). An emerging subset further emphasises (international) migration due to disasters and the expected impacts of global environmental change (e.g. Kolmannskog and Trebbi 2010). A particular focus therein is on the (legal) protection of these migrants (e.g. Cohen and Bradley 2010).

Diseases and other health emergencies constitute another major thematic cluster in the literature—as indicated by the strong contribution of medical sciences (see above). Within this cluster a lot of attention is given to the establishment and evaluation of early warning systems for cross-border spread of diseases (e.g. Dopson 2009; Grier et al. 2011). Secondly, a major focus is on the international institutional arrangements for combating disease outbreaks, covering technical medical assistance (e.g. Owens et al. 2005). The literature shows that a lot of global efforts have been undertaken over the last decades to improve international cooperation in response to disease outbreaks. However, the lite also highlights the many remaining challenges for effective collaboration, spanning institutional, political and legal domains (e.g. Fisher 2010). Interestingly, these barriers for collaboration are not only debated for low-income countries but also within the context of OECD members and high-income nations—for instance, in terms of the barriers for cross-border assistance between Belgium, Germany and the Netherlands (Post 2004). Particular emphasis is also placed on the need for improved ex ante contingency planning and training for medical assistance across borders (e.g. Jones et al. 2008).

Last but not least, an emerging body of literature focuses on theoretically and conceptually framing disasters and their governance (e.g. Lagadec 2009; Sapountzaki and Daskalakis 2015). In addition, a diverse range of other topics are discussed in the literature, illustrated in the bottom part of Table 4.1. These include, for instance, the analysis of the politics behind refusing international assistance in case of disaster (Nelson 2010) or the projection of power through involving military into cross-border disaster assistance in East Asia (Engstrom 2013).

4.2.2 Grey Literature

The analysis of grey literature was driven by two main components. Firstly, a bottom-up search in PreventionWeb was conducted. This is a comprehensive and widely used online platform in the field of disaster risk governance, with the explicit mandate for 'serving the needs of the disaster reduction community' (see UNISDR n.d.). An open word search for 'transboundary' within this portal resulted in 972 hits, out of which the largest subset (445) consisted of publications by international organisations, followed by policy plans and statements (198), conference announcements and reports (100) and other categories such as job announcements or news reports. A rough review of this list shows that major thematic clusters can be observed on transboundary flood risk management (around 170 hits) and drought risk management (130 hits), both increasingly debated under climate change perspectives. However, only 67 of the records are linked to the keyword of 'urban risk and planning'. In terms of regional focus, entries addressing disaster in Asia make for the largest contribution (295), followed by Europe (206) and Africa (160).

However, given the high number of records in the PreventionWeb portal, it proved impossible to apply a detailed review to the entire body of documents. The detailed analysis therefore, targeted specific reports which have been of high influence in the field of disaster risk reduction over the recent years. These include:

- The biannual Global Assessment Report (GAR) by the United Nations International Strategy for Disaster Reduction (UNISDR),
- The Special Report on Managing Extreme Events (SREX) by the Intergovernmental Panel on Climate Change (IPCC),
- The annual World Disaster Report by International Federation of Red Cross and Red Crescent Societies (IFRC),
- The annual World Risk Report by the Development Alliance Works and the United Nations University Institute for Environment and Human Security (UNU-EHS), and
- Special reports on disaster risk from the United Nations Economic and Social Commission for Asia and the Pacific (UN-ESCAP).

Overall, the coverage of cross-border disasters and their governance is strikingly shallow in all these reports. While many of the reports are published in series and feature an annual focal topic, none of them has in the past explicitly used cross-border or transboundary disasters as their overall feature topic. However, the reports frequently mention the importance to consider such types of disasters and the need to get a better handle on transboundary risk governance.

The 2013 GAR (UNISDR 2013), which is focused on '[t]he business case for disaster risk reduction', elaborates on the neglect of transboundary risks and their negative effects on economic sustainability. The report argues that '[r]isks are externalised or transferred across space and time to other locations and sectors' (p. 118) due to investments in hazard-prone areas. However, when a disaster materialises, the effects are likely to be transboundary, challenging economic performance at all ends of the system (ibid.). Looking ahead, the report further cautions that 'the more long term the perspective is on risk and uncertainty, the more it becomes an international and trans-boundary concern and less a national capacity issue' (p. 228). The GAR of 2011 (UNISDR 2011) also refers to transboundary disasters in a couple of places. In combination, these references open up an interesting tension. On the one hand, a high number of countries are engaged in transboundary risk governance projects and initiatives, run by a wide range of donors and organisations. On the other hand, however, the implementation of transboundary risk governance principles and actual procedures (i.e. the transfer from political agreements on paper to a real practice of transboundary risk governance) is often challenged. The 2011 GAR, for instance, elaborates on barriers around information sharing, political competition and lack of human and financial capacities in the case of transboundary risk governance in South Asia.

The IPCC's SREX report goes a great length in stressing the systemic nature of many risks related to extreme weather events as well as their risk management interventions, which both frequently 'cross national borders and transcend single nation policies and procedure' (IPCC 2012, p. 398). That is, the report explicitly recognises that hazards can 'apply to the contiguous zones of many countries, such as shared basins with associated flood risks' while other '[r]elationships and connections [that can be altered by hazard events] involving the movement of goods (trade), finance (capital flows and remittances), and people (displaced populations) can also have transboundary impacts' (p. 399). The report emphasises that such effects can particularly be triggered by extreme events. These should therefore always be considered as potentially transboundary in reach, irrespective of their original hazard extent. The report therefore calls for increased efforts towards transboundary risk governance. However, it does so by also emphasising the potential that can emerge from such cooperation:

> The interdependence of the global economy, the public good, and the transboundary nature of risk management, and the potential of regional risk pooling, can make international cooperation on disaster risk reduction and climate change adaptation more economically efficient than national or sub-national action alone. Notions of solidarity and equity motivate addressing disaster risk reduction and climate change adaptation at the international level in part because developing countries are more vulnerable to physical disasters. (p.396)

Referring to the Rio Declaration and the Charter of the United Nations Framework Convention on Climate Change (UNFCCC), the report further provides a reminder on key legal obligations (in terms of soft laws) around transboundary risk management:

> That states have a duty to prevent transboundary harm, provide notice of, and undertake consultations with respect to such potential harms is a soft law norm expressed under international environmental law. The more general duty to cooperate has evolved as a result of the inapplicability of the law of state responsibility to problems of multilateral concern, such as global environmental challenges. [...] From the duty to cooperate is deduced a duty to notify other states of potential environmental harm. This is reflected in Principles 18 and 19 of the Rio Declaration (a non-legal international instrument), that 'States shall immediately notify other States of any natural disasters or other emergencies that are likely to produce sudden harmful effects on the environment of those States' (Rio Principle 18) and 'States shall provide prior and timely notification and relevant information to potentially affected States on activities that may have a significant adverse transboundary environmental effect' (Rio Principle 19). (p. 402)

Thus the report supports its call for increased efforts towards transboundary risk governance by, raising awareness of, first, cross-border risk patterns; second, the opportunities emerging form transboundary risk governance; and third, the already existing obligations within ratified international agreements.

Interestingly, neither the World Disaster Report (annually published by the IFRC) nor the World Risk Report (published annually since 2011 by the Alliance Development Works and UNU-EHS) have in the past explicitly featured the topic of transboundary disaster risk in any of their reports. While the topic is mentioned in various places of these reports, the engagement does translate into a detailed analysis or conceptual framing.

One of the most explicit engagements with transboundary disasters can be found in UN-ESCAP's 2015 Asia-Pacific Disaster Report, entitled 'Disasters Without Borders—Regional Resilience for Sustainable Development' (UN-ESCAP 2015). As the title suggests, the report includes a strong plea for more stringent transboundary risk management in the region. The report stresses that the region is affected by a number of large-scale and cross-border hazard conditions, including earthquakes, tsunamis, tropical cyclones, transboundary floods, volcanic eruptions and droughts. Single hazard events in these categories can affect multiple countries in the region at the same time, necessitating transboundary efforts to reduce the risk of regional crises. However, a review of its problem analysis, particularly in the section on cross-border threats, also exemplifies one of the general problems in the field: that data for truly analysing the cross-border effects is largely lacking to date (see also the section below). Rather than exploring the cross-border effects in detail, this report has to make do with existing statistical data and rather lists the accumulation of disaster risk in the region. This is undoubtedly relevant but is analytically different from exploring the economic, social, informational and other linkages in cross-border disaster risk.

4.2.3 *Secondary Data*

In order to analyse whether and to what extent freely available secondary statistical data can capture cross-border disasters and their impacts, the main databanks of disaster information—EM-DAT and NatCatSERVICE—have been considered for the analysis.

The EM-DAT database, run since 1988 by the Centre for Research on the Epidemiology of Disasters (CRED), is compiled from various sources, including UN agencies, non-governmental organisations, insurance companies, research institutes and press agencies (CRED 2015). It lists the impacts from disasters related to natural hazards in terms of affected people and economic losses. However, despite being one the most comprehensive databases available, all data is prepared and provided with national resolution, due to the reporting mechanisms in the data sources it uses. Hence EM-DAT data does not allow for tracking cross-border effects of specific disasters in detail.

The NatCatSERVICE, run by Munich Re, provides annual reports on the largest disasters, measured in terms of their humanitarian and economic impact (Munich Re 2015). The figures on overall losses therefore include proxies for cross-border effects. However, given that Munich Re only publishes highly aggregated data out of their NatCatService, it is difficult to track cross-border impacts in specific countries in detail.

4.2.4 International Agreements in the Post-2015 Agenda

Disaster risk reduction is a major policy field in the so-called post-2015 arena. The year 2015 is of great importance as it saw four major international initiatives and agreements for fostering sustainable development and risk reduction: the passing of the Sustainable Development Goals (SDGs), the Addis Ababa Conference on Financing for Development, UNFCCC's 21st Conference of the Parties (with the goal of a new global climate agreement), and the 3rd UN World Conference on Disaster Risk Reduction. The latter, held in March in Sendai, has direct relevance for the topic of cross-boundary disaster risk governance. It led to a new intergovernmental agreement on disaster risk reduction, the so-called Sendai Framework for Disaster Risk Reduction (SFDRR) 2015–2030 (UNISDR 2015). Looking back to the lessons learned from the previous agreement, the Hyogo Framework for Action (UNISDR 2007), the Sendai Framework states that:

> International, regional, subregional and transboundary cooperation remains pivotal in supporting the efforts of States, their national and local authorities, as well as communities and businesses, to reduce disaster risk. Existing mechanisms may require strengthening in order to provide effective support and achieve better implementation. (p. 10)

Yet the Sendai Framework for Action goes beyond the Hyogo Framework for Action and emphasises international—and specifically transboundary—cooperation as one of the main responsibilities of national governments, anchored in the frameworks first guiding principle:

> Each State has the primary responsibility to prevent and reduce disaster risk, including through international, regional, subregional, transboundary and bilateral cooperation. (p. 13)

The framework further goes on to specify that:

> To guide action at the regional level through agreed regional and subregional strategies and mechanisms for cooperation for disaster risk reduction, as appropriate, in the light of the present Framework, in order to foster more efficient planning, create common information systems and exchange good practices and programmes for cooperation and capacity development, in particular to address common and transboundary disaster risks.' (p. 18)

While the Sendai Framework does not include binding principles on how to implement these requests, it is worth noting that the level of emphasis put on the topic is higher than in its predecessor framework. However, the below discussion will touch on the question of whether and how these claims on paper are likely to be implemented in action—and which further steps will be needed.

4.3 Conclusions: Achievements, Knowledge Gaps and Research Needs

In conclusion, the above review allows for synthesising the main achievements as well as remaining knowledge gaps and research needs in the field of cross-border and transboundary disasters. While the review can by no means be considered complete in covering all literature that has been published on the topic, it shows that considerable attention has been given to the issue. Scientific publications as well as major reports by international organisations working on disaster risk reduction stress the importance of recognising the transboundary and international reach of disasters and their impacts. They therefore agree on calling for strong efforts for improving transboundary and international risk governance efforts. However, the level of specification ranges from loose statements on wishes and visions to concrete suggestions for improved organisational mechanisms—for instance, within ASEAN (Lai et al. 2009). The plea for improved transboundary and international risk governance has also become a central element of the main intergovernmental agreement guiding disaster risk reduction in the post-2015 world, i.e. the Sendai Framework 2015–2030. This is a major advancement from the previous Hyogo Framework for Action 2005–2015, where the issue had received much less attention and was only vaguely mentioned—not transferred forcefully into a guiding principle as is the case in the Sendai Framework. However, whether and how these principles will be translated into action will need to be seen over the next years. The review suggests that despite the increased attention given to the topic, there are multiple barriers in terms of political will, institutional inertia, financial and human capital (reported especially in the scientific literature) and, last but not least, gaps in knowledge and data. The prospects for improved transboundary risk governance will therefore also depend on the whether and how existing knowledge gaps can be closed to allow for guiding policy and practice more effectively.

Along this line, the review brings to light a number of urgent research needs. First, there is a lack of strategic regional and global data collection to allow examination of the size, regional patterns, and sectoral patterns of cross-border disaster impact. The literature features a number of popular case studies for exploring cross-border disaster impacts. Amongst the most prominent examples are the 2011 floods in Bangkok and the Tohoku earthquake and tsunami, alluding to the strong relevance of transboundary disaster perspectives in urban—and urbanising—Asia. However, apart from heuristic analysis in singular case studies, a global approach to capturing the multiple cross-border effects of disasters is largely lacking to date. Neither the peer-reviewed literature nor the international reports provide such a comprehensive approach. In particular, the global databases on disaster impact statistics are to date not designed to capture such cross-border effects in a strategic and coherent manner—including those on migration, trade, economic production or any of the other effects frequently mentioned in the heuristic and conceptual literature on transboundary disasters. A more strategic collection of such data would not only enable a much more stringent and detailed scientific understanding of transboundary disasters, but would also help to focus political attention on the need for transboundary and cross-border risk governance to avoid these types of impacts and share the costs of their mitigation.

Second, cross-border disasters are often equated with a cross-border spread of hazards such as earthquakes or floods. Much less attention is given to the case where, in an increasingly complex world, even localised disasters such as urban floods can imply massive spillover effects across borders. The capture of these effects is much more complicated, but would be needed for truly understanding the disaster risk in integrated and highly connected systems of global production and social activity.

Third, the gap described in the above two points is particularly wide with regards to the soft impacts of disasters. These reach beyond the pure impacts on economic production, trade, currencies exchange rates, migration etc., which all could, in principle, be measured in hard numbers. Transboundary effects of disasters can also include issues such as teleconnected changes in risk management paradigms and practices. Again, while some illustrative examples are repeatedly reported in the literature (e.g. Germany's decision to exit nuclear power production following the Fukushima triple disaster) the literature lacks a more strategic assessment of these soft effects, including comparative studies or theoretical models for explaining such changes. Subtle transboundary effects at the interface of hard and soft might also include issues such as informal remittance flows or changes in investment decisions. These hybrid effects are also insufficiently captured and examined to date.

Fourth, next to the assessment of hard and soft impacts of transboundary disasters, there is a particular gap in examining and understanding the drivers and root causes of transboundary disaster risk. While most attention has to date been given to questions of transboundary and international cooperation in *ex post* disaster response, relief and recovery, a better understanding of the root causes and drivers is essential for any long-term and preventive risk reduction. This needs to include in particular a better deciphering of the causal structure and complexities driving disaster risk, including not only the currently predominant focus on assessing natural hazards (as one side of the risk equation) but also the risk contribution of socio-economic vulnerability and susceptibility.

Fifth, there is a significant need to account for future trends into the analysis of cross-border risk patterns. As indicated in the introduction, cross-border disaster risk is driven by—amongst other factors—the confluence of three megatrends: environmental and climate change; increasing regional integration and globalisation; and urbanisation. All three of these trends are going to continue or even intensify in the future, particularly in Asia and the Pacific region. However, scientific studies almost exclusively look backwards and examine past disasters. What is urgently needed is to complement these studies with scenario and other futurology approaches to identify and assess potential future trajectories in cross-border disaster risk. This should combine qualitative and quantitative approaches and should serve not only a scientific but especially a practical purpose, thereby leveraging increased action for transboundary risk governance today.

Sixth, there is a lack of institutional analysis on how to incentivise and improve cross-border disaster cooperation in contexts where the benefits might be delayed or uncertain. While all bodies of literature considered for this review agree on the normative call for improved transboundary risk governance, the current case studies on institutional and organisational enablers and barriers, despite being very insightful, only provide highly selective and context-specific accounts. Research over the next years should focus on synthesising and upscaling these accounts and on building abstract models and theoretical representations of the institutional dimensions of cross-border risk governance. Ideally, these can work to inform policymaking and designing institutions for transferring the conceptual claims made in the SFDRR and other documents into concrete policy and action.

Seventh, while the private sector is one of the main agents when thinking about cross-border disasters (both in terms of assessing negative impacts and ways of dealing with them), surprisingly little is known about how private sector actors make decisions to deal with cross-border risk. There is hence a need for an improved understanding of how these actors perceive and act upon the risk from cross-border disasters and whether and how they function as agents of change in wider risk governance constellations.

In sum, research and policymaking on transboundary and cross-border disaster risk reduction can build on a strong foundation and a widely shared set of goals. However, the field will need to receive increased academic and practical effort if it is to transform from a vision to practice.

References

Abolghasemi, H., Radfar, M. H., Khatami, M., Nia, M. S., Amid, A., & Briggs, S. M. (2006). International medical response to a natural disaster: Lessons learned from the Bam earthquake experience. *Prehospital and Disaster Medicine, 21*(3), 141–147. https://www.scopus.com/inward/record.uri?eid=2-s2.0-33748096810&partnerID=40&md5=c079c61468d3eb133f471548680de455.

Ansell, C., Boin, A., & Keller, A. (2010). Managing transboundary crises: Identifying the building blocks of an effective response system. *Journal of Contingencies and Crisis Management, 18*(4), 195–207.

Bakker, M. H. N. (2009). Transboundary river floods: Examining countries, international river basins and continents. *Water Policy, 11*(3), 269–288.

Bas, I., Zoicas, C., & Ionita, A. (2010). The management of large emergency situations – a best practices case study based on GIS for management of evacuation. *World Academy of Science, Engineering and Technology, 4*(6), 651–654.

Becker, G., Aerts, J., & Huitema, D. (2007). Transboundary flood management in the Rhine basin: Challenges for improved cooperation. *Water Science and Technology, 56*(4), 125–135.

Boin, A., Rhinard, M., & Ekengren, M. (2014). Managing transboundary crises: The emergence of European Union capacity. *Journal of Contingencies and Crisis Management, 22*(3), 131–142.

Brattberg, E., & Rhinard, M. (2013). Actorness and effectiveness in international disaster relief: The European Union and United States in comparative perspective. *International Relations, 27*(3), 356–374. https://www.scopus.com/inward/record.uri?eid=2-s2.0-84883384104&partnerID=40&md5=1cb4e63e78386b6c58d0fa687204e8e3.

Byard, R. W., & Winskog, C. (2010). Potential problems arising during international disaster victim identification (DVI) exercises. *Forensic Science, Medicine, and Pathology, 6*(1), 1–2. https://www.scopus.com/inward/record.uri?eid=2-s2.0-73949101827&partnerID=40&md5=b5960060f076a4ace89e2d99cdee2eaa.

Cao, L., Peng, B.-B., Wang, F., & Zhang, L.-Y. (2010). Medical rescue and relief work of the Chinese international rescue team in Pakistan flood disaster. *Chinese Journal of Emergency Medicine, 19*(11), 1143–1145. https://www.scopus.com/inward/record.uri?eid=2-s2.0-78650754504&partnerID=40&md5=2caf9d066b995769caf7c61bd6a52e6a.

Caron, D.D., Kelly, M.J., Telesetsky, A. (2014). The International Law of Disaster Relief, https://www.scopus.com/inward/record.uri?eid=2-s2.0-84953717325&partnerID=40&md5=b360ba4793b79d3b2f32e1ea583da788.

Cohen, R., & Bradley, M. (2010). Disasters and displacement: Gaps in protection. *Journal of International Humanitarian Legal Studies, 1*(1), 95–142. https://www.scopus.com/inward/record.uri?eid=2-s2.0-84906730020&partnerID=40&md5=18fa714097212a95520524551d422828.

CRED (Centre for Research on the Epidemiology of Disasters). (2015) EM-DAT International Disaster Database, viewed 29 April 2015, http://www.emdat.be.

Deebaj, R., Castrén, M., & Öhlén, G. (2011). Asia tsunami disaster 2004: Experience at three international airports. *Prehospital and Disaster Medicine, 26*(1), 71–75. https://www.scopus.com/inward/record.uri?eid=2-s2.0-80052383316&partnerID=40&md5=75457fa5f53cb9b387dd7e87c66ea6d3.

Diaz, J. O. P. (2008). Integrating psychosocial programs in multisector responses to international disasters. *American Psychologist, 63*(8), 820–827. https://www.scopus.com/inward/record.uri?eid=2-s2.0-56849096614&partnerID=40&md5=e88981e658a67e1ec7828cd27761f246.

Dopson, S. A. (2009). Early warning infectious disease surveillance. *Biosecurity and Bioterrorism, 7*(1), 55–60. https://www.scopus.com/inward/record.uri?eid=2-s2.0-65349087288&partnerID=40&md5=b3c4d0ffe94571d2fae43cdfe1242849.

Dudasik, S. (1982). Unanticipated repercussions of international disaster relief. *Disasters, 6*(1), 31–37.

Edwards, F. L. (2009). Effective disaster response in cross border events. *Journal of Contingencies and Crisis Management, 17*(4), 255–265. https://www.scopus.com/inward/record.uri?eid=2-s2.0-70849087035&partnerID=40&md5=4715447d49f03be1b78dcb86dcc0d186.

Einhaus, H. (1988). Emergency planning and management for disaster mitigation at the international level. *Regional Development Dialogue, 9*(1), 1–12. https://www.scopus.com/inward/record.uri?eid=2-s2.0-0024162586&partnerID=40&md5=bb2d666694f6a70062f8f3169d983d10.

Engstrom, J. (2013). Taking disaster seriously: East Asian military involvement in international disaster relief operations and the implications for force projection. *Asian Security, 9*(1), 38–61. https://www.scopus.com/inward/record.uri?eid=2-s2.0-84875304234&partnerID=40&md5=618bd0d0e395903f0fcb706aca77c399.

Fisher, D. (2010). Regulating the helping hand: Improving legal preparedness for cross-border disaster medicine. *Prehospital and Disaster Medicine, 25*(3), 208–212. https://www.scopus.com/inward/record.uri?eid=2-s2.0-77957001865&partnerID=40&md5=d4e19dd8ba83ba8f701a46bf63d85f62.

Gao, S., Mioc, D., Yi, X., Anton, F., Oldfield, E., Coleman, D.J. (2008). The Canadian geospatial data infrastructure and health mapping, CyberGeo 2008, https://www.scopus.com/inward/record.uri?eid=2-s2.0-67650911596&partnerID=40&md5=4521e1e5d27284a84beb5f68aa268a81.

Garschagen, M. (2014). *Risky change? Vulnerability and adaptation between climate change and transformation dynamics in Can Tho City, Vietnam*. Stuttgart: Steiner.

Gassebner, M., Keck, A., & Teh, R. (2010). Shaken, not stirred: The impact of disasters on international trade. *Review of International Economics, 18*(2), 351–368. https://www.scopus.com/

inward/record.uri?eid=2-s2.0-77954400835&partnerID=40&md5=60dc48cd304689cca66ed228c4e5aba7.

Grier, N. L., Homish, G. G., Rowe, D. W., & Barrick, C. (2011). Promoting information sharing for multijurisdictional public health emergency preparedness. *Journal of Public Health Management and Practice, 17*(1), 84–89. https://www.scopus.com/inward/record.uri?eid=2-s2.0-78650812097&partnerID=40&md5=0a1b031319cd4bda862dd5a7202e7bf3.

Habibzadeh, F., Yadollahie, M., & Kucheki, M. (2008). International aid in disaster zones: Help or headache? *The Lancet, 372*, 374.

IPCC (Intergovernmental Panel on Climate Change). (2012). In C. B. Field, V. Barros, T. F. Stocker, D. Qin, D. J. Dokken, K. L. Ebi, M. D. Mastrandrea, K. J. Mach, G.-K. Plattner, S. K. Allen, M. Tignor, & P. M. Midgley (Eds.), *Managing the risks of extreme events and disasters to advance climate change adaptation, A special report of working groups I and II of the Intergovernmental Panel on Climate Change*. New York: Cambridge University Press.

Ito, A., & Martinez, L. F. (2005). Issues in the implementation of the international charter on space and major disasters. *Space Policy, 21*(2), 141–149. https://www.scopus.com/inward/record.uri?eid=2-s2.0-20444427963&partnerID=40&md5=6f0fba88957d3b58c6c9c710a263c576.

James, L. (2008). International aid in disasters: A critique. In L. Dominelli (Ed.), *Revitalising communities in a Globalising world* (pp. 281–293). Burlington: Ashgate.

Janssen, J. A. E. B. (2008). On peaks and politics; governance analysis of flood risk management cooperation between Germany and the Netherlands. *International Journal of River Basin Management, 6*(4), 349–355. https://www.scopus.com/inward/record.uri?eid=2-s2.0-65849526443&partnerID=40&md5=0282900fffb9bd8f58743ceb54869b9d.

Jia'nan, L., Yan, S., & Qin, T. (2009). International disaster compensation fund: A new international financial aid mechanism. *Transition Studies Review, 16*(2), 479–483. https://www.scopus.com/inward/record.uri?eid=2-s2.0-70349998890&partnerID=40&md5=b237a7c87ef1476b7ae0bb38047aa963.

Jones, M., O'Carroll, P., Thompson, J., & D'Ambrosio, L. (2008). Assessing regional public health preparedness: A new tool for considering cross-border issues. *Journal of Public Health Management and Practice, 14*(5), E15–E22. https://www.scopus.com/inward/record.uri?eid=2-s2.0-56149125434&partnerID=40&md5=d7373353bdec0a30b07697866ecf3c96.

Kapucu, N. (2011). Collaborative governance in international disasters: Nargis cyclone in Myanmar and Sichuan earthquake in China cases. *International Journal of Emergency Management, 8*(1), 1–25. https://www.scopus.com/inward/record.uri?eid=2-s2.0-79957615528&partnerID=40&md5=ae0ed6ab4c922219a30233c8f3d07196.

Katel, O. N., Schmidt-Vogt, D., & Dendup, N. (2015). Transboundary water resources management in the context of global environmental change: The case of Bhutan Himalaya. In S. Shrestha, A. K. Anal, P. A. Salam, & M. van der Valk (Eds.), *Managing water resources under climate uncertainty: Examples from Asia, Europe, Latin America, and Australia* (pp. 269–290). Switzerland: Springer International Publishing.

Kent, R. C. (1987). *Anatomy of disaster relief: The international network in action*. New York: Pinter Publishers.

Kolmannskog, V., & Trebbi, L. (2010). Climate change, natural disasters and displacement: A multi-track approach to filling the protection gaps. *International Review of the Red Cross, 92*(879), 713–730. https://www.scopus.com/inward/record.uri?eid=2-s2.0-79956333612&partnerID=40&md5=9b39ecc6a4e7e388f42aa666855b2a57.

Lagadec, P. (2009). A new cosmology of risks and crises: Time for a radical shift in paradigm and practice. *Review of Policy Research, 26*(4), 473–486.

Lai, A. Y. H. (2012). Towards a collaborative cross-border disaster management: A comparative analysis of voluntary organizations in Taiwan and Singapore. *Journal of Comparative Policy Analysis: Research and Practice, 14*(3), 217–233.

Lai, A. Y. H., He, J. A., Tan, T. B., & Phua, K. H. (2009). A proposed ASEN disaster response, training and logistic centre enhancing regional governance in disaster management. *Transition Studies Review, 16*(2), 299–315.

Larsson, P. (2009). The crisis coordination arrangements (CCA). In S. Olsson (Ed.), *Crisis Management in the European Union: Cooperation in the face of emergencies* (pp. 127–138). Berlin: Springer. https://www.scopus.com/inward/record.uri?eid=2-s2.0-84892825167&partnerID=40&md5=122618d9ecec338b38d3bb0f0f475084.

Lau, G., Tan, W. F., & Tan, P. H. (2005). After the Indian Ocean tsunami: Singapore's contribution to the international disaster victim identification effort in Thailand. *Annals of the Academy of Medicine Singapore, 34*(5), 341–351. https://www.scopus.com/inward/record.uri?eid=2-s2.0-22544462223&partnerID=40&md5=f7cef9da949b6d5c9df534f44243cfc0.

Margesson, R. (2010). International crises and disasters: U.S. humanitarian assistance, budget trends, and issues for Congress. In J. C. Parish (Ed.), *U.S. Foreign assistance: Background, role and future* (pp. 217–234). Washington, DC: Nova Science Publishers. https://www.scopus.com/inward/record.uri?eid=2-s2.0-84892053861&partnerID=40&md5=1ce804eb6873aabca52e6503022e4b06.

McCann, D. G. C., & Cordi, H. P. (2011). Developing international standards for disaster preparedness and response: How do we get there? *World Medical and Health Policy, 3*(1), 1–4. https://www.scopus.com/inward/record.uri?eid=2-s2.0-79959564366&partnerID=40&md5=b22572122690c67f7cbda0c8744bc9c5.

Munich, Re. (2015). NatCatSERVICE, Munich Re, Munich, viewed 30 April 2015, http://www.munichre.com/de/reinsurance/business/non-life/georisks/natcatservice/default.aspx.

Nelson, T. (2010). Rejecting the gift horse: International politics of disaster aid refusal. *Conflict, Security and Development, 10*(3), 379–420. https://www.scopus.com/inward/record.uri?eid=2-s2.0-79956243743&partnerID=40&md5=197e3e8276671563317ca6345454b167.

Oh, C. H. (2015). How do natural and man-made disasters affect international trade? A country-level and industry-level analysis. *Journal of Risk Research, online first*. doi:10.1080/13669877.2015.1042496.

Oh, C. H., & Reuveny, R. (2010). Climatic natural disasters, political risk, and international trade. *Global Environmental Change, 20*(2), 243–254. https://www.scopus.com/inward/record.uri?eid=2-s2.0-77951255135&partnerID=40&md5=d4bc4ec21c68320a8048bd4a67e5a6dc.

Olivero, S., Migliorini, M., Stirano, F., Calandri, F., Fava, U. (2012). Cross-border strategic infrastructures: From risk assessment to identification of improvement priorities. The experience gained in PICRIT Project. In *Proceedings of the 4th International Disaster and Risk Conference: Integrative Risk Management in a Changing World—Pathways to a Resilient Society*, IDRC Davos 2012, pp. 539–541, https://www.scopus.com/inward/record.uri?eid=2-s2.0-84925017906&partnerID=40&md5=7670b3a841638c426562b1cff73b6c07.

Owens, P. J., Forgione, A., Jr., & Briggs, S. (2005). Challenges of international disaster relief: Use of a deployable rapid assembly shelter and surgical hospital. *Disaster Management and Response, 3*(1), 11–16. https://www.scopus.com/inward/record.uri?eid=2-s2.0-12144260518&partnerID=40&md5=7e8b0de4f9054011e04fd41e679eef22.

Park, K., & Reisinger, Y. (2010). Differences in the perceived influence of natural disasters and travel risk on international travel. *Tourism Geographies, 12*(1), 1–24. https://www.scopus.com/inward/record.uri?eid=2-s2.0-77949344871&partnerID=40&md5=629c293e581c0cd7d21359be06662a4e.

Parolai, S., Orunbaev, S., Bindi, D., Strollo, A., Usupaev, S., Picozzi, M., Di Giacomo, D., Augliera, P., D'Alema, E., Milkereit, C., Moldobekov, B., & Zschau, J. (2010). Site effects assessment in Bishkek (Kyrgyzstan) using earthquake and noise recording data. *Bulletin of the Seismological Society of America, 100*(6), 3068–3082. https://www.scopus.com/inward/record.uri?eid=2-s2.0-78650084514&partnerID=40&md5=a9e14c3c3960ab6f2f61fa5b2b5fe63f.

Petiteville, I., Ishida, C., Eddy, A., Frye, S., Steventon, M., Jones, B. (2014). International collaboration of space agencies to support disaster preparedness and response and country risk management. In *Proceedings of the 5th International Disaster and Risk Conference: Integrative Risk Management—The Role of Science, Technology and Practice*, IDRC Davos 2014, pp. 572–575, https://www.scopus.com/inward/record.uri?eid=2-s2.0-84924973551&partnerID=40&md5=37adc7349fc7715e60ae7dba2cd06c9d.

Pohl-Meuthen, U., Schlechtriemen, T., Gerigk, M., Schäfer, S., & Moecke, H. (2006). Cross-border emergency medical services: Hopes and reality. *Notfall und Rettungsmedizin, 9*(8), 679–684. https://www.scopus.com/inward/record.uri?eid=2-s2.0-33845327181&partnerID=40&md5=a1f67d07c6dff89186a3676248aedc29.

Post, G. B. (2004). Building the tower of Babel: Cross-border urgent medical assistance in Belgium, Germany and the Netherlands. *Prehospital and Disaster Medicine, 19*(3), 235–244. https://www.scopus.com/inward/record.uri?eid=2-s2.0-16544388613&partnerID=40&md5=b5272a0c6d33a423f7984db13fcaa810.

Pourhashemi, S. A., Khoshmaneshzadeh, B., Soltanieh, M., & Hermidasbavand, D. (2012). Analyzing the individual and social rights condition of climate refugees from the international environmental law perspective. *International journal of Environmental Science and Technology, 9*(1), 57–67. https://www.scopus.com/inward/record.uri?eid=2-s2.0-84857279467&partnerID=40&md5=d2936429bf3c99911d54f94d7f7f41b2.

Raschky, P. A., & Schwindt, M. (2012). On the channel and type of aid: The case of international disaster assistance. *European Journal of Political Economy, 28*(1), 119–131. https://www.scopus.com/inward/record.uri?eid=2-s2.0-84855195394&partnerID=40&md5=641eb1d8d8fa5ab17191d1f382890a37.

Rokach, A., Pinkert, M., Nemet, D., Goldberg, A., & Bar-Dayan, Y. (2008). Standards in collaborative international disaster drills: A case study of two international search and rescue drills. *Prehospital and Disaster Medicine, 23*(1), 60–62. https://www.scopus.com/inward/record.uri?eid=2-s2.0-46349101984&partnerID=40&md5=eeb5f940b5ffa6efbed1274e80dddb6b.

Rose, A., & Kustra, T. (2013). Economic considerations in designing emergency management institutions and policies for transboundary disasters. *Public Management Review, 15*(3), 446–462.

Rubert, A., & Beetlestone, P. (2014). Tools to improve the management of transboundary river basins for disaster risk reduction. *Water Science and Technology: Water Supply, 14*(4), 698–707.

Sapountzaki, K., & Daskalakis, I. (2015). Transboundary resilience: The case of social-hydrological systems facing water scarcity or drought. *Journal of Risk Research, pp.*, 1–18.

Stefanelli, J. N., & Williams, S. (2011). Disaster strikes: Regulatory barriers to the effective delivery of international disaster assistance within the EU. *Journal of International Humanitarian Legal Studies, 2*(1), 53–83. https://www.scopus.com/inward/record.uri?eid=2-s2.0-84905504404&partnerID=40&md5=c27c56b2017172bc66638ae302f55c0e.

Tapia, A.H., Maldonado, E., Tchouakeu, L-M.N., Maitland, C., Zhao, K., Bajpai, K. (2010). Crossing borders, organizations, levels and technologies: IS collaboration in humanitarian relief. In *16th Americas Conference on Information Systems 2010, AMCIS 2010*, vol. 1, pp. 301–311, https://www.scopus.com/inward/record.uri?eid=2-s2.0-84870384019&partnerID=40&md5=1acb824123279badf4cc5412abf43f70.

UN-ESCAP (United Nations Economic and Social Commission for Asia and the Pacific). (2015). *Disasters without Borders: Regional resilience for sustainable development*. Bangkok: UN-ESCAP.

UNISDR (United Nations International Strategy for Disaster Reduction). (2007). *Hyogo framework for action 2005–2015: Building the resilience of nations and communities to disasters*. Geneva: UNISDR.

UNISDR (United Nations International Strategy for Disaster Reduction). (2011). *Global assessment report on disaster risk reduction: Revealing risk, redefining development*. Geneva: UNISDR.

UNISDR (United Nations International Strategy for Disaster Reduction). (2013). *Global assessment report on disaster risk reduction: From shared risk to shared value—the business case for disaster risk reduction*. Geneva: UNISDR.

UNISDR (United Nations International Strategy for Disaster Reduction). (2015). *Sendai framework for disaster risk reduction 2015–2030*. Geneva: UNISDR.

UNISDR (United Nations International Strategy for Disaster Reduction). (n.d.). PreventionWeb. http://www.preventionweb.net.

Xu, J., Gong, A. D., & Li, J. (2008). An understanding of international cooperation in disaster reduction in view of great Wenchuan earthquake. *Journal of Natural Disasters, 17*(6), 139–141.

Yang, D. (2008). Coping with disaster: The impact of hurricanes on international financial flows, 1970–2002. *B.E. Journal of Economic Analysis and Policy, 8*(1), 1–43. https://www.scopus.com/inward/record.uri?eid=2-s2.0-46749086346&partnerID=40&md5=c933dff330391370b09e387ad295f4d2.

Yang, A., & Zhang, H. (2014). The study on evaluation system of transboundary disasters emergency management ability based on synergetic model. *International Journal of Earth Sciences and Engineering, 7*(6), 2400–2406. https://www.scopus.com/inward/record.uri?eid=2-s2.0-84938335429&partnerID=40&md5=e8c42f5452edbc7c2ba89bc33d0e515d.

Chapter 5
Disasters Across Borders: Borderlands as Spaces of Hope and Innovation in the Geopolitics of Environmental Disasters

John Hannigan

5.1 Introduction

Disasters defy borders. To be sure, this stubborn refusal to stay within neatly drawn geopolitical lines has been evident for a very long time. The Laki volcanic eruption in southeast Iceland in 1873 left a global trail of destruction that reached from Siberia to the Gulf of Mexico, eventually contributing to the death of six million people worldwide. Much of this was due to its cascading impacts, notably crop failures and livestock losses in Iceland, France and Japan. A decade later, another volcanic event, the Krakatoa eruption in the Sundra Strait between Sumatra and Java, bequeathed a massive footprint, including tsunamis that reached as far as South Africa and 5 years of global cooling. More recently, the 2004 Indian Ocean tsunami, a consequence of a powerful earthquake that occurred off the west coast of Sumatra, resulted in 200,000 casualties in 14 countries, extending as far as Somalia. Not only do disaster agents such as earthquakes, volcanoes, tsunamis, typhoons and floods not respect territorial boundaries, but how well we cope with them continues to be significantly constrained and problematised by the presence of national borders. It makes sense then to not only speak of *disasters without borders* (Hannigan 2012) but also of *disasters across borders.*

Cross-border environmental disasters can assume different forms. In the examples cited above, a common disaster agent typically impacts multiple political jurisdictions, either concurrently or sequentially. Little (2010, p. 29) distinguishes among three type of infrastructure failures that involve spillover in engineered urban systems (and which may also be applied to environmental disasters): *cascading failure* (a disruption in one infrastructure causes a disruption in one or more other infrastructures), *escalating failure* (a disruption in one infrastructure exacerbates an independent disruption of a second infrastructure) and *common cause failure* (a

J. Hannigan (✉)
Department of Sociology, University of Toronto,
725 Spadina Ave., Toronto, ON M5S 2J4, Canada
e-mail: john.hannigan@utoronto.ca

M.A. Miller et al. (eds.), *Crossing Borders*, DOI 10.1007/978-981-10-6126-4_5

disruption of two or more infrastructures at the same time because of a common cause, for example a natural disaster). In several of the other chapters in this volume, the authors report that megaprojects undertaken in one nation produced negative effects for states downwind or downriver. In the Mekong River basin, hydropower dam construction upstream occurs at the expense of ecosystems and livelihoods (fishing, rice production) tied to the river in countries situated at lower elevations. Similar problems occur internally. The issue of borders has been prominent along the Narmada River in western India, especially with the Sardar Sarovar dam: the benefits of the dam flow mostly to the state of Gujarat, while most of those whose land and homes are submerged live in Madhya Pradesh (Basu 2012, p. 100). Trans-boundary smoke or haze pollution attributable to Indonesian forest and palm oil plantation fires has become a recurring problem for the neighbouring nations of Singapore and Malaysia. Environmental disasters originating in distant continents may take decades to fully manifest themselves. For example, scientific experts deem as possible a complete drying out of Lake Reba (also known as the 'Pink Lake') on the Cap-Vert peninsula of West Africa within the next 80 years, in significant part as the result of environmental pressures linked to climate change—notably increasing evaporation produced by global heating, leading to salinisation and alkalinisation of the soil (Zen-Ruffinen and Pfeifer 2013).

The incidence of compounding cross-border disasters with multiple causalities and cascading impacts is widely thought to have been increasing, especially in Asia. Douglass (2016) links this to several types of urban transitions taking place across the continent: the agglomeration and formation of mega-urban regions, the spatial polarisation of urbanisation in high-risk zones such as coastal regions and major river deltas, new forms and magnitudes of vulnerability in urbanising settlements, and the expanding ecological reach and demands of cities into rural areas and across administrative borders. Asia's accelerated urban transition, Miller and Douglass (2016) assert, 'is both a major source and target of increasingly frequent and costly environmental disasters'. In similar fashion, Forman and White (2011, p. 6) note that rapid urbanisation in the Dominican Republic has put an increasing proportion of the population at risk from disasters. In Santo Domingo, the capital, the trend in urban development has been to 'build up', with only haphazard and inconsistent enforcement of building codes that mandate earthquake resistance. Santiago is built on a fault line and contains a hydraulic dam that threatens to overflow in the event of intense seismic activity. Much of the migration to Dominican cities is to coastal areas, and specifically to flood plains and other insecure areas that are not subject to proper urban planning.

In this chapter, I propose an alternative way of conceptualising 'disasters across borders'. At present, there are three main ways of framing the issue (see below). The disaster administration approach posits that cross-border problems result from faulty communication and incomplete planning. The solution lies in undertaking enhanced administrative measures such as universalised operating procedures and best practices. The governance approach focuses on disaster risks and vulnerabilities that differentially impact those living in poor neighbourhoods. Centralized, bureaucratic service delivery impedes an effective response, as do neoliberal policies that are geared more to slum clearance and urban redevelopment. This is compounded where grassroots communities under threat span national borders. Improved governance means strengthening decentralized, community-based responses to natural disasters. The legal and normative approach is directed primarily at the challenges faced by a growing number of cross-border, temporary migrants. Possessing virtually no citizenship or human rights in the host nations, they live in a type of liminal zone. Solutions are not easy to come by, but are more likely to be found in the province of soft law rather than in formal treaties and agreements.

Rather than take national borders as fixed and immutable, I suggest that they are fluid, contested and transitional. This is especially relevant to the case of cross-border migration during times of disaster. As the Finnish geographer Anssi Paasi (2011, p. 62) observes, 'Rather than neutral lines, borders are often pools of emotions, fears and memories that can be mobilised apace for both progressive and regressive purposes'. Bordering, then, both separates us and brings us together. It allows certain expressions of identity and memory, Paasi notes, while blocking others. By adopting this perspective, we can open up new ways of reimagining zones impacted by cross-border environmental disasters as constituting 'spaces of hope' with distinctive ecological and sociological identities that transcend sovereign constraints. The emphasis here is not on generating specific new initiatives, although some such as humanitarian corridors for cross-border disaster migrants can be suggested, but rather on re-focusing the discussion on borderlands and their inhabitants.

5.2 Borders and Border Spaces

The recent exodus of asylum-seekers from the Middle East toward Western Europe has reignited a global debate over the fixity and meaning of borders. This ongoing conversation features two conflicting viewpoints. One, formulated around the notion of the 'borderless world', draws on humanitarian discourse to assert a universal right to secure living arrangements and democratic freedom. By contrast, the doctrine of territoriality asserts the right of sovereign nations to exert exclusive force and law within a particular territory, and to enforce this by controlling access across their internationally recognised borders. In the case of the currently unfolding Syrian 'diaspora', some nations, notably Germany (with the support of Sweden, Austria and France) have drunk deeply (at least initially) from the well of compassion and moral conscience, opening their borders to hundreds of thousands of refugees and extending a 'welcome culture' (Anonymous 2015, 23–4). Others, notably Hungary and Macedonia, have rushed to seal their borders with barbed wire and walls, justifying this on the grounds that asylum-seekers are in fact economic migrants who, if admitted, would overwhelm the sanctity and robustness of traditional cultures and communities.

Kolossov (2005) compares and contrasts two theoretical perspectives, the political and the global, that have centrally informed border studies. There are two versions of the *political* approach. In the 'realistic' paradigm, the boundaries between states are interpreted as strict dividing lines protecting state sovereignty and national security. According to the 'liberal' paradigm, the principal function of state borders is to connect neighbours and to eliminate various international interactions; accordingly, the principal task at hand is 'to eliminate territorial disputes and border conflict and to develop a cross-border communication and infrastructure' (p. 612). More recently, a *global* approach has begun to garner considerable attention. This paradigm privileges the role played international networks. In this view, the global expansion of networks has transformed state boundaries into virtual lines, replacing them with multiple economic and cultural boundaries.

Asia has its own border issues. Introducing a special issue of the *Journal of Borderland Studies* on the topic of 'Asian borderlands', Willem van Schendel and Erik de Maaker (2014) remind us that many Asian borders were not demarcated until the second half of the twentieth century. Even today the exact location of these borders remains undecided, especially in regions whose terrain is characterised by mountains, rainforests, desert and marshland. Ganguly-Scrase (2012, p. 77) identifies the borders between nation states in South Asia as being 'largely artificial constructs arbitrarily drawn through ethnic, religious, cultural, and economic communities'. The fluidity of borders, which often isolates communities, is exemplified in situations where a river that often changes course is designated as a national border.

Nonetheless, state borders have taken on an increased importance in the face of unprecedented economic growth in the region. Even as Asian nations open their economies to the wider world, they have reinforced and militarised national borders. This echoes what has been happening recently within the EU, where Member States increasingly undercut the policy of free movement by imposing a stern regime of internal border controls. Border control, Barker (2016) observes, 'patrols the boundaries of belonging, sorting out who belongs and who does not'. This hardening of borders, van Schendel and Maacker (2014, pp. 3–4) note, requires increased demarcation and monitoring of cross-border mobility of people and goods, and applies to both land borders and maritime environments. Translating Cunningham's (2004) observations on the movement of people at the US-Mexico border to the Asia Pacific region, Ganguly-Scrase and Lahiri-Dutt (2012, p. 14) point out that borders are contradictory insofar as they are simultaneously sites of movement (of people, goods, capital) and sites of enclosure where 'rules are formulated, enacted, and negotiated concerning who has the right to mobility and who does not'.

5.3 Beyond Borders

By their very nature, borders are never immutable, that is, they are 'always in a state of becoming' (Mountz 2011, p. 65). The regions of Alsace-Lorraine on the French-German border, and Silesia, on the German-Polish-Czech border, constantly shifted nationalities throughout the nineteenth and twentieth centuries, depending on who won or lost militarily. Some land disputes between nations over border areas have gone on for centuries. For example, a coastal land dispute between Chile and Peru has ebbed and flowed since the Pacific War more than a century ago, while Argentina recently revived a long-time border dispute with Chile over ownership of 100 miles of contested land known as the Southern Icefields which contain the largest reserve of potable water in the world (Fendt 2010). The geopolitics of the Asia Pacific region have recently been dominated by an escalating string of confrontations between China and its neighbours over ownership of the Paracel and Spratly Islands.

Other ways of conceptualising borders as transnational spaces are possible here. Some of these are designed to enhance the securitisation of citizenship and the state. A decade ago the term 'the seam' was introduced into US military jargon. The seam refers to 'a zone between inside and outside national space, where old geopolitical divisions no longer hold; the border between police and military authority is blurred, and so too is the line between crime and terror' (Cowen 2010, p. 79). Maritime ports, for example, are increasingly becoming transformed into 'exceptional seam spaces', where the border represents a special 'transitional zone' rather than a two-dimensional, bifurcating line across absolute space (Cowen 2010, pp. 78–9). Kolossov (2005, p. 623) cites the concept of 'border space' that embraces not only the area along the boundary but also internally deep within state territory. Examples of this are international airports and special customs or free economic zones.

Other reconceptions of borders and bordering are more progressive. It is possible, for example, to reject the demarcation of space on purely political or jurisdictional grounds in favour of the idea of managing or governing a *commons*. Commons are collectively shared resource sites. Without regulation or cooperation, they will sooner or later be exploited and exhausted. The decade-long deliberations by the United Nations Law of the Sea Conference (UNCLOS III) highlights this. In order to protect untapped seabed resources from uncontrolled mining and oil drilling, UNCLOS redefined maritime space, designating the seafloor as lying beyond the regulatory geographies of state sovereignty (Hannigan 2016, p. 68) and under the supervision of the International Seabed Authority (ISA).

One promising cutting edge approach involves interrogating the very definition of a border. Chris Rumford (2011) has proposed that to understand borders fully we need to stop 'seeing like a state' and start 'seeing like a border'. In the first instance, this entails decoupling borders from fixed physical lines of demarcation. People can construct borders as local, national or transnational in scale, or reconfigure the border as a 'portal' (e.g. airport, maritime port, railway station). Second, bordering is not inevitably the business of the state. Rather, it could and should be the business of a variety of non-state actors ranging from NGOs and grassroots activists to ordinary citizens. Even agricultural producers who are striving to create a 'Protected Designation of Origin' status as a means of branding local produce such as Melton Mowbray pork pies and Stilton cheese are 'borderworkers', who in their own way 'are active in constructing, shifting and erasing borders' (p. 67). Third, the capacity to make or undo borders can become a source of political capital, that is, expanding constituencies to include non-citizens such as migrants and refugees who may coalesce at the borders. Fourth, borders and borderlands are fundamentally sites of contestation and claims-making. In sum, seeing like a border 'involves the recognition that borders are woven into the fabric of society and are the routine business of all concerned' (Rumford 2011, p. 68).

How then, can we start 'seeing like a border'? One way is to start thinking of borders as social and cultural spaces in their own right. Social scientists have labeled these as *borderlands*. Anthropologists Alexander Horstmann and Reed Wadley (2006) argue that rather than being dead zones, borderlands are vibrant sites of human agency. The authors draw a contrast between state borders that are characterised by essentialised tradition and community, and borderlands, which are complex social systems that question the nature of the state. The negotiations between populations and the state, they say, are particularly intense in the borderlands of Southeast Asia, where, with the possible exception of Singapore, the state's sovereignty in border regions may often be marginal.

Horstmann (2002) illustrates this empirically in a series of ethnographic case studies of networks of border people in between Thailand and Malaysia. The Patani Malays on the east coast of southern Thailand, the Sam Sam on the west coast, and the Kelantan Thais on the east coast of northeast Malaysia are ethnic and religious minorities (Thai-speaking Muslims in southern Thailand and Buddhist Thais from northern Malaysia) who are located in a 'diasporic trap'. On the one hand, they have been given citizenship, albeit a special category that withholds state resources and certain rights. At the same time, they embrace a form of dual citizenship, defined in part by holding double identity cards and partly through practice. Horstmann reports that they acquire multiple citizenship rights through various means: by registering the birth of their children just across the border, by marriage and by applying for naturalisation. Upon acquiring new citizenship rights, they carefully keep their existing identity cards. Life in these borderlands is anything but easy. For example, Thai fishermen from Ban Sarai in southern Thailand who fish illegally in Malaysian waters are quite vulnerable, facing the possibility of arrest and deportation by the police and economic exploitation by Malaysian middlemen. Nevertheless, trapped ethnic minorities who are accorded inferior positions in the space of the nation state are able to find ways to liberate and empower their lives through border-crossing practices.

5.4 Cross-Border Environmental Disasters

As noted in the introduction to this chapter, the incidence of compounding cross-border disasters with multiple causalities and cascading impacts is assumed to have been sharply increasing in recent years, although firm statistics on this are rather hard to come by. In discussing the impact of climate change on the cross-border displacement of people, Kälin and Schrepfer (2012) note that the number of people displaced by climate-related sudden-onset disasters is indeed very substantial—almost 95 million people between 2008 and 2010, according to the Norwegian Refugee Council—and will likely increase in the foreseeable future. Most, however, are displaced internally and do not appear to cross borders. There are some notable exceptions. The authors report that during a visit in Mozambique and South Africa in 2008 they heard anecdotal evidence that people from Mozambique and Malawi looked quite regularly for refuge in neighbouring countries when displaced by flooding (p. 14).

In the case of disaster, differing laws and norms across national borders constitute barriers to optimal relief and resettlement efforts. In his classic case study of the 1954 Rio Grande Flood, Clifford (1955) noted that core values such as *dignidad* (dignity, imbued with patriotic or racial pride) forced officials in the Mexican community of Piedras Niegras to refuse material goods from relief services agencies in the twin border city of Eagle Pass, Texas. In a subsequent study, Stoddard (1961, 1969) found that the same applied to the Mexican American population of Eagle Pass, who preferred receiving offers of inadequate housing and bare subsistence rations from kinfolk to the well-balanced nutritional meals, hospital aid, and physical comforts available to them through Anglo-dominated professional relief agencies. In cross-border disasters conflict can arise over differing expectations with regard to carrying weapons, the use of restricted pharmaceutical products, and standards of environmental protection. Edwards (2009) points out that a particularly difficult issue is the treatment of women responders and victims. Women disaster victims in traditional patriarchal societies may be required to receive food or other care only after the men have been served, or may be denied contact with male responders, thus potentially lengthening the time they must wait for care which can only be provided by women first responders.

Administrative solutions usually address and attempt to redress technical glitches such as those related to communicating and coordinating warnings about tsunamis, earthquakes or typhoons. Typically, cross-border disasters are conceptualised here as posing a host of organisational and managerial challenges, which can be largely met through better planning and coordination. Thus, Tricia Wachtendorf, whose MA thesis focused on cross-border (Canadian-American) interaction during the 1997 Red River Flood, describes disasters that extend beyond national borders as *trans-system social ruptures* (TSSRs). TSSRs, she says, 'have implications for empirical investigation, operational management and policy framing' (2009, p. 380). Such ruptures are most likely to be repaired in networked 'systems' with enhanced levels of dependence and interdependence. These networks can be physical (linked road systems, tightly coupled electrical grid lines and transformers) or social (interactions between organisational actors in the form of the supply and exchange of information, personnel and material resources). In the case of the Red River Flood, Wachtendorf found that routine, cross-border interaction during non-crisis times facilitated cooperation during the flood emergency: "Pre-established ties and pre-event mechanisms for interaction can work to correct a cross-border system failure more readily than if those social ties were not already established" (2009, p. 386). In the case of international trans-boundary disasters such as this, Edwards (2009) advises, 'a pre-disaster operational plan is essential to ensure that it is functional in the breadth needed'.

Miller and Douglass (2016) criticise this type of 'disaster administration' orientation on the grounds that it neglects or overlooks critical historical, social and cultural dimensions of environmental disasters in favour of a 'best practices' approach which spells out a standardised set of 'lessons learned' and universalised operating procedures. In the disaster administration approach, the prime responsibility for disaster response rests with public agencies, not at the grassroots level. If things go wrong, it is because of lack of trust or due to communication lapses between cross-border 'partners'. Consistent with the closed systems theories which dominated the study of organisations in the 1950s and 1960s, a steady state or equilibrium is assumed to be the normal state of affairs. Disasters are treated as 'ruptures' which must be repaired so things can be returned to a state of 'normality'. There does not seem to be much room here for significant innovation and change, although Kris Berse in his chapter in this book argues that a network of city-to-city cooperation in Asia that extends relief and technical assistance and transfers best practices, even where localities do not share a border, can 'provide a flexible mechanism for effecting changes at any point in the disaster cycle'.

In Asia and the Pacific, cross-border disaster risk management issues have increasingly been treated as a *governance* problem, with inclusiveness and cooperation as key determinants' (see Guilloux chapter in this volume). By placing applied research on disasters under the umbrella of governance rather than management, Miller and Douglass (2016) argue, we can improve our understanding of how closed and ineffective governance regimes exacerbate social inequality, placing the poor and the powerless at greater risk from disasters. Repairing system social ruptures is not sufficient. The rupture of a disaster often exposes underlying vulnerabilities in urban areas. In New Orleans, for example, the poor black parishes were inundated with water when Hurricane Katrina provoked a breach in the protective levees, while the middle-class, tourist-oriented city on higher ground endured only minor damage. In the 2011 floods in Bangkok, Thailand, the low-income suburbs were heavily flooded while the affluent inner-city area with its financial district was more or less protected (Chintraruck and Walsh 2016). In their chapter in this volume, Friend and Thinphanga focus on the Mekong region, showing that vulnerabilities and risks are distributed unevenly across different groups of people, the administrative boundaries of cities, and national boundaries.

Governance issues in cross-border disasters often revolve around the gap between centralised, bureaucratic service delivery and decentralised, community-based responses. Recent field studies in Indonesia and Thailand demonstrate that residents of marginalised flood-prone neighbourhoods have come to distrust official state programmes of flood management, relying instead upon 'their own independent coping mechanisms and acquired intergenerational knowledge to pursue their needs and interests outside official channels' (Miller and Douglass 2015, p. 2). This is compounded where these grassroots communities span national borders. As the classic sociological studies of flood response across the Rio Grande River indicate (see above), self-reliance trumps official disaster management delivery, even where the latter may be more effective.

A third approach to cross-border disasters treats them as essentially *legal-normative* challenges. Legal-normative issues are most often linked to the flow of people across national borders in environmental disasters such as floods and famine, where temporary migrants face a host of difficulties related to a lack of citizenship and human rights. Rather than signing formalised bilateral treaties, the most promising solutions here are said to reside in the domain of 'soft law' (Kälin and Schrepfer 2012).

One noteworthy attempt to theorise the protection of people displaced across international borders in the context of natural disasters and the adverse effects of climate change is the *Nansen Initiative*. Funded primarily by the governments of Norway and Switzerland, this three-year consultative process is organised around three 'pillars': international cooperation and solidarity; standards for the treatment of affected people regarding admission, stay and status; and operational responses including funding mechanisms and responsibilities of international humanitarian and development actors. Kälin and Schrepfer (2012, p. 61) outline a plan to recast cross-border displacement in environmental disasters under the auspices of the United Nations High Commissioner for Refugees that parallels that proposed in the Nansen Initiative. New instruments here, they specify, should contain four elements. People in need of protection (beneficiaries) should be defined as those whose return to their country of origin would be (i) legally impermissible (ii) not feasible (iii) unreasonable in terms of humanitarian considerations. Beneficiaries should be entitled to enter countries of refuge and remain there as long as the obstacles to their return exist. If, after a prolonged period of time, it becomes clear that return is impossible, permanent admission should be granted. Beneficiaries should be entitled to an array of status rights, including access to the labour market, housing, health services and education; protection against discrimination; and freedom of conscience, religion and opinion. Finally, the institutional arrangements governing the tenure of cross-border disaster migrants should be defined, notably those referring to financial and operational support.

The moral imperative behind approaches such as the Nansen Initiative dictates that fundamental human rights do not stop at the border. Unfortunately, enforcing this on the ground poses a formidable challenge. As Horstmann and his colleagues demonstrate in their ethnographic research on ethnic borderlands in Thailand (see above), even possessing double citizenship is no automatic guarantee of legal security and social acceptance on either of the dividing line. Kälin and Schrepfer (2012, p. 58) observe that there is seldom much understanding among government actors either of the rights related to admittance or of the status rights of environmental migrants living in foreign territories. Nor is there an established intergovernmental forum or process to address the issue in a consistent way. In the present context, they conclude,

> it is likely that negotiating a convention on cross-border movement of persons in the context of climate change would be very difficult because of largely incompatible interests of potential countries of origin and countries of destination. (p. 70)

A more appropriate approach lies in 'soft law'—for example, highlighting gaps in present international law and using these as a basis for future action. This course of action would mainly though not exclusively address the many gaps in present international law identified above (Kälin and Schrepfer 2012, pp. 71–2).

5.5 Disaster Cooperation and Disaster Diplomacy

Foreign offers of assistance during a disaster event are occasionally rebuffed altogether. Documented examples of this in Asia (Hannigan 2012) include: the refusal of international aid by the Chinese government during the 1976 Tangshan earthquake, in which as many as 700,000 residents may have perished; Japanese reluctance to accept international aid during the Great Hanshin Earthquake of 1995; the sealing of the border by the ruling military junta in Mynamar (Burma) after the 2004 Indian Ocean earthquake/tsunami, and again in 2008 after Cyclone Nargis; and North Korea's blocking of an offer from South Korea to transport emergency aid overland by rail to victims in a massive 2004 explosion in which a power line struck wagons of oil and chemicals. Reasons for this vary: a strategic move aimed at building domestic political support by demonstrating that a regime can cope alone without seeking external assistance; paranoia that foreign powers will use the occasion to invade; or simply a sense of profound isolationism. Having made its point, the recalcitrant state will then usually relent. North Korea permitted relief supplies to enter the country by sea, thereby avoiding visuals depicting lines of South Korean trucks crossing the demilitarised zone (but also delaying the arrival of aid by several days). China continued to refuse international aid but opened the disaster zone to earthquake engineering experts from the West.

More often than sealing the border and refusing aid, a nation will acknowledge that it needs help, but chafe at the heavy-handed intrusion of international aid workers and donors, something that has been magnified by recent efforts at politicising disaster management and humanitarian relief for foreign policy reasons. This was widespread in Indonesia after the 2004 Indian Ocean tsunami, where the United States and Australia took advantage of a dire situation to expand their position as regional political players. Sometimes, humanitarian assistance will collide with national laws and regulations. For example, during Hurricane Katrina food aid packages from NATO and the EU that included British beef were deemed not legally permitted because of an ongoing ban related to Mad Cow disease—the donations had to be warehoused and distributed to other nations whose food importation standards allowed them to accept the supplies (Edwards 2009).

On a more positive note, cooperation across borders can potentially provide an opportunity for the improvement of political relations. Despite years of mutual suspicion, in the immediate aftermath of the 2010 Haiti earthquake the Dominican Republic unconditionally opened its border; dispatched ambulances, helicopters and search and rescue teams; and facilitated the opening of a critical humanitarian corridor (see below) for the delivery of international assistance into devastated

Port-au-Prince (Forman and White 2011, p. 11). For a while, this new 'Good Samaritan' image re-framed Dominican-Haitian relations, leading to greater cooperation on issues of health, the border, security, climate change/the environment and cross island trade, although not human rights and migration (Kristensen and Wooding 2013, p. 3). Within a few years, however, this trend toward more soft diplomacy hardened, especially when the Dominican Republic began to step up the rate of repatriation of Haitians who had been displaced by the earthquake. Fulvio Attinà (2012, p. 29) has distinguished between disaster diplomacy and disaster cooperation. The former occurs primarily at the bilateral level and refers to situations where feuding countries temporarily suspend their differences in order to work together on disaster relief efforts. This is most likely to occur whether two nations share a land border or are located near one another. Research by Ilan Kelman (2011) indicates that disaster diplomacy does have the potential to reduce bilateral conflict, but only where a firm foundation for cooperation has been established prior to the disaster event.

These days, disaster cooperation must also take place at the global level, which Attinà defines as 'the wide set of relations and actions that are put in place by states and mostly international organisations to deal with disaster and everyday problems that cause large-scale shocks and setbacks' (Attinà 2012, p. 29). This has variously been called the 'international disaster relief system' (Green 1977; Cuny 1983); the 'international relief network' (Kent 1987); and the 'global policy field of natural disasters' (Hannigan 2012). Initially, disaster cooperation entailed sending money and supplies to stricken offshore populations, but today it is more likely to assume the face of scores of international non-governmental organisations descending upon the disaster area, often in competition with one another. For example, immediately after the October 2005 South Asia earthquake disaster, over 100 international organisations—United Nations agencies, INGOs (international nongovernmental organisations, European, NATO and bilateral partners—arrived in the earthquake zone to aid the relief effort (Hicks and Pappas 2006, p. 43).

With environmental disasters, which are larger and more complex, cooperation is likely to require a more complex form of diplomacy. Effective responses to those environmental problems that are said to magnify the probability and severity of disaster, especially in the nations of the South, demand global treaty-making efforts and other forms of collective action. Typically, global environmental treaty making and international cooperation that are needed to implement effective treaties

> require extensive consensus building, which in turn, requires effective ad hoc representation of all the stakeholders, face-to-face interaction among skilled representatives of the stakeholding interests, a real give-and-take aimed at maximizing joint gains, facilitation by appropriate neutral parties at various points in the process, informality that allows the parties to speak their minds, and extensive pre-negotiation that ensures opportunities for joint problem solving. (Susskind 1994, p. 61)

Even if this could be done, the international legal and policy framework for disaster management 'is insufficient to handle not only contemporary disaster threats but also the problems caused by the mega-disasters that will inevitably strike densely populated urban areas' (Haase 2010, p 227). This is especially the case where patchy legal structures and policy declarations inflate the potential for legal disputes and confusion during emergency situations between nations.

5.6 New Ways of Thinking About Cross-Border Environmental Disasters

The notion of borderlands as crucibles of social and cultural change raises the possibility they could function as 'spaces of hope' (to use the title of a well-known 2000 book by the geographer David Harvey) in environmental disasters. Rather than think of environmentally displaced people who move across borders during disasters as being simply helpless victims who need to be rescued and resettled by NGOs, it might make sense to consider how they improvise, cope, and employ knowledge accumulated over the course of similar disaster events in the past. This corresponds with an approach to urban disaster governance that endorses a strategy of grassroots self-reliance incorporating coping mechanisms and acquired intergenerational knowledge divorced from official political authority (Miller and Douglas 2015, p. 2). As the contributors to Hortstman and Wadley's book (2006) demonstrate, people are not only constrained by borders; the crossing of borders opens up new options of agency.

A borderland may have an *ecological* character that spans politically established boundaries. River basins, for example,

> represent closely integrated natural regions, while at the same time they constitute a basis of settlement and transportation systems and often determine boundaries between historically created territorial and cultural communities. (Kolossov 2005, p. 627)

As Alam observes in his chapter in this volume on trans-boundary disputes between India and Bangladesh, there is often a fundamental discordance between political and ecological boundaries that generates strong narratives and counter-narratives on both sides of the boundary. Any change in the ecosystem, for example the pollution of transnational rivers, produces adverse socio-economic impacts on the other side, notably in the form of habitat destruction, economic decline and population displacement. Disaster governance operates within an urban matrix that encompasses the ecological reach of cities into remote and rural areas and spills across borders.

This phenomenon has been recognised for a long time in the bioregionalism movement, whose central philosophical belief states that societies should be organised by commonality of place rather than by arbitrary political boundaries. A bioregion here refers both to a geographic terrain and a terrain of consciousness (Hannigan 2011, p. 328). While bioregionalism is unlikely to displace political boundaries as the formal structure for governance any time soon, it is more likely to gain traction in a form where it denotes 'a place or community linked to nature and with which residents identify in historical, cultural and material terms' (Lipschutz 1999, p. 101). This is made more difficult where a bioregion has become the site of competition for limited resources in a rapidly urbanising region. As Kelly and Adger's (2000) classic study in the Red River Delta in northern Vietnam demonstrates, the privatisation of state-managed cooperatives, the shift to aquaculture, and the enclosure and conversion to other agricultural uses of mangrove forests spelled disaster for poorer households, who were displaced from their traditional livelihoods in coastal and offshore fisheries and rendered more vulnerable to being battered by tropical cyclones or typhoons. Nonetheless, reconceptualising borderlands as distinct ecological units that span national boundaries may be one way of reworking how we think about cross-border environmental disasters.

Another possibility is to rework the concept of *humanitarian corridors*. To date, humanitarian corridors have mostly been applicable to conflict situations such as civil wars where INGOs are attempting to facilitate the safe transit of humanitarian aid in and refugees out of the crisis region. This has not been universally popular. Aid workers worry that there is a risk of humanitarian corridors becoming politicised, notably in the service of opposition forces who see them as a way of improving their strategic territorial position. Nonetheless, there are some scattered cases of humanitarian corridors being successfully implemented during and in the aftermath of natural disasters. For example, during the catastrophic 2010 earthquake in Haiti, a humanitarian corridor, both by land and sea, was opened between Port-au-Prince and Santo Domingo, the capital of the Dominican Republic. Santo Domingo served as an international logistics centre for cargo planes ferrying aid packages, which were then transported along the corridor, crossing the open border at Jimani. In reconceptualising such corridors within the context of cross-border environmental disasters, it might make sense to expand the emphasis from the transit of material aid such as food, water and medical supplies to the movement of people displaced by the disaster. A parallel situation is the Syrian diaspora; Giusy Nicolini, the mayor of Lampedusa, a small island at the southernmost tip of Italy that receives many fleeing from North Africa, has advocated the creation of humanitarian corridors to transport and distribute migrants across Europe (Palet 2016). In cross-border environmental disaster situations, such corridors might prove useful in moving those who have been displaced from their homes to temporary settlements in third nations more amenable to their ethnic, religious or linguistic identities.

5.7 Conclusion

As Miller and Douglass (2015, p. 1) have observed, compound environmental disasters produce disruptions that cannot easily be contained within existing political jurisdictions, necessitating the emergence of progressive trans-border networks, relationships and connections based upon common problems, ideas, knowledge and technologies within and among nation-states. In the immediate term, there are a number of specific measures that can be implemented in an effort to collaboratively deal with the effects of disasters that traverse sovereign territories: intercity aid networks, sub-national administrations in border regions, temporary protected status designation, and perhaps even the establishment of humanitarian corridors for environmentally displaced migrants. In the longer term, however, it is crucial that nations in Asia and the Pacific reconceptualise the meaning of borders and border populations. In particular, we need to visualise these in spatial, sociological, and ecological terms, imagining for example, the existence of borderlands as both distinct bioregions and liminal zones possessing both greater freedom and inclusivity.

In so doing, disaster researchers might profit from expanding their horizons to other allied fields of research, notably those related to diaspora, migration and refugee settlement. Increasingly millions of people displaced by war and internal strife reside in temporary camps, some of which have been in existence for decades. Most of these camps are situated well within national borders, but others can be found offshore (Australia) or at transit points between nations (Calais, France). Inner cities in South Africa are full of economic migrants from across Africa. Living a kind of 'tactical citizenship' and lacking a legible legal identity, they reside in a 'liminal' world, suspended somewhere between the formal and informal city, where applying 'good governance' principles is not a very effective policy option (Kihato 2017). In both of these cases, borders are as much a state of mind as a matter of legal and political jurisdiction. Natural disasters of the future, especially those that are environmentally linked, are likely to create massive numbers of temporary migrants, many of whom will find their way into cities. Coping with this will require transcending conventional disaster planning approaches and undertaking a more fundamental rethinking of the nature of boundaries, citizenship and livelihoods. As I have suggested in this chapter, one possibility here is to visualise borderlands as 'spaces of hope' rather than catchment areas for those trapped by disasters that defy borders.

References

Anonymous (2015). Germany! Germany!. *The Economist*, 12 September, pp. 23–4, http://www.economist.com/news/briefing/21664216-ordinary-germans-not-their-politicians-have-taken-lead-welcoming-syrias.

Attinà, F. (2012). Disaster and emerging policies at the international and social systems levels. In F. Attinà (Ed.), *The politics and policies of relief, aid and reconstruction* (pp. 21–41). Houndmills: Palgrave Macmillan.

Barker, V (2016). Reflections on Dutch border practices. *Border Criminologies blog*, University of Oxford, Faculty of Law, 18 March, https://www.law.ox.ac.uk/research-subject-groups/centre-criminology/centreborder-criminologies/blog/2016/02/reflections-dutch.

Basu, P. (2012). Aftermath of dams and displacement in India's Narmada River Valley: Linking compensation policies with experiences of resettlement. In R. Ganguly-Scrase & K. Lahiri-Dutt (Eds.), *Rethinking displacement: Asia Pacific perspectives* (pp. 95–114). Farnham: Ashgate.

Chintraruck, A., & Walsh, J. (2016). Bangkok and the floods of 2011: Urban governance and the struggle for democratisation. In M. A. Miller & M. Douglass (Eds.), *Disaster governance in urbanising Asia* (pp. 195–209). Singapore: Springer.

Clifford, R. A. (1955). *Informal group actions in the Rio Grande flood: A report to the committee on disaster studies*. Washington DC: National Research Council.

Cowen, D. (2010). Containing insecurity: Logistic space, U.S. port cities, and the 'war on terror'. In S. Graham (Ed.), *Disrupted cities: When infrastructure fails* (pp. 69–83). London: Routledge.

Cunningham, H. (2004). Nations rebound. Crossing borders in a gated globe. *Identities, 11*(3), 329–350.

Cuny, F. C. (1983). *Disasters and development*. New York: Oxford University Press/Oxfam America.

Douglass, M. (2016). The urban transition of disaster governance in Asia. In M. A. Miller & M. Douglass (Eds.), *Disaster governance in urbanising Asia* (pp. 13–43). Singapore: Springer.

Edwards, F. L. (2009). Effective disaster response in cross-border events. *Journal of Contingencies and Crisis Management, 17*(4), 255–265.

Fendt, L. (2010). Argentina revives longtime border dispute with Chile in the Patagonian ice fields. *Santiago Times*, 20 May.

Forman, J. M., & White, S. (2011). *The Dominican response to the Haiti earthquake: A neighbor's journey*. Washington, DC: Center for Strategic & International Studies.

Ganguly-Scrase, R. (2012). Neoliberal development and displacement: Women's experiences of flight and settlement in South Asia. In R. Ganguly-Scrase & K. Lahiri-Dutt (Eds.), *Rethinking displacement: Asia Pacific perspectives*. Farnham: Ashgate.

Ganguly-Scrase, R., & Lahiri-Dutt, K. (2012). Dispossession, placelessness, home and belonging: An outline of a research agenda. In R. Ganguly-Scrase & K. Lahiri-Dutt (Eds.), *Rethinking displacement: Asia Pacific perspectives* (pp. 3–29). Farnham: Ashgate.

Green, S. (1977). *International disaster relief: Towards a responsive system*. New York: McGraw Hill/Council on Foreign Relations.

Haase, T. (2010). International disaster resilience: Preparing for transnational disaster. In L. K. Comfort, A. Boin, & C. C. Demchak (Eds.), *Designing resilience: Preparing for extreme events* (pp. 220–243). Pittsburgh: University of Pittsburgh Press.

Hannigan, J. (2011). Implacable foes or strange bedfellows? The promise and pitfalls of econationalism in a globalized world. In T. W. Harrison & S. Drakulic (Eds.), *Against orthodoxy: Studies in nationalism* (pp. 314–332). Vancouver: UBC Press.

Hannigan, J. (2012). *Disasters without borders: The international politics of natural disasters*. Cambridge, UK: Polity Press.

Hannigan, J. (2016). *The geopolitics of deep oceans*. Cambridge, UK: Polity Press.

Hicks, E., & Pappas, G. (2006). Coordinating disaster relief after the south Asian earthquake. *Society, 43*(5), 42–50.

Horstmann, A. (2002). Dual ethnic minorities and the local reworking of citizenship at the Thailand-Malaysian border. CIBR Working Papers in Border Studies, no. CIBR/WP02–3, http://www.qub.ac.uk/cibr/WPpdffiles/CIBRwp2002_3_rev.pdf.

Hortsmann, A., & Wadley, R. L. (2006). Introduction. In A. Hortsmann & R. L. Wadley (Eds.), *Centering the margin: Agency and narrative in Asian borderlands* (pp. 1–26). New York: Bergdahn Books.

Kalin, W & Schrepfer, N 2012, *Protecting people crossing borders in the context of climate change. Normative gaps and possible approaches*. United Nations High Commission for Refugees, Geneva. February, http://www.unher.org/protect.

Kelly, P. M., & Adger, W. N. (2000). Theory and practice in assessing vulnerability to climate change and facilitating adaptation. *Climatic Change, 45*, 5–17.

Kelman, I. (2011). *Disaster diplomacy: How disasters affect peace and conflict*. London: Routledge.

Kent, R. C. (1987). *Anatomy of disaster relief: The international network in action*. London: Pinter Publishers.

Kihato, C. W. (2017). The liminal city: Gender, mobility and governance in a twenty-first century African city. In J. Hannigan & G. Richards (Eds.), *The handbook of new urban studies*. London: Sage.

Kolossov, V. (2005). Border studies: Changing perspectives and theoretical approaches. *Geopolitics, 10*(4), 606–632.

Kristensen, K., & Wooding, B. (2013). *Haiti/Dominican Republic: Upholding the rights of immigrants and their descendants.* Oslo: Norwegian Peacebuilding Resource Centre.

Lipschutz, R. (1999). Bioregionalism, civil society and global environmental governance. In M. V. McGinnis (Ed.), *Bioregionalism* (pp. 101–119). London: Routledge.

Little, R. G. (2010). Managing the risk of cascading failure in complex urban infrastructures. In S. Graham (Ed.), *Disrupted cities: When infrastructure fails* (pp. 27–39). London: Routledge.

Miller, M. A., & Douglass, M. (2015). Introduction: Decentralizing disaster governance in urbanizing Asia. *Habitat International, 52*, 1–4.

Miller, M. A., & Douglass, M. (2016). Disaster governance in an urbanising world region. In M. A. Miller & M. Douglass (Eds.), *Disaster governance in urbanising Asia* (pp. 1–12). Singapore: Springer.

Mountz, A (2011). Border politics: Spatial provision and geographical precision. In: Johnson, C., Jones, R., Paasi, A., Amoore, L., Mountz, A., Salter, M., & Rumford, C. (eds.), *Interve ntions on rethinking 'the border' in border studies. Political Geography*, vol. 30, pp. 65–6.

Paasi, A. (2011). Borders, theory and the challenge of relational thinking. In: Johnson, C., Jones, R., Paasi, A., Amoore, L., Mountz, A., Salter, M., & Rumford, C. (Eds.), *Interventions on rethinking 'the border' in border studies. Political Geography*, vol. 30, pp. 62–3.

Palet, L.S. (2016) Giusy Nicolini: Governing at the world's most dangerous border, OZY, 12 March, http://www.ozy.com/rising-stars/giusy-nicolini-governing-at-the-worlds-most-dangerous-border/60218.

Rumford, C 2011, Seeing like a border. In: Johnson, C., Jones, R., Paasi, A., Amoore, L., Mountz, A., Salter, M., & Rumford, C., *Interventions on rethinking 'the border' in border studies. Political Geography*, vol. 30, 67–8.

Stoddard, ER (1961). *Catastrophe and crisis in a flooded border community: An analytical approach to disaster emergence.* Ph.D. thesis, Department of Sociology and Anthropology, Michigan State University.

Stoddard, E. R. (1969). The U.S. Mexican border as a research laboratory. *Journal of Inter-American Studies, 11*(3), 477–488.

Susskind, L. E. (1994). *Environmental diplomacy: Negotiating more effective global agreements.* New York: Oxford University Press.

Van Schendel, W., & De Maaker, E. (2014). Introduction to the special issue: Asian borderlands: Introducing their permeability, strategic uses and meanings. *Journal of Borderlands Studies, 29*(1), 3–9.

Wachtendorf, T. (2009). Trans-system social ruptures: Exploring issues of vulnerability and resiliency. *Review of Policy Research, 26*(4), 379–393.

Zen-Ruffinen, B. & Pfeifer, H.R. (2013) Coastal waters in arid climate: a 'natural chemical risk' for sustainable land management and public health. Paper delivered at the Workshop, Linking Sustainable Development, Global Migration, Climate Change Adaptation and Disaster Risk Reduction, Faculty of Geosciences and Environment, The University of Lausanne, Switzerland, http://www3.unil.ch/wpmu/dixansfgse/workshop.

Part II
Transboundary Governance in Riparian Regions

Chapter 6
Urban Transformations Across Borders: The Interwoven Influence of Regionalisation, Urbanisation and Climate Change in the Mekong Region

Richard Friend and Pakamas Thinphanga

6.1 Introduction

There is growing global interest in urbanisation and urban risks, particularly around climate change and disasters, and the corresponding need to build resilience. Much of the struggle to avoid climate catastrophe will be played out in the urban arena. Urbanisation is a major contributor to greenhouse gas (GHG) emissions and global climate change, while patterns of urbanisation place a higher concentration of people and economic assets in vulnerable locations. Equally, global policy debates acknowledge the growing awareness of the leading role that cities can play, and the potential of city governments for filling the vacuum of inaction over global environmental challenges.

Much of the effort around disaster risk reduction and climate change adaptation has focused on local and community dimensions (Cutter et al. 2008). Related literature has also highlighted the 'interlinked disasters' (Shimizu and Clark 2015) and the ways in which interlinkages across people and places through increasingly globalised systems, structures and processes influence how disasters occur, and how their impacts cascade beyond specific locations (Adger et al. 2009). These approaches have informed our understanding of global health crises as well as economic crises, while also recognising the role of cities as nodes of transmission within these broader networks. However, there has been less consideration of the role that urbanisation plays as a transformative process in reshaping and redistributing vulnerability and risk, and of the role that regional economic integration plays in transferring vulnerability and risk across national borders.

R. Friend (✉)
Environment Department, University of York, York, North Yorkshire, UK
e-mail: richard.friend@york.ac.uk

P. Thinphanga
Thailand Environment Institute Foundation (TEI), Bangkok, Bangkok, Thailand

M.A. Miller et al. (eds.), *Crossing Borders*, DOI 10.1007/978-981-10-6126-4_6

In this chapter we address how the phenomena of globalised and regionalised cities and urban regions create new patterns of risk and vulnerability, and how the ways in which urban regions are linked also create fault lines through which the impacts of shocks, crises and stresses cascade beyond the site of specific events. Our focus on environmental disasters is largely on climate-related shocks and crises, while also recognising that disasters are rarely attributable to one set of factors alone, whether climate, environmental or man-made. Yet climate change presents important context for discussion here given the clear connections between urbanisation and global environmental change, and the widely noted significance of urban areas as locations of climate vulnerability.

Urbanisation and global environmental change must also be approached from a political economy perspective. Recent literature has highlighted the tension between capitalism and global environmental change (Klein 2014; Pelling et al. 2012). Viewing contemporary urbanisation as both a product and necessary feature of global capitalism focuses attention on a fundamental clash of rationalities between capital and ecology, and the way in which capital investments in land that underpin urbanisation deal with the risks and vulnerabilities that such investment creates. Urbanisation in Asia illustrates a critical tension between the investment logic of filling wetlands and land transformations, with ecological imperatives to maintain natural water sources and flood protection. The vulnerabilities that emerge through capital's need for a spatial fix through investment in land (Harvey 2001) is offset by the redistribution of vulnerabilities through localised infrastructure solutions, combined with a hedging approach to risk management in which investment portfolios are spread across multiple locations and borders.

This chapter emerges from several years' involvement in city level implementation and action research projects focusing on urban climate resilience. It is part of an attempt to develop a coherent theoretical approach to our understanding of urbanization in the Mekong region as being partly driven by regional economic integration, and how such forces reshape climate and environmental vulnerabilities and risks as being both local and regional. The chapter draws on our own empirical data garnered from engagement in a number of projects; as members of research teams and as actors in facilitating multi-stakeholder dialogue, and as partners in collaborative research efforts (Friend et al. 2016). The chapter is also supplemented by literature reviews, and a reading of a recent regional disaster, the 2011 floods that affected much of Thailand and ongoing controversy around industrial estates, that itself influenced how local actors in secondary cities approached the challenges of urbanization, regionalization and climate change.

6.2 Moving from a Place-Based Approach to Climate Disasters

The starting point for assessing urban climate vulnerability within much of the emerging research on urban climate resilience has been focused on hazards of space—hazardous zones (Cutter et al. 2008) and vulnerable spaces. Projections of future climate change have been used to identify the locations that might be directly vulnerable to a range of anticipated climate variables. From this assessment, the locations of vulnerability and risk can be mapped, and the implications for people, or spheres of economic activity that are located in these vulnerable spaces, can be calculated.

Of course the calculation of climate vulnerability is more sophisticated, taking on board the combined influences of exposure, sensitivity and adaptive capacity, while compensating for a range of future development scenarios. Not all locations or people that are similarly exposed display the same degree of vulnerability, and in many cases issues of adaptive capacity are the most significant in determining overall vulnerability. Adaptive capacity itself is a function of a range of factors related to assets, capabilities, power and knowledge, often defined in shorthand in terms of poverty and wealth. This broad approach has led to a proliferation of projections with efforts to determine the likely impacts of climate change on specific locations, through mapping vulnerable space and people who are located in this space, and site-based interventions to reduce vulnerability.

Taking climate projections as the starting point for assessing urban vulnerability becomes problematic due to the very nature of urbanisation across Asia (Friend et al. 2015; Institute for Social and Environmental Transition–International et al. 2014). The city in Asia itself is a moving target. The pace of change is so intense that it is increasingly difficult to define the limits of the urban space, whether from the perspective of administrative, physical or ecological boundaries.

Patterns of urbanisation create complex linkages between and across territories, sectors and—most importantly—between and among people. Urbanisation brings people together in new patterns of settlement and employment that are often highly mobile, and in ways that transcend traditional categories of urban and rural space. With greater diversity in urban areas, patterns of shared identity and community are less clear, often multi-scaled and overlapping, and generally not directly associated with territory or with location of residence, but more closely associated with ethnicity, language, religion, class and employment.

The understanding of disasters has increasingly adopted concepts of 'interlinked disasters' (Shimizu and Clark 2015) and cascading impacts of shocks and crises across territories. Disasters are increasingly marked by the degree of intensity and uncertainty, but also the complexity with which disasters have repercussions and ramifications both spatially and temporally. The networks that bind locations together can thus also be fault lines by which disasters are transmitted across disparate territories. The concept of teleconnections (Adger et al. 2009) draws attention to nested relationships, and the synergistic and interdependent nature of social-ecological relationships; globalisation is making such interdependencies critical determinants of local vulnerability (p. 151). As urbanisation becomes part of regional and global phenomena, with cities increasingly linked by chains of markets, production, transport and communication—a kind of 'pan-urbanism' (Moris 2014)—it is the nature of these linkages that increasingly determines what occurs in a specific urban locality.

Moving from looking at disasters as being bound by space, and from site-specific to systemic and networked dimensions of disasters, illuminates the increasingly interdependent global systems on which urbanisation depends—water, food, energy, transport, waste and information. However, even this perspective tends to take the urban itself to be spatially bound, and representative of the local (Shimizu and Clark 2015). Notions of the 'local' become problematic in parts of the world where urbanisation is accelerating at a pace and in ways that are difficult to predict, and that are being brought together in regional and global configurations (Friend et al. 2015). This has been referred to as the 'double exposure' whereby 'regions, sectors, ecosystems and social groups will be confronted both by the impacts of climate change and by the consequences of globalisation' (O'Brien and Leichenko 2000, p. 222).

With such a rapid transformation unfolding, part of the challenge is in understanding the drivers of urbanisation and the ways in which risk is calculated, created and reinvented as a function of the transformations of land use and values, infrastructure and production systems, and exchange mechanisms. The scale of investment in urbanisation, and more specifically in real estate markets as well as in industrial development, reminds us of the dependence on capital flows, and the dependence of capital on creating new opportunities for accumulation (Isono 2010).

Patterns of urbanisation in the Mekong reveal some limitations of a territorially focused approach to vulnerability. The Mekong is now emerging as a region, or rather a region of regions that is increasingly linked to the global scale. Patterns of regionalisation themselves influence the ways in which urbanisation unfolds, how locations are linked across national boundaries, and how vulnerability and risk is calculated and manufactured.

6.3 Interlinked and Interlocked Urban Systems

A critical but often overlooked dimension of regionalisation is the way urban systems expand across territories, allowing for agglomeration and accumulation in specific sites, while also creating new sets of linkages (Harvey 2001). The focus on such systems, with its theoretical grounding in complex social-ecological systems (Leach et al. 2007), takes the concept of double exposure (O'Brien and Leichenko 2000) further in order to consider the significance of urban systems and their inherent fragility, and the ways in which they are interlinked and interconnected.

A critical aspect of contemporary urbanisation is the dependence on systems for food, water, energy, communications and transport that extend way beyond the physical, geographic space of the city (Elmqvist 2014; Tyler and Moench 2012). The dependence on complex systems of technology and infrastructure allow for the scale and reach of urbanisation in its contemporary form (Graham and Marvin 2001). The ability of financial markets and production systems to operate across diverse territories and move rapidly from one location to another allows for the scale of investment, agglomeration of assets and capital accumulation.

Such urban systems are a combination of natural ecologies, physical infrastructure and technology, and agents (people, individuals, households, communities and organisations) and institutions (rules, norms and practice). For example, water is a natural resource and part of a complex ecological cycle. Its extraction, distribution and use is shaped by man-made physical infrastructure and technology, that is itself managed and accessed by interactions between individuals, organisations, companies and government. The access to and distribution of water is also shaped by rules, norms, laws, policies and crucially practice. Water can thus be seen as a complex system comprising infrastructure and technology, agents and institutions bringing together biophysical and social dimensions, yet still dependent on and influencing ecological processes (Tyler and Moench 2012).

These specific characteristics of urbanisation—from both the complex social-ecological systems and the political economy perspective—have not been adequately addressed in the literature on urban disasters (Friend et al. 2015). Similarly, the influence of urban systems on shaping vulnerability and poverty has also received limited attention. Accessing urban systems is essential for ensuring access to critical services (health, nutrition, employment and shelter) and the benefits that they deliver, and thus for ensuring wellbeing and reducing poverty and vulnerability. In the majority of rapidly growing cities, a significant proportion of people lack adequate access to reliable, high-quality systems and services (Friend and Moench 2013).

Moreover, there is something in the nature of infrastructure and technology systems that opens up the potential for failure (Ahern 2011; Tyler and Moench 2012). As recent economic and weather-related shocks have demonstrated in dramatic fashion, even the best of urban infrastructure and technology systems have some element of fragility; a disturbance in any one part in such a system can have cascading impacts on other parts, both through people and places. In many ways, the greater the dependence on these systems, the more dramatic and far-reaching are the consequences of failures.

The ways in which urban systems operate further blurs the boundaries between the city and the non-city and between the social and physical dimensions of systems. Cutting across different geographies, systems linkages cross regional and global scales as resources, capital, labour and information move on transport and communication infrastructure between and across urban and rural areas (Friend and Moench 2013). The goods and services on which urban populations depend are part of increasingly global production and distribution networks, themselves sustained by global transport and communications technology and infrastructure, and of course globalised capital (Graham and Marvin 2001; Parnell and Robinson 2012). Cities are not only linked to their immediate rural hinterland; through these increasingly multi-scale and complex interlinkages, they are increasingly networked across regions, and across the globe (Sassen 2005).

In such a globalised economy, there is an inherent competition between cities: between those that are well connected and thus become centres of investment and growth, and those that are more marginalised, with cities competing for various forms of investment and economic growth (McCann 2004). These networks and patterns of investment, wealth and power between cities across the globe create new centres and peripheries of development and dependency, and axes of competition, reminding us of earlier work on world systems (Wallerstein 2004). Yet against earlier expectations of dependency theory, cities in the former periphery of the Global South are emerging as global centres linked to other such urban centres, while also exhibiting harsh extremes between wealth and deprivation (Roy 2009). Mumbai stands out as an example of these extremes, being home to both some of the most expensive real estate on the planet, and also the most extreme absolute poverty. Similarly, some urban centres prosper while being surrounded by a hinterland of urban peripheries across broader geographies: the non-resilient and failed cities. The prosperity and security of the centre requires the impoverishment and vulnerability of the periphery.

As a transformative process, urbanisation creates not only a new vulnerability context for people in both urban and rural areas, but also a specific urban ecology that derives its character from changing land use and urban design and the influences these have on natural features like hydrology, temperature, and air quality (Parnell et al. 2007; Pelling and Manuel-Navarrete 2011). Urbanisation is itself associated with a range of environmental problems. In monsoonal Asia the phenomenon of urban heat islands is argued to increase urban temperatures by several degrees in comparison with less built-up environments in the perimeter (Srivanit et al. 2012). The combination of higher temperatures and the high humidity increase the heat index, which is directly related to how the human body is able to cool, especially during the night. Major heat waves in cities in Pakistan and India have demonstrated the potential impacts of such a high heat index in cities in which many working people are exposed to high temperatures during the day due to the nature of their work, and have only limited cooling in their homes (Ammann et al. 2014). Problems of heat in Asian cities are further compounded by poor air quality and high levels of pollution.

The interlinkages between different locations and economies are also evident in changing patterns of livelihoods, employment and migration. Rural livelihoods are increasingly dependent on off-farm employment, or what Winkels (2011) has referred to as 'stretched livelihoods'. Migration between rural and urban economies is increasingly significant, with members of farming households often employed at certain times of the year (or more permanently) in urban economic sectors, and remittances flowing between the two locations (McKay 2005).

Much of the urban economy is informal and inherently vulnerable, with people engaged in labour relations that are poorly regulated and unpredictable, in which they have limited rights, often suffering abuse and exploitation while earning low wages. The informality of cities also encompasses the ways in which people access services and urban systems, going through institutional mechanisms that are shaped by corruption, criminality and patronage, yet often paying prices that are higher than the formal market, while enjoying a quality of service that is far below market standards. Urban people's wellbeing, whether as individuals, households, or communities and neighbourhoods, is related not just to their ability to access systems and services, but also to complex institutional arrangements (Friend and Moench 2015).

Across this story is a degree of dependence on systems and structures over which individuals, households and even neighbourhoods have at best only limited influence, and in most cases, extremely limited control. Increasingly it is through these systems, and their inherent fragilities, that impacts of disasters becomes manifest beyond specific locations. As we discuss below, such systems are themselves shaped by regional and global political and economic forces.

6.4 Regionalisation: Reshaping Borders, Linkages and Dependencies

Regionalisation is a reconfiguring of nation states and national borders, not necessarily in the ways in which these are mapped, but in some of the core functions that govern how resources, goods and services are transferred between locations. Indeed, the very purpose of regional economic integration is to reconfigure the economic function of borders. Rather than restricting movements of people, goods, and capital national borders are being reconfigured to facilitate movement across locations, creating trans-boundary markets and efficiencies of resource access and distribution.

There is thus an unavoidably regional dimension to the reconfiguring of vulnerabilities and risks. Economic integration creates new linkages across different regional assemblages, and new patterns of urbanisation both in terms of human settlement and in terms of industrial production. These patterns of regional integration shape the territorial reach of urban risks and vulnerabilities, creating new transmission lines through which impacts of shocks and crises cascade across different locations. Yet at the same time, underpinning investment and physical infrastructure is a calculation of vulnerability and risk that increasingly is regional and global in nature. The implications of these shifts for environmental and climate-related shocks requires closer conceptual and empirical scrutiny.

Central to urbanisation is the transformation of landscapes, in terms of their values and uses, and the flows of resources between different locations. Urbanisation in the Mekong (and similar regions) is very much driven by the logic of capital—the expansion of markets, and the integration of different locations and people into an increasingly regionalised and globalised economy. Urbanisation requires the agglomeration of economic assets, and the transformation of land—both in how it is utilised and in its production and exchange value. There is an environmental dimension to how risk and returns are calculated. Across the region, urbanising areas target low-value land where the immediate capital returns on investment are the highest. Yet from an environmental perspective, low-value land in this region tends to be land that is hazardous, being prone to seasonal flooding and very often in the agricultural floodplains or along the coastal strips.

Alongside patterns of investment come calculations of risk. One of the puzzles in this story is the way in which space that is identified as hazardous is targeted for investment. This is partly due to an incomplete appreciation of risks, particularly those associated with uncertain future climate risks. But it is also due to inherent characteristics of capital investment targeting low value land. This is calculated on a principle of hedging, and in this way distributing risk across locations, building redundancy into the portfolio of investments. The risk of a disaster in one location is balanced by the assessment of low level of risk for more than one location facing a disaster at the same point in time.

Such a discussion inevitably takes us into a consideration of issues of governance. In many ways, the way that urbanisation has unfolded in Asia is a consequence of failures of governance. The basic tools of urban policy and planning in the region are notoriously weak. Land use planning is little more than painting-by-numbers: a retrospective mapping of land use changes that have occurred on the ground, rather than a tool for strategic long-term planning (Ribiero 2005). Similarly, environmental governance through Environmental Impact Assessment (EIA) in the region occurs at the latter stages of project development, and rarely has the independence or legal authority to influence project development in any significant way. These failings are largely a product of the collusion of capital and state, with the state playing both the regulatory role, and the role of investor. With such a tension between these competing roles, considerations for environmental and social concerns are easily over-ridden. Additionally, in urban areas the public goods of systems and services, whether they are water, energy, waste or transport are largely privatised. There is very little public space in urban Asia. But perhaps even more significantly the public policy process is itself privatised, with private companies take on planning roles that would normally be the realm of the public sphere. Rather than the privatisation of projects, we are witnessing the privatisation of the complete urban project (Shatkin 2007).

6.5 Regionalisation: The Specific Case of the Greater Mekong Sub-region (GMS)

Urbanisation is of course a global phenomenon, one that is accelerating and intensifying in many parts of the world but especially in Africa and Asia. The specific drivers and patterns of urbanisation are shaped by local context and circumstances, by histories of colonialism and integration into globalised markets and production patterns. While there are universal characteristics, there is also a growing need for theoretical explanations that are grounded in the particular experience of the Global

South, differentiated further across specific countries (Parnell and Robinson 2012; Patel 2014; Roy 2009).

Mainland Southeast Asia illustrates the changing urban landscape and the relationship with environmental disasters. While cities have a long history in Southeast Asia, the current drive of urbanisation has its roots in patterns of globalisation and regionalisation that are political and economic in nature. The pace of urbanisation in the region is among the highest in the world. In little more than a generation this part of the world has witnessed a dramatic transformation from a largely subsistence agricultural economy, distanced from the globalised economy, to one of the main drivers of global economic activity.

The region of the Mekong is itself a recent construct, and one that has not yet come to fruition. It is born of a history of shifting lines of national authority and allegiance that have emerged through conflicts between colonial powers, ruling elites and changing notions of identity, and that have created networks of power and dependency with shifting centres and peripheries. National boundaries were rarely mapped in the pre-colonial period, with state authority created and reinforced through lines of tribute (Winichakul 1994). Similarly, ethnicity was mixed, and many of the larger ethnic groups of the region—the Karen, Cham, Mon and Hmong—were left out in the post-colonial creation of nation states.

The boundaries of the 'Mekong region' can be defined in different ways: according to the ecological boundaries of the Mekong river basin itself, or more broadly to encompass the countries of Cambodia, Lao PDR, Myanmar, Thailand and Vietnam as well as parts of China. The central place of the Mekong in these notions of a region is as much a legacy of the imagining of the French colonial powers, as any sense of shared identity or common interest among its people or governments.

In using the term Mekong to describe a rather fluid region, we also recognise that it is just being born as a region, and indeed is one of many regions that create an increasingly fluid landscape beyond borders of nation states. A number of different regions are currently in operation—the Greater Mekong Sub-region (GMS), initially funded by the Asian Development Bank (ADB) through support to transport and energy infrastructure systems, brings together the countries of Cambodia, Myanmar, Thailand, Vietnam and southern provinces of China. Despite its name, it brings together a number of major river basins beyond the Mekong with the majority of people within the region having little direct connection with the river itself.

The countries of the Mekong are also members of alternative constellations that are designed to counterbalance the power of China, while creating economic opportunities. The Ayerwady-Chao Phraya-Mekong Economic Cooperation Strategy (or ACMECS), established in 2003 on the edge of an ASEAN summit, brings together Myanmar, Cambodia, Laos, Thailand and Vietnam in a similar commitment to foster economic cooperation between the countries. Each of these countries is also a member of ASEAN and APEC, broader alliances of security and economic interest.

A less ambitious scope of regional cooperation has been established for the lower Mekong River basin with the creation of the Mekong River Commission (MRC), focusing on cooperation in sharing of the water resources of the basin, and enshrined in the 1995 Mekong Agreement. Much of the efforts under the MRC have been focused on cooperation in the development of the hydropower sector, most recently along the mainstream of the river.

Much of the motivation in regionalising these countries is in the creation of market opportunities both as labour and as consumers, applying former Thai Prime Minister Chatchai Choonhavan's motto of "turning battlefields into market places". Here of course, urbanisation plays an important role in creating a market for consumption and a whole set of urban values in which consumption patterns, habitat and lifestyle are intertwined (Friend and Thinphanga 2018).

6.6 The Expansion of Infrastructure and Technology

Infrastructure and technology allow for current patterns of urbanisation and regionalization—for what Sassen (2005, p. 2) refers to as the 'the geographic dispersal of economic activities that marks globalisation, along with the simultaneous integration of such geographically dispersed activities'. This itself requires, and opens, opportunities for investment in the physical infrastructure that brings locations and economies together in ways that allow for greater efficiencies. But equally it is through this networked infrastructure—that allows for movement of information, capital, goods and services—that the impacts of shocks in one location reverberate across the globe.

The extent of this transformation was difficult to predict only 20 years ago (or less), but the way it is unfolding clearly relates back to earlier theoretical arguments around urbanisation in the region. The problem of defining the boundaries of the urban has been a persistent area of interest in urban studies of Southeast Asia. This has led to theoretical approaches that have talked of mega-urban regions (McGee 1991) that bring together urban and rural centres within a specific national region. Similarly, debates around the 'desakota' continuum have highlighted the interlinkages and dependencies between rural and urban areas. A slightly more refined version of this notion of rural-urban linkages, and the blurring of boundaries, can be found in the use of the term 'rurban'. Urbanisation is a transformation of rural space, economy and society as much as it is a transformation of the city itself. Most significantly, it is the linkages between the two that stand out. As Douglass (1995, p. 64) predicted over 20 years ago,

> spatial development is more characterized by expanding networks of rural-urban linkages that defy simple models of spatial structure. They also present new issues and problems for urban and regional planning and management.

The transformation is not necessarily a one-way, evolutionary transition from rural to urban, or from agricultural to industrial, as we can see from the diverse experience of countries such as Laos and Myanmar, as well as the experience of India and South Asia. Across these different theoretical debates is the recognition of the murkiness of categories of rural and urban. However, they have generally been applied to notions of regions within national boundaries. Similarly, vulnerabilities are not solely determined by location, but more by the nature of the linkages between locations, their inherent fragilities and failures, and the ways in which risk and impacts of specific shocks are transferred between the two.

6.7 Emerging Patterns of Cross-Border Vulnerability and Risk

As in other parts of the world, much of the most rapid urbanisation in the GMS is driven by trans-boundary trade and production patterns and economic ambitions, with 'trans-border urban regions' and 'urban corridors' that follow the core transport (and energy) infrastructure that has been put in place across the GMS.

The initial investment that underpins the GMS has been around the construction and expansion of core regional infrastructure: transport and energy. A grid of roads—referred to as corridors—cut across the region from North-South and East-West linking key trade and communications centres, and supporting the development of industrial complexes. Alongside these transport investments, the energy sector has also seen investment with the vision of an integrated regional energy grid. Much of this energy is to be generated by hydropower, with the region seeing a massive expansion of dams across the tributaries, and within recent years, along the mainstream of the Mekong River. These investments have also been hugely controversial, with concerns for environmental and social impacts at site level, but also across the region (Molle et al. 2009; Haefner in this volume).

While the initial investment for the Greater Mekong Subregion (GMS) came from the Asian Development Bank (ADB) this is now dwarfed by the private sector investments, with much of the capital coming from the region. Rather than the global capital from the North, the national capital from within the Mekong as well as from East Asia (China, Korea and Japan) is increasingly influential, less guided by the social and environmental safeguards of the ADB and other IFIs. This can also be seen as an expansion of existing industrial production to multiple locations in the region taking advantage of preferential labour costs, and of opening new markets. The level of investment in the region is also a product of the shocks in global financial markets. With loose capital struggling to find a base that provides reliable revenue streams, the spatial fix of multi-scale investment in real estate, both urban and industrial, is all the more attractive.

Part of the focus of the GMS is on overcoming the constraints of the previously established borders. These constraints are partly institutional but location of Special Economic Zones and related growth triangles in the GMS is also dependent on broader trans-border transport and production systems; their viability is not just tied to their location relative to borders and their ability to overcome the borders at which they are located, but also to their ability to benefit from broader multinational linkages.

One of the most striking features of this scale of investment is the emergence of urban regions and urban corridors, beyond the boundaries of specific cities or specific industrial parks, straddling national borders in growth triangles, and across coasts through transport corridors. As Scott et al. (2002, p. 12) observe of a global phenomenon of global cities and city regions:

> Whereas most metropolitan regions in the past were focused mainly on one or perhaps two clearly-defined central cities, the city-regions of today are becoming increasingly polycentric or multi-clustered agglomerations.

Moreover, these city regions exist beyond contiguous locations.

City regions, and indeed growth triangles are no longer necessarily territorially contiguous. For example, the case of Map Tha Phut and the expansion of industrial zones in Dawei (Myanmar) and Quy Nhon (Vietnam) illustrates attempts at the regional reconfiguring of industrial production and trade, and the redistribution of patterns of risk and vulnerability.

One of the key areas of linkage and commonality is in the form of the investment capital that supports these developments. Two of the major Thai companies are involved in both Quy Nhon and Dawei, demonstrating the regional expansion of Thai capital. PTT Public Company Limited was the lead investor in the Non Hoi Oil Refinery Complex, with an initial total budget of US$28 billion dollars while Ital-Thai is one of the leading investors in Dawei (Bangkok Post 2015).

The motivation for these projects is partly around expanding markets and increasing efficiencies. Yet there is a more sinister motivation behind these investments. In many ways they represent a response to existing environmental and social impacts that constrain the ability of Thai capital to invest within Thailand's own national boundaries, and the expansion of industries that have been mired in controversy in Thailand. The combination of a history of conflict along with strengthened environmental legislation has been identified as a constraining influence on the petrochemical industry in Thailand. In 2010 the then-Prime Minister was quoted as arguing the case for exporting such industries.

> 'Some industries are not suitable to be located in Thailand,' Abhisit Vejjajiva, the Thai prime minister, said in explaining the project to viewers of his weekly television address recently. 'This is why they decided to set up there,' he said, referring to Dawei (International Herald Tribune 2010)

This was a position taken up by successive governments and apparently endorsed in the National Economic and Social Development Board's (NESDB) 11th five-year plan that was passed in 2011 (National Economic and Social Development Board 2011). In some ways this can be seen as a progressive decision in recognising the impact of industrial development, and the failings of environmental governance and land use planning. Yet even this acknowledgement of past failings has recently been overruled, with the military junta government passing a law that allows it to use special powers to push through the development of Special Economic Zones (SEZs), and promotion of coal-fire power stations. Recently it has been announced that Map Tha Phut will be expanded, and a high-speed rail link established to the north-east and onwards to China.

When considered as stand-alone investment projects, there are considerable risks directly related to their location. For example, from a climate perspective the multi-billion dollar oil refinery project at Quy Nhon is located on a hazardous part of the Vietnamese coast, vulnerable to storms and floods (DiGregorio 2015). But this is only one project in a global investment portfolio. While the risk of a climate event in Quy Nhon might be considered to be high, the investor's portfolio is spread across different locations. The risk of a similar event striking more than one of these investments at the same time would be considered to be extremely low. This diversification of risk—or rather, a hedging approach to climate risk—can be considered as building the resilience of the global investment, but of course does not necessarily address the implications of a specific disaster. Ultimately the project was cancelled due to volatility of global crude oil markets, but with no apparent concern for environmental risk.[1]

With the Thai (and global) economy facing a dramatic downturn, the expansion of investment in industrial zones, particularly SEZ's, and major public infrastructure plays a central role in national economic strategy. It is a strategy that also aims to increase domestic demand, and as such is centred on development of the urban economy, and the consumer demand that is associated with urban life (Siam Commercial Bank 2011). At the same time, the imperative to push ahead with such investments has put additional pressure on environmental governance. The military junta has announced that it will use its own extraordinary powers under Section 44 to accelerate the EIA process, arguing that 'there are too many delays caused by land use issues, environmental impact assessments (EIA) and protests' (Wangkiat 2016).

6.8 The Thailand Floods of 2011 and Emerging Drought

The ways in which the causes and impacts of disasters spread beyond specific locations became apparent in the floods that Thailand experienced in 2011 and the drought crisis that is now unfolding in 2016. During the floods of 2011, the urgency of protecting the industrial base of the country and the urban centre of Bangkok lead to frantic efforts to divert floodwaters away from the city centre to rural areas (Marks 2015). Farmlands were flooded in order to protect urban and industrial centres. Of course, managing a flood of such a scale proved challenging and many urban and industrial centres were also flooded. These dramatic events are widely referred to as the Bangkok Floods—privileging the story of the capital city over the rest of country and neighbouring provinces, thus framing the event in the imagination as being place-specific. Yet the causes and the impacts were not confined to Bangkok, and indeed, the enormous effort dedicated to protecting the economic, administrative

[1] At the time of writing this project had been approved by the Vietnamese government. However, this decision was reversed in 2016 with the collapse in global crude oil prices (Vietnam Economic Times 2016).

and political heart of Bangkok contributed to the way in which impacts were created. The flood itself crossed administrative boundaries of provinces, as the waters flowed downstream from the Upper North. Significantly, the flood in the lower part of the basin was due to this upstream runoff rather than to localised rainfall.

The vulnerability to flooding of Bangkok and much of the Chao Praya basin that was so crudely exposed in 2011 has deep historical roots that illustrate the ways in which Thailand's incorporation into a globalising industrial economy also influenced patterns of vulnerability. Part of the history of changing flood vulnerability across the basin lies in the transformation of land use in the lower basin, and extensive deforestation of the upper watershed of the Chao Phraya (Roachanakanan 2012; Srisawalak-Nabangchang and Wonghanchao 2000). The history of the expansion of the country's industrial base also provides some context to the way in which the crisis of 2011 unfolded. For the lower part of the Chao Phraya basin around the Central Plains, this history also exposes in dramatic fashion the clash of rationalities between the logic of capital investment in land speculation and transformation on one hand, and the ecological dimensions of risk and vulnerability on the other. Much of the Central Plains of Thailand that became the industrial and urban base of the country had previously been designated agricultural land, in land use plans that also recognised the flood protection functions that the plains provided to the city of Bangkok (Bello et al. 1998; Roachanakanan 2012). However, these plans were overturned as foreign investment sought to locate the new industrial base of the country in affordable locations, close to major transport infrastructure at the Laem Chabang deep-sea port, and Don Muang International Airport.

With investment from overseas, largely led by Japan, the deep- sea port of Laem Chabang and the petrochemical plant of Map Tha Phut were opened along the Eastern Seaboard. With a need for locating additional factories and warehouse facilities within convenient communications reach of these industrial centres and transport centres, the lower parts of the Chao Praya basin were targeted for additional industrial estates (Roachanakanan 2012; Shatkin 2004). Much of the investment of the 1970s and 1980s that drove this land conversion was in response to the oil crises of the time, and the shift of capital investment into land speculation and away from manufacturing (Bello et al. 1998). Similar patterns of speculative capital investment in real estate have emerged as a result of the 2008 financial crisis (Harvey 2012).

This pattern of investment went directly against earlier land use plans that had designated these areas as green space, for agricultural uses and flood protection. Indeed, this earlier land use planning was itself informed by the natural ecology and hydrology of the basin; historically these areas had always been prone to a natural, annual cycle of flooding. With this level of investment, the lower Chao Phraya basin witnessed a dramatic refashioning, with further investment in residential housing estates and road infrastructure (Bello et al. 1998).

The most extreme manifestation of this pattern of investment targeting low-value, flood-prone agricultural land is illustrated by the case of Suvannabhumi International Airport, the most important air link between Thailand and the global economy. Prior to the construction of the airport the area was better known as King Cobra Swamp—a low-lying rice-producing area that was widely recognised as providing drainage and flood protection to the ever-expanding eastern Bangkok. Despite warnings from environmentalists of the risk of locating the airport in the swamp, the investment went ahead (Hutanuwatr et al. 2015). Suvannabhumi provides a neat, but not unique example, of the competing rationalities of capital investment versus risk protection, with the consequences becoming apparent during the 2011 floods.

This refashioning of the lower basin has created additional vulnerability to flooding, altering natural hydrology but requiring enormous, often flawed, state efforts to prevent flooding in areas where the waters would naturally flow, in order to divert the flood waters against the natural flow (Marks 2015). There was a clear international dimension to this effort. Foreign investors demanded that their assets be protected, with the threat of relocating the investment that the Thai economy had by now come to depend on.

The impacts of the 2011 flood crossed administrative boundaries within the country, pitting provincial and district authorities against each other. The impacts also went beyond national boundaries of Thailand. The economic costs of the floods have been widely reported. Since Thailand had become the centre for a globalised production chain, particularly of hard drives, computer chips and also car parts, the impacts of flooded warehouses and shutdown production had rippling effects across the globe (Fuller 2011). Production in the USA and Japan that depended on these inputs ground to a halt, with implications for globalised markets, and of course, employment security for workers involved in these industries.

There is another side to this story that has not yet been investigated adequately. Thailand is not only a regionalised economy in terms of production but also in terms of absorbing labour. Whereas Thailand was exporting labour to other parts of Southeast Asia and the Gulf through the 1990s, Thailand has now become labour short, and an importer of labour. Much of the economy depends on migrant labour from Myanmar, Cambodia and increasingly from Vietnam and Laos. Statistics for the levels of such migrant labour are inconsistent and generally thought to be significant underestimates, Official figures suggest a total number of migrant workers of around 3.25 million (Huguet 2014, p. 1), but suggestions that there are about three million workers from Myanmar and over one million from Cambodia seem plausible.

These workers send remittances back to their rural homes, and thus provide critical financial resources to struggling rural society in the region. In many ways, migration of rural labour can be seen as evidence of increasingly stretched livelihoods (cf. Winkels 2011). One of the questions as yet unanswered regards the implications of loss of income and other impacts of the floods on migrant workers—and the implications for their homes, particularly in rural Cambodia and Myanmar. An emerging concern relates to the importance of remittances in ensuring household viability, but also that rural livelihoods that depend on these remittance flows are highly precarious, susceptible to a wide range of shocks and crises.

The ways in which migration patterns and climate-related shocks are interlinked is further revealed in the case of migrant workers in Phuket. Migrant workers from Myanmar in Phuket live a precarious existence. With a complex process of legal registration, and an employment market that is dominated by a network of well-connected agents and brokers, interviews with local migrant rights activists suggest that only one third have access to all the labour rights and social services to which they would normally be entitled.

In addition to the violations of basic rights, and the informality of their employment status, the vulnerability of these migrant workers can also be seen through the ways in which they access critical urban systems. Workers either live in camps established by the construction firms for whom they work, or in informal communities that are located in marginal land—for example, along a former mangrove forest that has been cleared by Thai investors in order to locate rental accommodation for the migrant workers. In addition to the problems migrant workers have in accessing basic social services of education and health, access to water and electricity is outside of formal systems, with service costs disproportionately high, and of poor quality.

A further twist to this story of migration has emerged recently from interviews with stakeholders in southern Thailand. Government clampdowns on migrant workers throughout 2015 combined with a downturn in the economy has pushed many migrant workers from Myanmar out of Thailand. However, the floods that have hit many parts of Myanmar during the rainy season of 2015 have pushed an additional group of labourers across the border into Thailand. This provides an interesting dimension in which vulnerability created by precarious employment status and rights becomes linked to disaster migration. Of course, the migrant workers who have moved as a response to flood impacts in their own country also find themselves living in vulnerable locations, with poor access to urban systems and services and limited rights. The story comes full circle.

With El Niño taking effect in 2015/2016, an intense drought brought reservoirs to all-time lows and raised concerns that water supply for urban centres and industrial parks could not be met from available supplies. The first response across the country was to restrict irrigation supplies to rice farmers—once again, relocating risk from urban to rural locations. Additionally, the drought revitalised old infrastructure plans, with large river basin diversion schemes from the Mekong mainstream into the Northeastern river basins of Chi-Mun being brought back to the policy table, and negotiations for China to release more water from their upstream dams to provide water for downstream countries. Again responses to crisis are shaped by the path dependency of previous investments and agendas.

6.9 Conclusion

This chapter has sought to lay out some exploratory thoughts on how to approach the increasingly cross-border dimensions of climate vulnerability that emerge from viewing urbanisation from the perspective of regional economic integration, and from the perspective of dependence on critical systems of infrastructure and technology. In this final section, we aim to point to the potential implications of such approaches both for fields of research and academia, and also for policy.

Both the regional dimensions of urbanisation and the complex systems perspectives steers us towards the need for understanding vulnerability to shocks and crises that will be multi-scale, interlinked and interlocked, with impacts that cascade across locations and people, and that precipitates actions that also then creates new rounds of potential shocks and crises.

Underpinning patterns of urbanisation across the Mekong region is the tension between the rationality of capital's investment in transforming land values, and the ecological functions of a part of the world that is largely characterised by wetlands and floodplains. The main response to emerging vulnerabilities to flood and drought that this investment creates is to redistribute and disperse risk—through construction of site-specific infrastructure, extraction of resources from a more distant hinterland, and hedging a wide portfolio of investments in multiple locations.

This is not to say that the interest in spatial characteristics of vulnerability, risks and hazards should be abandoned, but it will be increasingly important to factor in the ways in which the impacts of specific events cascade across locations. This requires a more global analysis of the ways that regions are being created through means of economics and of infrastructure and technology, and a political economy grounded analysis of the drivers of urbanisation that contribute to the ways in which vulnerability and risk are calculated, created and distributed. An analysis of disaster risk that goes beyond spatially bounded analysis and action also points to the need to rethink policy and practice around social protection, and social and environmental safeguards for infrastructure development.

The consequences of cascading impacts of shocks and crises across different locations raises enormous problems for current governance mechanisms, whether in terms of environmental governance or in terms of disaster risk reduction and social protection. With such high rates of labour migration in the GMS, much of the rural economy of the region is dependent on remittances that come from urban and industrial areas that are themselves increasingly vulnerable to shocks and crises. Yet the mechanisms for social protection, and in particular emergency relief in the face of a specific disaster, are focused on the location of the event rather than where the impacts are felt most acutely. Yet it may well be that the most significant economic and social impact of an event will be quite distant from its actual location. Of course this raises problems for social protection: understanding the vulnerability of a Cambodian rural household is directly influenced by potential disruptions to remittance flows, and that these in turn may be influenced by climate-related shocks located far away. Again, there is a need to rethink these approaches with similar challenges of needing to act across administrative jurisdictions.

The cross-border dimensions of disaster risk also raise challenges for the governance of infrastructure investments that underpin this regional integration. The remit of local government is territorially defined, with no jurisdiction or mechanisms in place to address broader infrastructure and technology systems. Additionally, much of the investment in and ultimately ownership of the infrastructure and technology systems is in the hands of the private sector. The rationality of private-sector calculations of risk is partly around reducing direct impacts, but also mitigating the severity of impacts by hedging investments in different locations. Even at this site level, the governance mechanisms to reduce risk—whether land use planning, Environmental Impact Assessment (EIA) or regular environmental monitoring—are rarely functioning in any meaningful way, and are increasingly under threat from repressive governments.

Spatially oriented approaches to climate impacts only make sense if cities are seen in isolation from the broader processes of urbanisation and transformation outlined above. Alternatively, applying a conceptual approach that combines complex-systems and actor-oriented approaches to urbanisation within the context of global climate change debate, creates opportunities for rethinking issues of risk, vulnerability, impacts of disasters and moreover, of adaptation, mitigation and resilience.

References

Adger, W. N., Eakin, H., & Winkels, A. (2009). Nested and teleconnected vulnerabilities to environmental change. *Frontiers in Ecology and the Environment, 7*(3), 150–157.

Ahern, J. (2011). From fail-safe to safe-to-fail: Sustainability and resilience in the new urban world. *Landscape and Urban Planning, 100*(4), 341–343.

Ammann, C., Ikeda, K., & MacClune, M. (2014). Projecting the likely rise of future heat impacts under climate change for selected urban locations in south and Southeast Asia. *The Sheltering Series, 9*, 1–23.

Bangkok Post. (2015). Deal puts Dawei on fast track: Gas-fired power plant will lead development (http://www.bangkokpost.com/print/648464/).

Bello, W., Cunningham, S., & Poh, L. K. (1998). *A Siamese tragedy: Development and disintegration in modern Thailand*. London/New York: Zed Books.

Cutter, S. L., Barnes, L., Berry, M., Burton, C., Evans, E., Tate, E., & Webb, J. (2008). A place-based model for understanding community resilience to natural disasters. *Global Environmental Change, 18*(4), 598–606.

DiGregorio, M. (2015). Bargaining with disaster: Flooding, climate change, and urban growth ambitions in Quy Nhon City, Vietnam. *Pacific Affairs, 88*(3), 577–598.

Douglass, M. (1995). Global interdependence and urbanisation: Planning for the Bangkok mega-urban region. In T. G. McGee & I. M. Robinson (Eds.), *The mega-urban regions of Southeast Asia*. Vancouver: University of British Columbia Press.

Elmqvist, T. (2014). Urban resilience thinking. *Solutions, 5*(5), 26–30. http://www.thesolutionsjournal.org/node/237196?page=16.

Friend, R. M., & Thinphanga, P. (2018). Urbanisation, climate change and regional integration in the Mekong Region. In K. Archer, & K. Bezdecny (Ed.), *Handbook of cities and the environment*. London: Routledge.

Friend, R. M., & Moench, M. (2013). What is the purpose of urban climate resilience? Implications for addressing poverty and vulnerability. *Urban Climate, 6*, 98–113.

Friend, R., & Moench, M. (2015). Rights to urban climate resilience: Moving beyond poverty and vulnerability. *Wiley Interdisciplinary Reviews: Climate Change, 6*(6), 643–651.

Friend, R. M., Thinphanga, P., MacClune, K., Tran, P., & Henceroth, J. (2015). Understanding urban transformations and changing patterns of local risk: Lessons from the Mekong Region. *International Journal of Disaster Resilience in the Built Environment, 6*(1), 30–43.

Friend, R. M., Choosuk, C., Hutanuwatr, K., Inmuong, Y., Kittitornkool, J., Lambregts, B., Promphakping, B., Roachanakanan, T., Thiengburanathum, P., Thinphanga, P., &

Siriwattanaphaiboon, S. (2016). *Urbanising Thailand implications for climate vulnerability assessments, Asian Cities Climate Change Resilience Network Working Paper Series 30*. London: IIED.
Fuller, T. (2011). Thailand flooding cripples hard-drive suppliers. *New York Times*, 6 November, http://www.nytimes.com/2011/11/07/business/global/07iht-floods07.html?pagewanted=all&_r=0
Graham, S., & Marvin, S. (2001). *Splintering urbanism: Networked infrastructures, technological mobilities and the urban condition*. London: Routledge.
Harvey, D. (2001). Globalization and the spatial fix. *Geographische Revue, 2*, 23–30.
Harvey, D. (2012). *Rebel cities from the right to the city to the urban revolution*. London: Verso.
Huguet, J. W. (2014). Chapter Thailand migration profile. *Thailand Migration Report 2014*, p. 1.
Hutanuwatr, K., Krisanapan, A., Charoentrakulpeeti, W., Knobnob, N., & Huong, L. T. T. (2015). *Urbanisation, poverty and vulnerability of Bueng Bua community in Lad Krabang Development Area*, TEI Working Paper 6 (in Thai).
Institute for Social and Environmental Transition–International, Thailand Environment Institute, & Vietnam National Institute for Science and Technology Policy and Strategy Studies. (2014). *Urban vulnerability in southeast Asia: Summary of vulnerability assessments in Mekong-Building Climate Resilience in Asian Cities (M-BRACE)*. Bangkok: Institute for Social and Environmental Transition–International.
International Herald Tribune. (2010). An industrial project that could change Myanmar. *New York Times*, 26 November, http://www.nytimes.com/2010/11/27/world/asia/27iht-myanmar.html?_r=0
Isono, I. (2010). Economic impacts of the economic corridor development in Mekong Region. In M. Ishida (Ed.), *Investment climate of major cities in CLMV countries, BRC Research Report* (Vol. 4, pp. 330–353). Bangkok: Bangkok Research Centre, IDE-JETRO. http://www.ide.go.jp/English/Publish/Download/Brc/pdf/04_chapter9.pdf.
Klein, N. (2014). *This changes everything: Capitalism versus the climate*. London: Simon and Schuster.
Leach, M., Bloom, G., Ely, A., Nightingale, P., Scoones, I., Shah, E., & Smith, A. (2007). *Understanding governance: Pathways to sustainability*. Brighton: STEPS Centre.
Marks, D. (2015). The urban political ecology of the 2011 floods in Bangkok: The creation of uneven vulnerabilities. *Pacific Affairs, 8*(3), 623–653.
McCann, E. J. (2004). Urban political economy beyond the 'global city'. *Urban Studies, 41*(12), 2315–2333.
McGee, T. G. (1991). The emergence of desakota regions in Asia: Expanding a hypothesis. In N. S. Ginsburg, B. Koppel, & T. G. McGee (Eds.), *The extended metropolis: Settlement transition in Asia* (pp. 3–25). Honolulu: University of Hawaii Press.
McKay, D. (2005). Reading remittance landscapes: Female migration and agricultural transition in the Philippines. *Geografisk Tidsskrift–Danish Journal of Geography, 105*(1), 89–99.
Molle, F., Foran, T., & Kakonen, M. (Eds.). (2009). *Contested waterscapes in the Mekong: Hydropower, livelihoods and governance*. London: Routledge.
Moris, J. (2014). *Reimagining development 3.0 for a changing planet* (Institute of Development Studies (IDS) Working Papers, vol. 2014, no. 435, pp. 1–49). Brighton: IDS.
National Economic and Social Development Board. (2011). *The eleventh national social and economic development plan, 2012–2016*. Bangkok: NESDB, Office of the Prime Minister.
O'Brien, K. L., & Leichenko, R. M. (2000). Double exposure: Assessing the impacts of climate change within the context of economic globalisation. *Global Environmental Change, 10*(3), 221–232.
Parnell, S., & Robinson, J. (2012). (Re)theorizing cities from the global south: Looking beyond neoliberalism. *Urban Geography, 33*(4), 593–617.
Parnell, S., Simon, D., & Vogel, C. (2007). Global environmental change: Conceptualising the growing challenge for cities in poor countries. *Area, 39*(3), 357–369.
Patel, S. (2014). Is there a 'south' perspective to urban studies? In S. Parnell & S. Oldfield (Eds.), *Routledge handbook on cities of the global south* (pp. 37–47). London: Routledge.

Pelling, M., & Manuel-Navarrete, D. (2011). From resilience to transformation: The adaptive cycle in two Mexican urban centers. *Ecology and Society, 16*(2), 1–11.
Pelling, M., Manuel-Navarrete, D., & Redclift, M. (Eds.). (2012). *Climate change and the crisis of capitalism: A chance to reclaim self, society and nature*. London: Routledge.
Ribiero, G. (2005). Research into urban development and cognitive capital in Thailand. *Journal of Transdisciplinary Environmental Studies, 4*(1), 1–5.
Roachanakanan, T. (2012). *'Floodways and flood prevention in Thailand: Reflections on the great flood in 2011', paper delivered at the world flood protection, response, recovery and drawing up of flood risk management conference*. Thailand: Bangkok.
Roy, A. (2009). The 21st-century metropolis: New geographies of theory. *Regional Studies, 43*(6), 819–830.
Sassen, S. (2005). The global city: Introducing a concept. *Brown Journal of World Affairs, 11*(2), 27–43.
Scott, A. J., Agnew, J., Soja, E. W., & Storper, M. (2002). Global city-regions. In A. J. Scott (Ed.), *Global city-regions: Trends, theory, policy*. Oxford: Oxford University Press.
Shatkin, G. (2004). Globalization and local leadership: Growth, power and politics in Thailand's eastern seaboard. *International Journal of Urban and Regional Research, 28*(1), 11–26.
Shatkin, G. (2007). The city and the bottom line: Urban megaprojects and the privatization of planning in Southeast Asia. *Environment and Planning, 40*(2), 383–401.
Shimizu, M., & Clark, A. L. (2015). Interconnected risks, cascading disasters and disaster management policy: a gap analysis. *Planet@Risk, 3*(2), 1–4. Global Risk Forum GRF, Davos.
Siam Commercial Bank. (2011). *Looking beyond Bangkok: The urban consumer and urbanisation in Thailand*. Thailand: Insight Economic Intelligence Centre.
Srisawalak-Nabangchang, O., & Wonghanchao, W. (2000). Evolution of land-use in urban-rural fringe area: The case of Pathum Thani Province. In *Proceedings of the International Conference: The Chao Phraya Delta: Historical Development, Dynamics and Challenges of Thailand's Rice Bowl*. IRD (Institut de Recherche pour le Developpement), Kasetsart University, Chulalongkorn University/Kyoto University.
Srivanit, M., Hokao, K., & Phonekeo, V. (2012). Assessing the impact of urbanisation on urban thermal environment: A case study of Bangkok metropolitan. *International Journal of Applied Science and Technology, 2*(7), 243–256.
Tyler, S., & Moench, M. (2012). A framework for urban climate resilience. *Climate and Development, 4*(4), 311–326.
Vietnam Economic Times. (2016). Non Hoi refinery project abandoned. (http://vneconomictimes.com/article/business/nhon-hoi-refinery-project-abandoned).
Wallerstein, I. (2004). *World systems analysis: An introduction*. Durham: Duke University Press.
Wangkiat, P. (2016). Laying down the 'dictator law' for money. *Bangkok Post*, 20 March, http://www.bangkokpost.com/news/special-reports/903648/laying-down-the-dictator-law-for-money
Winichakul, T. (1994). *Siam mapped: A history of the geo-body of a nation*. Honolulu: University of Hawaii Press.
Winkels, A. (2011). Stretched livelihoods – The social and economic connections between the Red River Delta and the central highlands. In T. Sikor, P. T. Nghiem, J. Sowerwine, & J. Romm (Eds.), *Upland transformations: Opening boundaries in Vietnam*. Singapore: National University of Singapore Press.

Chapter 7
Environmental Disasters in the Mekong Subregion: Looking Beyond State Boundaries

Andrea Haefner

7.1 Introduction

There are 261 transboundary river and lake basins worldwide, which cover nearly half of the earth's land surface (Wolf 1998). Freshwater resources are scarce and different nations, actors and users compete for limited sources in transboundary river basins, often conflicting with each other. For instance, in the Mekong River Basin, current hydropower and navigation developments in certain countries impact on traditional sources of income such as fisheries and rice production in downstream countries. This chapter focuses on the Mekong River and especially the challenge of handling environmental disruptions and disasters including floods, pollution spills and hydropower dam constructions not sorely as national concerns but as regional challenges which need to be dealt with on several government levels including a role for Intergovernmental Organisations and NGOs. This chapter will address interlinkages of environmental disasters and urbanisation in a cross-border environment analyzing their interdependence and their impacts on the livelihoods, governments and the region as a whole using the Xayabouri Dam as a case study. This is increasingly important as resources are limited, populations increase and people move from rural to urban areas further increasing the demands on limited resources. Key research questions include: What are the key challenges of water (in) security in regards to environmental disasters? How are these transboundary challenges governed?

In answering these central questions, this chapter draws on qualitative research by the author in the Mekong region in 2012 and 2013 which was part of a larger project and included 40 interviews with government representatives, NGOs, aid agencies, international organisations and researchers (see also Haefner 2013a, b, 2014). In addition to the interviews, the analysis draws on articles and reports published in scholarly journals and newspapers supplemented by government,

A. Haefner (✉)
Griffith Asia Institute, 7/228 Vulture Street, Brisbane, QLD 4101, Australia
e-mail: a.haefner@griffith.edu.au

M.A. Miller et al. (eds.), *Crossing Borders*, DOI 10.1007/978-981-10-6126-4_7

international organisations and international aid agency reports. The chapter begins with an overview of the importance of water governance and its interlinkages before introducing the Mekong River and the main case study the Xayabouri Dam. The chapter illustrates the interlinkages of water, energy and food in an urbanising context that is and will be increasingly affected by environmental disasters. Following this section, a governance lens will be used to look at the role of regional institutions, mainly focusing on the Mekong River Commission (MRC), and the role of NGOs. Overall, this chapter argues that it is important to look beyond the state boundaries, achieving a regional perspective as this is the only way to tackle transboundary challenges and increase resilience to environmental disasters as linkages between the global, regional and national level and between different levels of governance shape how disasters occur and how their impact is distributed.

7.2 Setting the Scene: Understanding Linkages

Conflict between and within states over freshwater resources poses an interesting challenge for scholars and policymakers due to competing claims over water and the concept of territorial sovereignty that is broken by transnational rivers. Worldwide, transboundary river and lake basins cover nearly half of the earth's land surface which account for approximately 60% of the global freshwater flow, and over 90% of the world's population lives in a country that shares a river basin (Task Force on Transboundary Waters 2008). Most of the world's shared water resources remain outside transboundary agreements between all riparian countries, and some have no agreements in place. While freshwater resources are in theory renewable, in reality they are finite, poorly distributed and often controlled by one state or group (Gleick 1993). Additionally, while fresh water is fundamental for all ecological and societal activities, such as energy and food production, industrial development, transportation, health and employment, it is often unequally distributed within a region or even within one country (Gleick 1993; Frey 1993). In China, for example, south and southeastern parts of the country have sufficient water, while water scarcity is a major problem in the northern and eastern parts. As a result, the future development of many countries depends on the successful management and allocation of these natural resources, including successful cooperation between riparian states in regards to transnational water management.

Some key problems related to water management include water shortages, inefficient water use, inadequate sanitation, destruction of the ecosystem, waterlogging, salinisation, water pollution, and flooding of cultivated, urban or industrial areas. Not all of these major problems are relevant in all transboundary river basins; challenges vary (Schultz and Uhlenbrook 2007). On the Mekong River, water governance is closely related to environmental effects, especially regarding water quality and water flow. For instance, damming on the upstream river not only reduces the water flow, creating problems of water scarcity or allocation in the downstream countries, but also influences the water and soil quality (Lowi 1999). This is interlinked with growing energy demands, which is another major aspect of trans-boundary governance in the Mekong River basin. Problems arise when hydropower developments create environmental problems. Furthermore, food and energy depend on access to water, which adds to the pressures facing transnational river basins like the Mekong River—especially where urbanisation is increasing (Schneider 2011).

Water resource management and allocation plays a crucial part in potential conflict between riparian states in most, if not all, regions in the world. This is why it is important to focus on water resource management and related issues: development increasingly impacts not only water security but also energy and food security, problems that can only be tackled while using a comprehensive approach including water, energy and food. This is even more important today, as challenges in urban Asia are rapidly increasing due to growing resource demands, a larger urban population and increasing environmental disasters affecting the region. Current trends continue to show movement of migrants from the rural Mekong Delta towards urban centres, and due to Asia's rapid urbanisation more and more people are living in locations that are highly exposed to floods, cyclones, tropical storms and tsunamis (Dun 2011; Miller and Douglass 2015). This is especially prominent on the Mekong as several capitals and large cities are located along the river shore, in delta areas or on coastlines which will be increasingly affected by floods due to rising river water levels, melting of the Tibetan-Himalayan glaciers or sea level rise which all will be further impacted by climate change. These threats are more prominent in urban areas as urbanisation depends on water, food, energy, transport, shelter, waste and communications systems which are interlinked and can have effects in different places and different times (Friend et al. 2014). For instance, much of the food production in Vietnam comes from the main rice-growing areas in the Mekong Delta, which produces around 50% of the country's staple food and 60% of the fish-shrimp production (Dun 2011). It is also the region of the country most threatened by floods and typhoons, with modeling indicating that a sea level rise of 1 m would directly inundate 39% of the Mekong Delta (Garschangen 2015). Environmental disaster in the Mekong Delta would impact the national prices of rice in Vietnam, but also regional and global rice prices: in 2008 Thailand and Vietnam produced 51% of the world's rice exports and exports of fishery products of the four lower Mekong countries reached 5.6 billion USD in the same year (Schaffer and Li 2012).

However, hydropower developments directly affect food, energy and water securities as hydropower dams increase energy security, but at the same time often impact water quantity and reduce seasonal water flows. Food security is also affected, as hydropower dams reduce fish stocks and threaten river bank agriculture by regulating water flow, thereby reducing sediment flow. Generally, on one side, if dams are built and designed sustainably, they can create reservoirs for multiple purposes, including water storage and managed release, flood protection, electricity generation and irrigation. However, on the other side, dams interrupt water flows and fish migration, impact siltation and displace people and ecosystems (Martin-Nagle et al. 2011).

Although linkages of food, energy and water have always been present, this is especially important in an urbanising Asia as resources are limited, populations increase and people move from rural to urban areas, further increasing the demands on limited resources; this significantly reinforces the nexus approach[1] as a means for achieving sustainable development and a peaceful region. This is even more important against the backdrop of increasing floods, droughts and other disasters which have in recent years influenced the livelihoods of those living directly on the river. This can be seen, for instance, in the flooding in China and the Mekong region in 2010 or the landslides in Thailand in 2009, which had devastating impacts on the region and especially the local population. Similarly, the 2011 floods along the Mekong resulted in dire humanitarian and agricultural losses, killing 141 people in Cambodia, 30 people in Laos (while 400,000 were affected) and destroying 5000 hectares of rice fields in the Mekong Delta (Neo 2012). Overall, in the last three decades, Asia has experienced 40% of total floods worldwide, and 90% of people who are vulnerable to floods live in Asia (Miller and Douglass 2015). Cambodia, and especially the Tonle Sap area, is also increasingly affected in both frequency and intensity of flooding since 1999, with floods considered a major disaster in 2000, 2001, 2009 and 2011.

These events can be exacerbated by large infrastructure projects including hydropower dams, but also by increasing navigation or other impacts on the river basin (Goh 2006). Also, climate change increasingly affects the region—especially due to melting of the Himalayan glaciers which alters the seasonal access, exaggerating droughts, floods, landslides and other localised disasters, in turn affecting water and food security.

7.3 The Mekong River: A Threatened Lifeline for Millions

The Mekong is an important subregion in the Asia-Pacific as more than 70 million people live directly on its river banks and its significance extends to the wider region (Fig. 7.1). Additionally, the river hosts a unique and significant ecological system with some of the world's highest diversity of fish and snails, including the endangered Irrawaddy dolphin and the giant Mekong catfish (Campbell 2009).

In the past, the Mekong River has been at the centre of human conflict and wars for several hundred years. In future decades, the Mekong will have even more strategic importance as it is the centre for food, accommodation, and employment, and plays a crucial part in the future development of the six riparian countries: China, Myanmar, Thailand, Cambodia, Vietnam and Laos. Some of these countries are the poorest in East Asia, with Laos and Cambodia having an estimated per capita GDP of 5000 USD and 3300 USD, respectively, in 2014 (Central Intelligence Agency

[1] The nexus approach has gained momentum in the last 5 years especially after the Bonn 2011 Conference and the Nexus 2014 conference and has attracted significant interest from international organizations, the private sector and other global players as a way of tackling the interdependencies between water, energy and food security. The nexus approach focuses on the urgent need for joined initiatives to policy and practice which lies at the heart of the nexus approach.

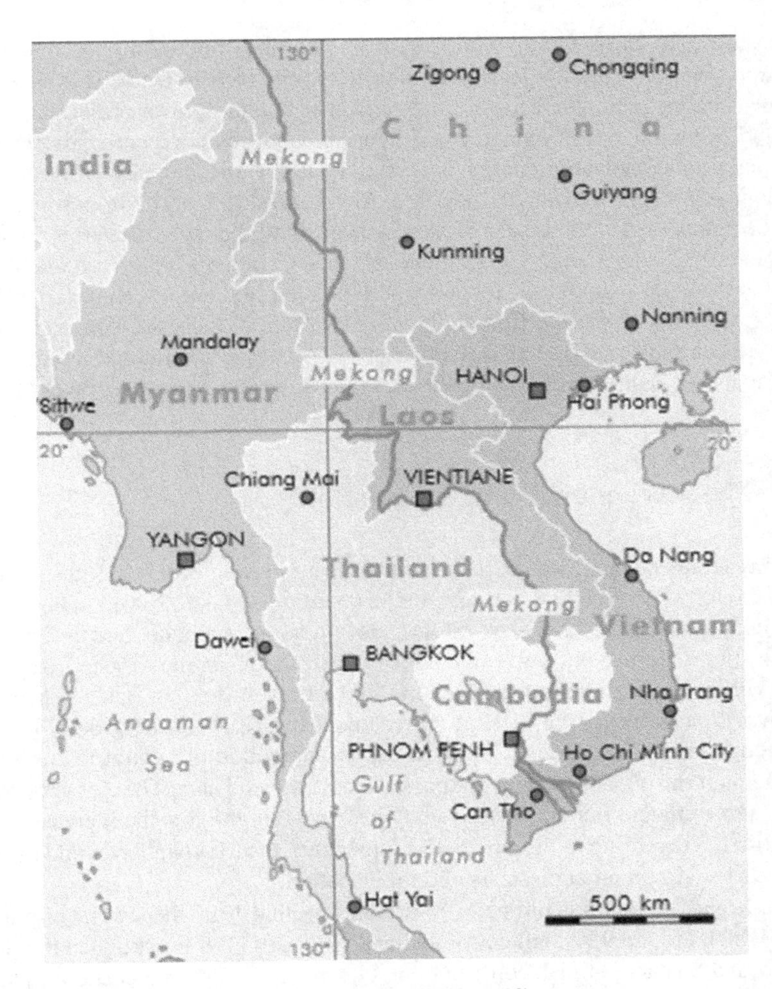

Fig. 7.1 The Mekong Subregion (Source: Haefner 2013, p. 19)

2015a, b; Mehtonen et al. 2008). Transnational challenges and environmental degradation[2] affect East Asian countries in different aspects and locations, and are impairing the Mekong subregion in particular. Floods, droughts, famines, soil erosion and deforestation are influencing the livelihoods of those living directly on the river (Emmers et al. 2006). Often these incidences are connected as floods can cause landslides, contaminate water supply or affect river bank agriculture, impacting both rural and urban areas. Although the majority of the population (72%) in the Lower Mekong Basin (LMB) still lives in rural environments, urbanisation plays an

[2] Environmental degradation is here defined according to the UNISRD as 'the reduction of the capacity of the environment to meet social and ecological objectives and needs'.

increasing role as environmental disasters including regular floods can trigger migration, which is predominantly from rural to urban areas (Dun 2011; Neo 2012).

The Mekong region is facing a steady increase in hydropower projects as a result of rapid economic development of the riparian countries in recent years. The need for cheap and renewable energy is rising, fostering the increase in hydropower development where logistically possible. As of mid-2013, 35 hydropower projects have been finalised in the Lower Mekong Basin, all of them on tributaries, including 15 in Laos, 12 in Vietnam, 7 in Thailand and 1 in Cambodia. Although the Mekong River is the central source of food, accommodation and employment for millions, each riparian country has different plans for its share of the river, which can sometimes be contradictory (see also Haefner 2013a). This is highlighted by the case of the Xayabouri Dam, which will be the first mainstream dam on the Lower Mekong.

7.4 The Xayabouri Dam: A Case Study

The Xayabouri Dam is currently one of the most pressing issues in Laos related to the Mekong; it receives international media attention and is followed intensively by regional governments, activists, NGOs, development agencies and international organisations. This dam, currently under construction in the northern region of Laos, will be the first dam of the Mekong in the Lower Mekong Basin (LMB). The dam was first proposed by the Lao government in 2010 and the ground-breaking ceremony was held at the end of 2012 despite neighbouring countries Cambodia and Vietnam raising concerns about possible negative impacts. The Xayabouri Dam is an interesting and important case study because it highlights the increasing difficult situation regarding hydropower development of the Mekong River and its direct linkage to environmental disasters and urbanisation.

The Xayabouri Dam is projected to produce around 1200 megawatts of electricity, of which around 95% will go to Thailand (IRN 2011). It is estimated that it will take around 8 years to finish construction, with an estimated total cost of 3.8 billion USD (Schmidiger and Sierotzki 2015). According to a high-level official with the MRC in Vientiane, Laos, the Xayabouri Dam is expected to bring 130 million USD a year in royalties and taxes (personal communication, 6 June 2013). Key developers of the project include Thailand's Ch. Karnchang Public Company, the Electricity Generating Authority of Thailand (EGAT),[3] the Lao government, and Ratchaburi Electricity Generating Holding Company. A Memorandum of Understanding (MoU) with Thailand's Ch. Karnchang Company as developer was signed in late 2008, and was followed by a Power Purchasing Agreement with EGAT and the Lao government in mid-2010. This agreement allows EGAT to purchase electricity, to be transferred through a 200 km long transmission line to northern Thailand, at a cost of 2.159 Thai Baht per kilowatt-hour (IRN 2011).

[3] EGAT is a Thai state enterprise that owns and manages the majority of Thailand's electricity generation capacity, as well as the nation's transmission network.

Because Thailand provides the market for the produced electricity, Thailand's Ministry of Energy and Electricity and EGAT play key roles in the development process. This also includes Ch. Karnchang, which is Thailand's second-largest publicly traded construction company and therefore has an impact on the country. As Thailand has already developed most of its domestic hydropower potential, it seeks to import from the neighbouring countries to secure its growing energy demand. Also, previous strong local opposition to dam projects including the Pak Mun Dam further supports the increasing involvement in hydropower development abroad, mainly in Laos, where civil society is weaker (IRN 2013a; Matthews 2012). Often the Xayabouri Dam is seen as a Thai project, since Thailand provides the financing, the labour force, the construction equipment and material, and buys the electricity (Varchol 2012). Estimated impacts of the Xayabouri Dam include the loss of biodiversity, changes in the ecosystem, and the reduction of fish (Radio Free Asia 2013). A highly controversial point is the use of fish ladders on the Xayabouri Dam, as proposed in the first draft of the dam. Fish experts argue that these ladders would not work due to the high diversity of fish in the Mekong, including different sizes and high number of fish, and the height of the dam (IRN 2011). It is estimated that around 2100 people would be resettled by the project and that more than 202,000 people living on the river banks in close proximity to the dam site would be affected through loss of income and food security, including loss of agricultural land, loss of opportunities of gold panning[4] and lack of access to forest products (*ibid.*).

Besides the impact on biodiversity, fish stocks and food security, another problem of the proposed dam is that the actual cost of the already existing revenues from a 'healthy' Mekong River is not counted, and social and environmental damages—as well as the necessary money for mitigation and compensation—are missing from any calculations (IRN 2011). This means that the actual cost is unknown and cannot be weighed against the income generated through the dam investment. Another key issue is that the costs and benefits will be unevenly distributed between the people within Laos, as well as between the riparian countries in the Lower Mekong. This means that the main beneficiaries in Laos will be the elite and politicians, whereas the costs will be at the local level where income is reduced due to decreased fish stocks, reduced agricultural productivity and hardship due to relocations. Looking at the distribution across borders, benefiting countries will include Thailand as a result of increasing energy supply or Laos through increased income revenue through the project. However, negative impacts will be felt mostly in downstream countries Cambodia and Vietnam due to changes in water flow, water quantity and water quality, which will directly affect food security due to reduced fish catches and other income sources. The construction will also impact water security as the water flow is restricted and further controlled.

While Cambodia and Vietnam were outspoken against the project until the end of 2012, since then the rhetoric from both countries has waned. Some experts argue that the redesign of the Xayabouri Dam was shared bilaterally with Vietnam—

[4] Gold panning is still a big source of income in Laos, especially for women living around Luang Prabang.

although the MRC itself did not receive the documents until July 2013. However, Vietnam's changed attitude was probably not due to the change of the design of Xayabouri; as a hydropower expert with the MRC said, 'I don't think that the new design, if they got it, played a major role' (personal communication, 14 June 2013). Others argue that a bilateral deal between both countries was achieved, which could include other areas such as rubber plantation or logging along the border between Laos and Vietnam or the building of the Luang Prabang Dam by a Vietnamese developer. In the Vietnamese media, pressure against hydropower development still exists, however it does not specify the Xayabouri Dam anymore. Further, Vietnam progresses with its studies in the Mekong delta focusing on impacts from upstream dams. An expert in water management in the Mekong Basin said that overall, Vietnam does not have a strong standpoint against dam building because around 50% of the Mekong River flow which comes from the Vietnamese Highlands has been blocked by dams built in Vietnam (personal communication, 24 May 2013).

The Lao government perspective—which Thailand supports—is that the engineers altered the Xayabouri Dam's design to focus on the fish passages, which meant that the project could proceed. Since international pressure focused on fish passages, as opposed to resettlement or water flow, it was easy for the government to adjust this one area. Further, the Lao government aims to overcome the country's situation of being one of the least developed in the world, and hopes to stimulate economic development and move from being a landlocked country to a land-linked one. As stated by the Lao Vice-Minister for Energy and Mines, Dr. Viraphonh Viravong, hydropower is important because it is clean, cheap and renewable, further affirming that 'hydropower contributes something like 33% to the natural capital of the wealth of Laos. And if Laos wants to leave behind its least developed country status by 2020, this is our only choice' (Varchol 2012). Moreover, decision-makers in the country will benefit sufficiently from large construction projects. Similarly, large constructions are still seen as a sign of power and a symbol of a growing and influential Laos.

In this context, it is also important to remember that the projects that proceed in the Mekong region are often the result of not only technical but also political decisions. Similarly, government departments in the countries do not have one position; often, there are several drivers behind the decision-making process. Further, controversies regarding the Xayabouri Dam include the low fixed price Thailand pays for electricity generated from it. The price is already relatively low by today's standards; according to a hydropower expert with the MRC, this anomaly will be even more prominent over time when energy prices increase (personal communication, 14 June 2013). The Finnish engineering company Pöyry's[5] role in the process was also an unexpected move by the Lao government. The company carried out the Environmental Impact Assessment (EIA) and then became the lead engineer of the project (expert in water management in the Mekong Basin, personal communication, 24 May 2013). In a manner similar to other hydropower developments in the

[5] Pöyry is a Finnish-based global consulting and engineering firm focusing on energy, forestry, infrastructure and environment.

region, this creates a conflict of interest: the company has an advantage by making certain statements and approving projects independently of possible social and environmental impacts (IRN 2013b). In general, hydropower development concentrates wealth and distributes negative impacts; a high-level MRC member stated that banks make profits, while the ways the government uses the income from revenues is often unclear (personal communication, 6 June 2013). This is even more important as the economic value of the natural capital in the region is substantial. Besides the contribution to fisheries, river-dependent agriculture yields around US$4.6 billion in rice paddy growth and up to US$574 million in riverbank gardens per year (Ponte 2012).

Hydropower developments in general, and specifically those on the Mekong River, are an explicit example of the need to focus beyond water security and include the challenges of energy and food security in the equation. The case study of the Xayabouri Dam further demonstrates that increasing energy security, for instance due to more hydropower developments in Laos, has clear tradeoffs and is linked to increasing food insecurity in downstream countries, especially Cambodia and Vietnam. Cambodia is concerned about the Tonle Sap Lake, the biggest lake in Southeast Asia, which will be affected most as it is the primary source of food and livelihoods for 1.6 million people and the source of approximately 10% of the current national GDP (Vannarith 2012). The lake depends on the Mekong River's enormous water flows in the wet season in order to reverse the direction in the dry season, carrying high amounts of fish and other animals that spawn in the lake during the wet season. This accounts for at least 60% of the annual protein intake of Cambodia's population, and is therefore crucial for food security in the country. Cambodia wants to keep the seasonal flow and the flood pulse system of the Mekong, as opposed to a regulated flow through dams, in order to protect the unique ecosystem of Tonle Sap Lake. On the other hand, Cambodia also plans to build dams within its territory, which could have a devastating impact on Cambodia and the region—seen, for instance, in the planned Lower Sesan II hydropower dam. Hydropower is expected to account for 77% of Cambodia's total electricity generation by 2030, in contrast to less than 4% in 2007 (Daily Fusion 2013).

Vietnam would also be affected strongly by increasing hydropower developments since around 50% of its annual rice crop is produced in the Mekong delta, which shows increasing salinisation. Because of the Mekong River's characteristics, Vietnamese farmers can grow three rice crops per year in the Mekong Basin, resulting in Vietnam being the world's second-largest rice exporter in 2012 (Fernquest 2012). This again shows the close interlinkage of food and water security, as around 3000 l of water are necessary to produce 1 kg of rice. Globally, it is estimated that a quarter to a third of fresh water is used for rice production (Mueller 2007). The Vietnamese goal of sufficient dry season flow is one of the main issues regarding damming on the Upper Mekong, because this influences the cultivation of rice and aquatic production and could have an impact on rice prices on a national, regional and even global level. However, Vietnam also needs to satisfy its growing energy demand and relies on energy imports from neighbouring countries such as Laos. Additionally, whenever the geographical conditions allow, Vietnam builds hydropower dams, mainly on the Dong Nai River and Sesan River, to support the growing national energy demand (IRN 2001). This again demonstrates the close interlinkages between water, energy and food in the region. Besides sufficient dry season flow, increasing floods due to climate change are another challenge: although rice uses more water than most other grains, most rice types cannot sustain being flooded for more than 4 days (Mueller 2007). This is especially critical in the Mekong delta, and will become more of a concern in the future due to increasing floods and droughts affecting rice production and as a result food security.

Furthermore, the Xayabouri Dam and similar projects demonstrate tightening rural-urban linkages: the construction of dams can lead to sudden downstream flooding due to heavy rain or seasonal glacier melt, when dams release water to prevent collapse in rural and remote regions, whereas the dam construction in the first place serves urban energy and water needs (Miller and Douglass 2015). Also, besides hydropower developments, there are other water users in a fast-developing region including water and waste water for population and industry, mining and agro-industry and tourism which further enhance the need for an interaction of various stakeholders.

7.5 Multilevel Governance: Regional Institutions, National Units and NGOs

The Mekong River Commission (MRC) is the regional organisation dealing with the Lower Mekong, which has existed in its current form since 1995 and includes the four lower Mekong countries: Laos, Thailand, Cambodia and Vietnam. The MRC has a limited mandate and does not include the two upstream countries Myanmar and China (Haefner 2013b). In September 2010, the Lao government submitted the proposal of the Xayabouri Dam to the other regional countries through the 'Procedures for Notification, Prior Consultation and Agreement' (PNPCA) of the MRC. This process is based on Article 5 of the 1995 Mekong River Agreement, which defines 'reasonable and equitable utilisation' of the river (MRC 1995). Section B 2a of the same article requires that countries notify, consult and reach agreement with the member riparian states if they utilise the waters on the mainstream of the Mekong River for intra-basin use during the dry season, which is the case in the Xayabouri Dam (MRC 1995).

The PNPCA process of the MRC was used for the Xayabouri Dam for the first time and was seen as a test. At the same time, the MRC released a major Strategic Environment Assessment (SEA) report stating that mainstream dams on the Mekong should be deferred for another 10 years due to massive possible risks and unknown impacts on the river (International Center for Environmental Management 2010), which further demonstrates the difficult position of the MRC. The Xayabouri Dam is going ahead despite concerns of the other riparian countries, Cambodia and Vietnam, regarding the PNPCA—which means that the process was unsuccessful and did not fulfill its purpose. As a local expert in water management in the Mekong Basin stated that 'within its mandate, the MRC handled the PNPCA very well. The MRC didn't do anything wrong but the PNPCA ruined the reputation of the MRC' (personal communication, 24 April 2013).

Besides intergovernmental institutions, international and national NGOs are also active in the region. However, this varies significantly from country to county in the Mekong Region. For instance, Thai and Cambodian NGOs are more active and outspoken against hydropower developments, whereas this is rather limited in Laos. One initiative was the Save the Mekong campaign, which included a coalition of civil society organisations focusing on socio-economic and environmental implications of dam development. Other civil society and NGO networks active in Cambodia—such as the River Coalition of Cambodia (RCC) or the NGO Forum (NGOF)—advocate their standpoint and try to influence the Cambodian government in order to change policies. However, sometimes this influence is very limited and weak: an expert with the RCC reported that members cannot achieve consensus. Some members want to oppose the construction of the Xayabouri Dam, while other members do not oppose it but instead want to secure a fair and decent compensation and proper relocation arrangements (personal communication, 6 December 2012). These different views make it difficult to voice one opinion, which is important in order to achieve anything in a context where civil society already has a difficult standpoint and public participation is often limited or non-existent. This is even more evident if projects have transboundary effects and not only national implications.

The Xayabouri case shows the problematic nature of large infrastructure developments in the region and their impact on the local population and the uneven distribution of wealth, especially among the poorer rural communities. As the first mainstream dam on the lower Mekong River, the Xayabouri Dam will foster future mainstream dam projects since a precedent will have been set. Future developers will argue that the first dam has the biggest impact and that if the Xayabouri Dam goes ahead, other dams are fine as well. This seems the likely development on the Mekong River as there are several other proposed mainstream dams, one being the Don Sahong Dam in southern Laos in close proximity to the Cambodian border.[6] However, this view would neglect the problem of cumulative impacts on the river and sediment flow through several dam projects—an issue that is often not appropriately addressed in single projects, which ignore the problem of transboundary effects downstream. Furthermore, the Xayabouri case is a unique case as the PNPCA was used for the first time. However, most would argue that this was unsuccessful as the dam construction went ahead, although the MRC recommended further studies on the impacts of mainstream dams. Similarly, the downstream countries Cambodia and Vietnam still do not support the project and maintain that the PNPCA process was not finished, with the main critical comments by the two downstream

[6] Construction on the Don Sahong Dam started in 2014. Until today, it is disputed by many if the dam is a mainstream or tributary dam due to its location.

countries focusing on retention of sediment and fish migration (Schmidiger and Sierotzki 2015).

Besides hydropower developments and recent initiatives such as the MRC Initiative on Sustainable Hydropower and the Rapid Basin-wide Hydropower Sustainability Assessment Tool, in regards to flood management and environmental disaster management, the four lower Mekong countries signed the Flood Management and Mitigation Agreement in 2001. This was followed by an agreement signed between China and the MRC in 2002 which included the report of daily water levels and twice daily rainfall information from the upper Mekong basin from 2003 in order to forecast floods (Lebel et al. 2013). Since early 2002, China has sent technical daily water level and rainfall data to the MRC during the rainy season in order to forecast possible flooding downstream (Hensengerth 2009). However, reliable and continuous data exchange and water release schedules from Chinese reservoirs during the dry period still do not exist (although in the 2010 dry season China began to provide data from the Jinghong Power Station on the mainstream Mekong and the Man'an tributary for downstream countries). Data sharing is important in order to reduce the risk of insufficient water availability during droughts, especially in the delta region in Vietnam (Viet Nam News 2015). Similarly, the Greater Mekong Subregion (GMS) also established a program titled Food Control and Water Resources Management which is linked to the MRC program (Haefner 2013b). However, as can be seen in recent flood events, further regional improvement are necessary to increase resilience, adjust to increasing environmental disasters, and ensure water, energy and food security. This is especially important as 80% of the rainfall falls within just 6 months, causing flooding and landslides, followed by 6 months of drought and falling water levels in the Mekong (Guttman 2012). As the case study shows, the challenges in the fast growing Mekong region are large and a multi-stakeholder and multi-level governance approach is necessary to be successful. In addition to the national governments and the Mekong River Commission, this should also include NGOs and private companies.

Besides hydropower development, improved navigation can also become a more contested topic in the future due to more pollution (especially on the Upper Mekong) and an amplified risk of accidents impacting on the environment. The impact will be more local (in comparison to large hydropower projects)—for instance, through spills of dangerous goods—but are still visible and add to the cumulative effects. Increased navigation can lead to unusual water fluctuation patterns, an intensification of riverbank erosion and a decrease in sediment load impacting on the ecological function of the Mekong (Lazarus et al. 2006). Besides the impact on tropical freshwater fish and rising water levels, the changes will also impact the local villages collecting Mekong water weed, which is an important source of food and protein (Lazarus et al. 2006; Southeast Asia River Network 2004). The unusual rapid water fluctuation also affects the rich riverine ecosystems, local livelihoods and fishing communities; for instance, on the Thai-Lao border, fish catch declined by 50% due to water fluctuation as a result of rock blasting to improve navigation (Southeast Asia River Network 2004). Furthermore, the increased presence of large vessels presents hazards for small fishing boats, which can be affected directly or through the waves on the river (Japan Environmental Council 2005). In the long run, the water flow to the sea will be faster as a result of navigation due to a decrease in resistance, which will lead to an increased risk of floods including flash floods in the Mekong delta area and the urban centres (Roberts 2001). Combined with hydropower developments, navigation could further increase the risks of floods and erosion in the region and as a result further impact on the livelihoods in the region.

7.6 Conclusion

Freshwater resources are scarce and different nations, actors and users often come into conflict with each other as they compete for their use in transboundary river basins. Across Asia, deltas such as the Mekong Delta are home to cities and city regions that are major economic hubs, with many millions of people who face significant flood risk which will further increase in the future (Chan et al. 2012). This chapter focused on the Mekong River, and especially the challenge of handling environmental disruptions and disasters including floods and hydropower dam construction not solely as national concerns, but as a regional challenge which needs to be dealt with on several levels (including roles for intergovernmental organisations and NGOs). A special focus was on looking beyond the state boundaries and taking a comprehensive approach including water, energy and food: this is increasingly important as resources are limited, populations increase and people move from rural to urban areas, further increasing the demands on limited resources. This is especially important as over the last two decades, there have been unusually large flooding events in the Mekong Delta region which have adversely impacted the people in the region, resulting in a rural-urban migration.

Given the current environment of hydropower expansion on the Mekong River, challenges will increase. This was demonstrated by the case study of the Xayabouri Dam and future dams, which will further increase the risks of environmental disasters such as floods and droughts in the region and affect the millions of people who depend on the river for food, accommodation and employment. Combined with increasing frequency and intensity of environmental disaster (especially in urban areas), the approach needs to be comprehensive and include various sectors and stakeholder focusing on water, energy and food security. Further, it is important to look beyond the state boundaries, achieving a regional perspective as this is the only way to tackle transboundary challenges and increase resilience to environmental disasters. This is especially important in an interconnected world where linkages between the global, regional and national level and between different levels of governance shape how disasters occur and how their impact is distributed. This includes on the one side the responsible regional institutions such as the MRC but increasingly also the role of non-state actors such as NGOs and local units which are affected by environmental disasters.

References

Campbell, I. C. (2009). Chapter 1: Introduction. In I. C. Campbell (Ed.), *The Mekong: Biophysical environment of an International River basin* (pp. 1–11). New York: Elsevier.
Central Intelligence Agency (2015a). 'Cambodia', in The World Factbook, CIA, Washington, DC, viewed 23 November 2015, https://www.cia.gov/library/publications/the-world-factbook/geos/cb.html.
Central Intelligence Agency (2015b). 'Laos', in the world Factbook, CIA, Washington, DC, viewed 23 November 2015, https://www.cia.gov/library/publications/the-world-factbook/geos/la.html.
Chan, F. K. S., Mitchell, G., Adekola, O., & McDonald, A. (2012). Flood risk in Asia's urban mega-deltas: Drivers, impacts and response. *Environment and Urbanization ASIA, 3*(1), 41–61.
Daily Fusion (2013). China becomes 'key investor' in Southeast Asia's hydropower. Daily Fusion, 19 August.
Dun, O. (2011). Migration and displacement triggered by floods in the Mekong Delta. *International Migration, 49*(S1), 200–223.
Emmers, R., Greener-Barcham, B., & Thomas, N. (2006). Institutional arrangements to counter human trafficking in the Asia-Pacific. *Contemporary Southeast Asia, 28*(3), 490–511.
Fernquest, J (2012). Rice exports: Fall from 1st to 3rd. Bangkok Post, 25 October.
Frey, F. W. (1993). The political context of conflict and co-operation over international river basins. *Water International, 18*(1), 5468.
Friend, R. M., Thinphanga, P., MacClune, K., Henceroth, J., Tran, P. V. G., & Nghiem, T. P. (2014). Urban transformations and changing patterns of local risk: lessons from the Mekong Region. *International Journal of Disaster Resilience in the Built Environment, 6*(1), 3043.
Garschangen, M. (2015). Risky change? Vietnam's urban flood risk governance between climate dynamics and transformation. *Pacific Affairs, 88*(3), 599–621.
Gleick, P. H. (1993). Water and conflict: Fresh water resources and international security. *International Security, 18*(1), 79–112.
Goh, E. (2006). China in the Mekong River basin: The regional security implications of resource development on the Lancang Jiang. In M. Caballero-Anthony, R. Emmers, & A. Acharya (Eds.), *Non-traditional security in Asia*. Hampshire: Ashgate
Guttman, H (2012). 'Keynote: Water, food and energy security in the Mekong Basin', delivered at Mekong2Rio International Conference on Transboundary River Basin Management, Phuket, Thailand, 1 May.
Haefner, A. (2013a). Non-traditional security: The case of environmental challenges in the Mekong subregion. *Pacific Geographies, 40*, 17–22.
Haefner, A. (2013b). Regional environmental security: Cooperation and challenges in the Mekong subregion. *Global Change, Peace and Security, 25*(1), 27–41.
Haefner, A (2014). Cooperation and conflict along transnational rivers: The role of regional institutions, paper delivered at the Oceanic Conference for International Studies, Melbourne, Australia, July.
Hensengerth, O. (2009). *Money and security: China's strategic interests in the Mekong River basin, Chatham House Briefing Paper*. London: The Royal Institute of International Affairs.
International Center for Environmental Management (2010). Environmental assessment of hydropower on the Mekong mainstream: Summary of the Final Report. Prepared for the Mekong River Commission.
IRN. (2001). *Planned dams in Vietnam, report*. Berkeley: International River Network.
IRN. (2011). *The Xayaburi dam: A looming threat to the Mekong river, fact sheet*. Berkerley: International River Network.
IRN (2013a). The Mekong dams dispute: Four trends to watch. Viewed 29 April 2016, https://www.internationalrivers.org/blogs/267/the-mekong-dams-dispute-four-trends-to-watch.
IRN. (2013b). *'Pöyry's role in the Xayabouri dam controversy', factsheet*. Berkerley: International Rivers Network.

Japan Environmental Council. (2005). The Mekong region: Incorporating the views of regional civil society. In Japan Environmental Council (Ed.), *The state of the environment in Asia* (pp. 131–150). Tokyo: Springer.

Lazarus, K., Dubeau, P., Bambaradeniya, R., Friend, R., & Sylavong, L. (2006). *An uncertain future: Biodiversity and livelihoods along the Mekong River in northern Lao PDR*. Vientiane: The World Conservation Union (IUCN) Lao PDR Country Office.

Lebel, L, Sinh, BT, Garden, P, Bui, VH, Subsin, N, Le, AT, Nguyen, TPVV (2013) Risk reduction or redistribution? Flood management in the Mekong region. ResearchGate, December.

Lowi, M. R. (1999). Water and conflict in the Middle East and south Asia: Are environmetnal issues and security issues linked? *Journal of Environment and Development, 8*(4), 376–396.

Martin-Nagle, R, Howard, E, Wiltse, A Duncan, D (2011). *Conference Synopsis: Bonn2011 Conference: the Water, Energy and Food Security Nexus: Solutions for the Green Economy*, Federal Ministry for the Environment, Nature Conservation and Nuclear Safety (BMU), Federal Ministry for Economic Cooperation and Development (BMZ) and OOSKAnews, Inc., Bonn, Germany.

Matthews, N. (2012). Water grabbing in the Mekong Basin: An analysis of the winners and losers of Thailand's hydropower development in Lao PDR. *Water Alternatives, 5*(2), 392–411.

Mehtonen, K., Keskinen, M., & Varis, O. (2008). The Mekong: Iwrm and institutions. In O. Varis, A. K. Biswas, & C. Tortajada (Eds.), *Management of transboundary rivers and lakes*. Berlin: Springer.

Mekong River Commission (1995). *Agreement on the cooperation for the sustainable development of the Mekong River basin*. Chiang Rai, Thailand: Mekong River Commission.

Miller, M. A., & Douglass, M. (2015). Governing flooding in Asia's urban transition. *Pacific Affairs, 88*(3), 499–515.

Mueller, H (2007). In Feist, S. (Ed.), *Hoffnung Auf Das Kleine Korn: in Den Philippinen Waechst Der Reis Der Zukunft (Hope for the little corn: The Philipines grow rice for the future.* Weltmacht Wasser (Worldpower Water), Munich, Germany: Herbig.

Neo, L. (2012). Climate change impacts, vulnerabilities and adaptation measures in the lower Mekong Basin. *Asian Journal of Environment and Disaster Management, 4*(4), 355–378.

Ponte, E. (2012). *Preliminary analysis of water security in the Mekong River basin, UNU-CRIS working papers 12*. Brugge: United Nations University Institute on Comparative Regional Integration Studie.

Radio Free Asia (2013). Laos confirms 'preparatory' work on controversial dam project. Radio Free Asia, viewed 10 August 2013, http://www.rfa.org/english/news/laos/confirms-07312013151754.html.

Roberts, TR (2001). Downstream ecological implications of China's Lancang hydropower and Mekong navigation project. Paper delivered at the International Symposium on Biodiversity Management and Sustainable Development in the Lancang-Mekong River Basin, Xishuangbanna, China.

Schaffer, A., & Li, D. (2012). Strenghtening climate adaptation in the lower Mekong River basin through a regional adaption action network. *Asian Journal of Environment and Disaster Management, 4*(4), 527–547.

Schmidiger, R., & Sierotzki, K. (2015). *Xayaburi hydroelectric power project*. Vientiane: Pöry.

Schneider, K (2011). Water challenges Asia's rising powers—part I. Yale Global Online, viewed 29 April 2016, http://yaleglobal.yale.edu/content/water-challenges-asia-powers-part-i.

Schultz, B., & Uhlenbrook, S. (2007). *Water security: What does it mean, what may it imply?', discussion draft paper for session on water security*. Delft: UNESCO-IHE (Institute for Water Education).

Southeast Asia River Network. (2004). *Downstream impacts of hydropower and development of an interantional river: A case study of Lancang-Mekong*. Chiang Mai: Southeast Asia River Network (SEARIN).

Task Force on Transboundary Waters. (2008). *Transboundary waters: Sharing benefits, sharing responsibilitites*. Zaragoza: UN-Water Thematic Paper.

Vannarith, C. (2012). *'A Cambodian perspective on Mekong River water security', blog post.* Phnom Penh: Cambodian Institute for Cooperation and Peace (CICP).

Varchol, D (2012). The Mekong [film].

Viet Nam News (2015). Mekong faces water shortage. Viet Nam News, 9 October 2015. http://vietnamnews.vn/society/276846/mekong-faces-water-shortage.html.

Wolf, A. T. (1998). Conflict and cooperation along international waterways. *Water Policy, 1*(2), 251–265.

Chapter 8
Cross-Boundary Disaster Risk Governance: Lessons from the Pearl River Delta

Alain Guilloux

8.1 Introduction

Tackling disaster risk and related vulnerabilities has increasingly been conceptualised as a governance issue, with inclusiveness and cooperation as key determinants. Developed countries and territories are often regarded as better equipped to limit the damaging consequences of natural hazards and thus reduce casualties, but recent disasters including the heat wave in Europe in the summer of 2003, Hurricane Katrina in 2005 and the Great East Japan earthquake and tsunami in 2011 have clearly shown that even the most developed countries are not immune and can be seriously affected.

While reducing disaster risk within a single country is challenging, in particular where multiple jurisdictions and governance levels overlap, handling disaster risk across borders may be an even more complex task. This chapter addresses disaster risk governance in China's Greater Pearl River Delta (GPRD). Keeping in line with the theme of the book, the focus of enquiry is on the governance of disaster risk across the Hong Kong-Guangdong border. It is claimed here that this border can and should be conceptualised as a quasi-international border. The chapter argues that the evolution of disaster risk governance in Hong Kong, in Guangdong and across the border has generally been reactive, rather than informed by strategic thinking. The region is underprepared for future risks in the light of rising factors of vulnerability, including breakneck urbanisation, demographic trends and climate change.

Key data were gathered from official government sources from both sides of the border. In addition, clarifications were sought directly from Hong Kong government officers by e-mail. Given the difficulty to identify relevant informants in Guangdong province, Hong Kong-based journalists and NGO workers were interviewed to provide additional information.

A. Guilloux (✉)
Independent Scholar, 2/F 12 B Tai Yuen Village, Yung Shue Wan, Lamma Island, Hong Kong
e-mail: alguil@yahoo.com

M.A. Miller et al. (eds.), *Crossing Borders*, DOI 10.1007/978-981-10-6126-4_8

In the first part, transboundary governance is explored with reference to environmental hazards in an urbanising world. The relevant academic literature is reviewed and governance challenges specific to rapidly expanding megacities explored. A key finding is that studies of transboundary environmental disasters focus more on the sharing of freshwater resources than on rapid-onset, life-threatening crises. Besides, breakneck urbanisation challenges traditional, hierarchical public administration and governance models. The second part explores the nature and dimensions of the borders that criss-cross the Greater Pearl River Delta region, with a specific focus on the hierarchical order that organises its various entities. The institutional arrangements and constraints relevant to disaster risk governance and response across the Pearl River Delta are then examined in the third section. Their evolution over time is mapped for two types of environmental disasters: tropical storms and epidemic outbreaks.

Finally, future risks are assessed for the region in the light of climate change and unbridled urbanisation.

8.2 Governing Transboundary Environmental Disasters in an Urbanizing World

In this section, the linkages between environmental disasters, transboundary governance and urbanisation are examined. An environmental disaster is defined here as 'one that destroys important environmental amenities or one that causes harm to human interests via an environmental change' (Farber 2011, p. 1785). Boin, Rhinard and Ekengren have suggested that a transboundary crisis occurs 'when the life-sustaining systems or critical infrastructures of multiple member states are acutely threatened' (Boin et al. 2014, p. 131). Environmental disasters that cut across national borders may be addressed more effectively where regional institutions, such as the European Union, are in a position to establish common rules. The lack of similar rule-based intergovernmental institutions across the Asia-Pacific region has led to sub-optimal results, as evidenced with the recurring haze episodes affecting Singapore, Indonesia and Malaysia. ASEAN approaches were found to have limited effectiveness even with regard to regular, annual episodes such as the transboundary haze pollution resulting from forest fires (Nurhidayah et al. 2015). Where applicable, customary international law may help foster solutions to intractable transboundary issues. Treaties governing international waters, which include third-party dispute settlement, were found to be more effective than river basin organisations to prevent militarised conflict between riparian claimants (Mitchell and Zawahri 2015). Sharing knowledge and skills relevant to transboundary issues also helps improve mutual understanding and build relevant partnerships with a view to devise workable solutions (Stacey et al. 2015).

Calls have been made to engage informal civil society networks, including scientists or policy makers, to help tackle transboundary governance of water resources, be they fresh water or sea water, fraught with numerous difficulties (Armitage et al. 2015). The potential of transboundary conservation as a way to foster peace across international borders has been questioned, as in the case of Nicaragua-Costa Rica relations (Barquet 2015). Nonetheless, transboundary protected areas are more likely to be established by countries with a history of militarised disputes, provided these disputes are low-intensity and do not lead to fatalities (Barquet et al. 2014). Riparian states are sometimes at odds over the sharing of water resources, although the case was made that environmental and geopolitical security concerns can lead to harmonious, win-win agreements (He et al. 2014). Demand rather than supply was shown to be a key determinant across the Mediterranean, Middle East and Sahel regions, thus pointing to population pressure and development needs, rather than climate variability, as key determinants of conflict (Böhmelt et al. 2014; Selby and Hoffmann 2014).

Research on transboundary governance tends to focus more on conservation issues and the sharing of resources (primarily water) than on disasters that potentially threaten human lives. Nonetheless, this literature helps make sense of the values, institutions and processes conducive to transboundary problem-solving. In particular, the structure of incentives that drive the behaviour of various stakeholders has to be ascertained (Dallimer and Strange 2015).

While the European Union failed to handle the Chernobyl nuclear disaster and mad cow disease episode effectively, its capacity to respond to transboundary life-threatening crises has been repeatedly upgraded since the 9/11 attacks (Boin et al. 2014). The Lisbon Treaty was a key step in integrating the external policy of the European Union, but in terms of transboundary responses to disasters, 'the most important development may well be the frantic development of 'sense-making tools', aimed at the collection, analysing, and sharing of information on the causes, dynamics, and effects of transboundary threats' (ibid., p. 134).

Disaster risk governance is understood here as the set of policies, institutional arrangements, programmes and activities designed to help prevent or minimise disaster risk, mitigate the impact of, or respond to disasters. Since the late 1980s, the international community has increasingly focused on disaster risk reduction. Following the decision of the United Nations General Assembly to declare the 1990s as the International Decade for Disaster Risk Reduction, Japan hosted a series of World Conferences on Disaster Risk Reduction in Yokohama (1994), Hyogo (2005) and Sendai (2015).

In December 1999 the UN General Assembly endorsed the International Strategy for Disaster Reduction to focus the attention of all member states on the urgent need to minimise the human casualties and economic losses which result from disasters. The UN General Assembly also established the United Nations International Strategy for Disaster Reduction (UNISDR), with the mandate

> to serve as the focal point in the United Nations system for the coordination of disaster reduction and to ensure synergies among the disaster reduction activities of the United Nations system and regional organizations and activities in socio-economic and humanitarian fields. (United Nations General Assembly 2001, p. 2)

The involvement of businesses and nonprofit actors, as well as communities and local governments, is seen as important to the success of the UN-led intergovernmental normative, institutional and programmatic agenda. In recent years, many voices have called for tighter alignment of the disaster risk reduction, climate change adaptation and development agenda, arguing that development has been repeatedly hampered by recurring disasters, including extreme weather events induced by anthropogenic climate change (see, for instance, Intergovernmental Panel on Climate Change 2012). Synergies between policy measures and related investments have been noted across the development, climate change adaptation and disaster risk reduction communities, and calls been issued, by aid recipients and donors alike, to streamline the allocation of limited ODA resources that cut across all three sets of goals (World Conference on Disaster Risk Reduction 2015). Conference participants also noted a broad consensus around key concepts such as vulnerability and resilience.

While the United Nations International Strategy for Disaster Reduction (UNISDR) takes note of such convergences and synergies, the term 'environmental disaster' is not included in its detailed and extensive glossary of terms. Instead, UNISDR defines environmental degradation as 'the reduction of the capacity of the environment to meet social and ecological objectives and needs' and notes that

> The types of human-induced degradation are varied and include land misuse, soil erosion and loss, desertification, wildland fires, loss of biodiversity, deforestation, mangrove destruction, land, water and air pollution, climate change, sea level rise and ozone depletion. (United Nations International Strategy for Disaster Reduction 2009, p. 14)

In its glossary UNISDR also lists 'environmental impact assessments' and states that they 'should include detailed risk assessments and provide alternatives, solutions or options to deal with identified problems' (United Nations International Strategy for Disaster Reduction 2009, p. 15). Beyond these semantic issues and attempts at delineating various types of disasters neatly, compound disasters are becoming an increasingly pressing concern.

Compound disasters, as understood here, 'can be multiple independent events, sequential or simultaneous events in large disasters, progressive failures of infrastructure, catastrophic infrastructure collapses or disruption of communications, transportation or energy networks' (Eisner 2015, p. 139). More specifically, the interlinkages between the consequences of synchronous or sequential events have contributed to refocusing the attention of crisis managers on the need to 'prevent unmanageable cascading and debilitating events' (Lagadec 2006, p. 489). Macro trends including population growth, urbanisation (including in areas at risk of technological hazards), the expansion of critical infrastructure into high-risk locations and anthropogenic climate change point to the urgency of reconceptualising existing paradigms and disaster risk governance arrangements. Widespread vulnerability and accelerated change across the developing world only add to the urgency. As developing countries industrialise, mapping technological hazards across an ever-expanding array of nuclear and chemical plants (plus related sites for storage or processing of toxic waste, and the critical infrastructure used to transport it) is a daunting task. At the same time, anthropogenic climate change may cause more frequent extreme weather events such as tropical storms. The combination of rising temperatures, urbanisation and breakneck industrialisation is likely to increase the frequency and intensity of compound disasters, such as the March 2011 East Japan Tohoku earthquake and tsunami followed by the Fukushima nuclear meltdown (Eisner 2015).

UNISDR has adopted the concept of ecosystem services from the Millennium Ecosystem Assessment, which categorises the benefits ecosystems provide to people and communities as regulating, provisioning, supporting and cultural services. The focus is here on disasters directly linked to the regulating services of ecosystems, such as floods, drought, land degradation and disease. Ecosystem-based disaster risk reduction calls for the integration of ecosystem management and disaster risk governance. While this approach, based on investments in sound and sustainable ecosystem management, may help reduce the vulnerability of communities to disasters, a larger range of infrastructure projects may be regarded as cost-effective in the urban areas of developed countries than in those of many developing ones. This is a reflection of the ways in which the global political economy works. In particular, 'disaster risk reduction imposes higher opportunity costs in poorer countries because it can take money away from other priorities that may result in more lives saved at lower cost' (Williams 2011, p. 12).

As hubs that underpin and disseminate global economic growth, megacities are the hallmark of the global, contemporary urban age. Megacities are generally understood as continuous urban areas with a resident population in excess of ten million. Connectivity is of particular importance, as megacities are key centres of transport and communications infrastructure and networks. Since megacities in the developed world are growing at a slower pace, nearly all future growth in the urban population, and of megacities in particular, will happen in the subtropical areas of the developing world. As the share of the global population living within 100 km of a coastline is increasing, the focus naturally turns to subtropical coastal megacities when considering current and future environmental and disaster risk governance challenges. The rapid rise in the population of megacities, especially in the coastal, subtropical areas of developing countries, is associated with increased vulnerability. In particular, informal settlements house a growing proportion of urban dwellers, most at risk of being affected by weather-related, geophysical or technological disasters. In addition to single, weather-related, geophysical or technological disasters, megacities located in coastal, subtropical areas are also more likely to be vulnerable to compound disasters in the future.

Rapid industrialisation, population density and migration from rural areas to cities have shaped urbanisation in East Asia, with polycentric growth often leading to the merging of multiple urban areas into one. Researchers have long struggled to establish valid criteria for defining urban areas and making international comparisons, as individual countries use different criteria. Chomitz et al. (2005) suggested (1) population density and (2) distance to cities with populations of over 100,000. Elaborating on these criteria, Uchida and Nelson (2010) suggested an agglomeration index based on three factors: minimum population size, minimum population density, and maximum travel time by road. This is not a panacea, however, as it was found that 'when applied to Indonesia in unmodified form, it was found to define nearly all of the island of Java (home to 140 million people) as one large urban area' (World Bank 2015, p. 5).

While no megacity straddles international borders, a large number of urban areas, in particular in East Asia, face critical governance challenges as institutional arrangements fail to keep up with the pace and forms of breakneck urbanisation. As noted in a World Bank report,

> Hundreds of urban areas in the region now cross local administrative boundaries. About 350 urban areas in East Asia spill over local administrative boundaries. In 135 of these urban areas, no single jurisdiction encompasses even half of the total urban area. (World Bank 2015, p. xx)

China is in a league of its own with regard to urban expansion, which consumed 23,600 km^2 between 2000 and 2010, as measured by comparing satellite data (World Bank 2015, p. 11). The Central Government has promoted urbanisation 'by decree' as a way to speed up modernization. The Greater Beijing Area project, or Beijing-Tianjin-Hebei conurbation, mulled over for more than a decade, has yet to realise the vision of a network of cities linking Beijing, Tianjin and satellite towns in the surrounding Hebei Province. A key reason is that '[c]onflicts of interest between the three local governments abound, as they are unable to agree on their respective roles' (*The Economist*, 13 December 2013). This is not an isolated example:

> Of the 81 urban areas in China of more than 1 million people, 57 are 'fragmented' urban areas, that is, no one county or district in the urban area has more than 50 percent of the overall urban land. Some 17 are 'spillover' urban areas that are at least 50 percent contained within one county but urban expansion has spilled over into others. Only seven are contained within a single county. (World Bank 2015, p. 74)

The report concludes that:

> The future prosperity of East Asia's urban areas will depend in large measure on tackling the challenge of governing multijurisdictional urban regions effectively. Many of the region's urban areas cannot be effectively served by local governments acting independently. (World Bank 2015, p. xxi)

The pace of urbanisation in East Asia in general and China in particular is unprecedented. An important consequence is the jurisdictional fragmentation, which affects urban governance throughout the region. These issues are now examined in the case of the Greater Pearl River Delta region.

8.3 Borders and Boundaries in the Greater Pearl River Delta Region

The Greater Pearl River Delta/Pearl River Estuary Bay Area (GPRD/PRE) is the largest and most populated megacity worldwide. Also called Zhujiang Delta or Zhusanjiao, the urban region extends far beyond the banks of the Pearl River, including counties and areas far removed from the Pearl River Estuary Bay Area. The Greater Pearl River Delta has a land area of 39,380 km^2 (15,200 mi^2), of which 7000 km^2 were regarded as urban land in 2010 (World Bank 2015, p. 69).

One of the key hubs of the global economy, in particular in terms of logistics, manufacturing and finance, the GPRD/PRE has been flagged as a high-risk urban region, first and foremost with regard to hydro-meteorological hazards (Swiss Re 2014). At the same time, it has significantly upgraded disaster preparedness and cross-boundary coordination over the past three decades and is rated as 'very well prepared' to respond to storms (Epstein 2013).

While the largest part of the region falls directly under the jurisdiction of the People's Republic of China, an entirely different set of laws and institutions, largely inherited from British colonial rule, applies to the Hong Kong Special Administrative Region (Ghai 1997). Following Deng Xiaoping's suggestion during his 1992 tour of South China, the Pearl River Delta Economic Zone was set up in 1994 'as a pioneering demonstration area of modernisation for Guangdong province' (Ma 2012, p. 99) but left out both Hong Kong and Macao. Ma (2012) argues that the Greater Pearl River Delta (GPRD), which consists of the Pearl River Delta (PRD), Hong Kong and Macao 'has become a global city-region' (Ma 2012, p. 98).

The Greater Pearl River Delta consists of 11 different entities ranked at various levels. The Hong Kong Special Administrative Region and Macao Special Administrative Region, set up in 1997 and 1999 respectively, stand at the First or Provincial level, on par with the Guangdong Provincial Government.

Nine of Guangdong Province's 19 prefectures are included in the Pearl River Delta region. At the sub-provincial level, Guangzhou and Shenzhen are not under the administrative control of the provincial government. As Special Economic Zones (SEZ), Shenzhen and Zhuhai enjoy 'the relaxation of authority, and the freedom to experiment with new policies and measures' (Yeung et al. 2009). One rank below, Huizhou, Jiangmen, Foshan and Zhaoqing are prefecture-level cities with authority over county governments. Lastly, Dongguan and Zhongshan, although regarded as prefecture-level cities, do not include county-level divisions. In short, various hierarchical levels characterise the 11 entities that constitute the Greater Pearl River Delta (China Internet Information Center n.d.).

Although the Hong Kong SAR and Guangdong Province are in principle on par, 'Hong Kong's de facto semi-state status under the OCTS [One Country Two Systems] makes cross-border integration in the Greater PRD much more complicated than other city-regions of the country' (Yang and Li 2013, p. 33). The Central Government's main priority following the return to Chinese rule of Hong Kong in 1997 and Macao in 1999 was not to integrate both territories within the wider Pearl River Delta region, but to ensure the successful implementation of 'One Country Two Systems.' Provincial authorities in Guangzhou were therefore largely ignored or bypassed.

Weakened by decentralisation in the 1990s, Guangdong authorities regained some initiative to drive economic integration through negotiations with the Hong Kong SAR Government. While business and government elites in Hong Kong have long insisted on the benefits of integration with the PRD, Hong Kong residents have in recent years been increasingly divided over its benefits. More generally, Peter Cheung argues that 'the intergovernmental dynamics inside the Greater Pearl River Delta (PRD) can be interpreted as the policy advocacy and coordination efforts of entrepreneurial local state actors in a fragmented authoritarian polity' (2012, p. 21).

However the key issue, to keep in line with the theme of this volume, is the nature of the border that separates Hong Kong from the mainland. A key reason why this border is not just a sub-national, administrative division is that it was created, and later modified, by a series of international treaties between Britain and China. The Sino-British Joint Declaration of 1984 is the last of these treaties. Britain agreed to return Hong Kong to China in 1997 on the understanding that individual rights and liberties, institutions and lifestyle in the territory would remain unchanged for 50 years. The provisions of the Basic Law, Hong Kong's mini-constitution drawn up in 1990, mirror those of the Joint Declaration (Ghai 1997).

The Hong Kong SAR is an international actor on its own, a member of a number of international organisations, and has the power to make treaties. Both the European Union and the United States have endorsed the Joint Declaration and the Basic Law and expressed support for Hong Kong's autonomy, emphasising their many interests in Hong Kong's future with regard to trade, investment, the rule of law, and large expatriate communities residing in the territory. In 1992 the US Congress enacted the United States-Hong Kong Policy Act, establishing itself as a guarantor of Hong Kong's autonomy in line with the Joint Declaration and the Basic Law (United States Congress 1992).

Regardless of sovereignty issues, borders may present various levels of intensity. What is meant here is the level of control over the circulation of goods, people and ideas. At one extreme, the DMZ that separates both Koreas is a strong barrier. At the other end, European countries signatories to the Schengen agreement have relaxed border controls among themselves to facilitate near-seamless exchanges. The Hong Kong-Guangdong border lies somewhere in-between. The Hong Kong SAR issues its own passports and its own currency, pegged to the US$ and not the Yuan. It has its own visa regime and controls immigration at border points. It is a separate customs territory and a full member of the World Trade Organisation. Established by international treaties, under close watch by both the United States and the European Union and characterised by strict immigration and border controls, the Hong Kong-Guangdong border separates territories ruled by different principles and laws. For the purpose of this study, it should be regarded as a quasi-international border.

Thus, the Greater Pearl River Delta region is a complex array of entities, including Special Administrative Regions, Special Economic Zones and prefectures ranked in several categories, each endowed with different levels of authority. For the purpose of this study, in line with the overarching theme of this volume, the focus is hereafter on disaster risk governance arrangements across the Hong Kong-Guangdong border.

8.4 Disaster Risk Governance Across the Greater Pearl River Delta Region

How does the complex institutional design of the Greater Pearl River Delta region affect disaster risk governance and the disaster risk reduction agenda? Relevant policies, institutions and governance arrangements in China and Hong Kong are examined before turning to the cross-border handling of two types of environmental disasters: tropical storms and outbreaks of infectious diseases.

China has actively supported the post-Hyogo process and enacted a sustainability policy (Yi et al. 2012), and has emphasised in various intergovernmental fora the need to better align sustainable development, climate change adaptation and disaster risk reduction policies. China issued the *Disaster Reduction Action Plan of the Peoples' Republic of China (2006–2015)* (Office of the China National Committee for Disaster Reduction 2006), updating an earlier version in line with the *Hyogo Framework for Action 2005–2015: Building the Resilience of Nations and Communities to Disasters* (United Nations Office for Disaster Risk Reduction (UNISDR) 2007). The revised plan tightly follows the gap analysis and priorities for action included in the final report of the World Conference on Disaster Reduction.

The Post-Hyogo Framework recommends 'the creation and strengthening of national integrated disaster risk reduction mechanisms' (World Conference on Disaster Reduction 2005, p. 6), but the architecture of China's vast disaster management apparatus lacks integration. Unlike in Japan, the USA or many other countries, there is no central authority tasked with all aspects of disaster risk governance. In the PRC, these functions are disseminated across numerous ministries and government departments (Yi et al. 2012, p. 303). An array of single-disaster response plans have been issued and 'more than 100 laws and regulations related to disaster management … such as Flood Control Law, Earthquake Disaster Mitigation Act, Fire Prevention Law, Meteorology Law' enacted (Yi et al. 2012, p. 304). What is missing is a core disaster management law that specifies the respective roles, duties and responsibilities of national, provincial, prefecture-level and local authorities with regard to disaster risk reduction, disaster prevention, preparedness, mitigation, early warning, rescue, recovery and reconstruction. The *Disaster Reduction Action Plan of the People's Republic of China (2006–2015)* lacks specifics in this regard and simply spells out the need to

> intensify the function of comprehensive disaster reduction coordination of the China National Committee for Disaster Reduction, strengthen the coordination and cooperation between relevant national departments and establish the communication and sharing mechanism on disaster reduction information between the departments. (Office of the China National Committee for Disaster Reduction 2006, p. 6)

Disaster management is organised hierarchically from the national to the provincial and prefecture level. Provincial-level disaster governance arrangements reflect the large number of national-level authorities involved, and single-disaster specific laws and regulations. Thus, the Guangdong Emergency Management Office consists of a Guangdong Emergency Management Committee (with a Director General and three Deputy Director Generals) and no fewer than 68 units and seven Specialized Emergency Management Institutions (Emergency Management Office, the People's Government of Guangdong Province n.d.). The ways in which the Guangdong Emergency Management Committee oversees and coordinates the activities of such a large number of relevant units and institutions is an area for further research.

Kapucu's extensive literature review on collaborative and multi-level disaster management systems in urban areas concluded with the observation that 'collaborative and networked relationships constitute the core of the urban/metropolitan emergency management' and emphasised the 'leadership and political will that would create, foster and enhance an environment of cooperation rather than competition' (2012, p. S43). An empirical study of local governments in Florida found 'a link between the magnitude of local disaster vulnerabilities and local officials' acceptance of regionalism as an organizational approach and as a promoter of emergency preparedness' (Caruson and MacManus 2008, p. 299). While these findings may apply to like units such as Florida counties, extending them across the uneven institutional architecture of the Greater Pearl River Delta region would be hazardous. There are no clearly unified levels in the Chinese administrative system, as units which are in principle on par may enjoy different administrative powers. Unevenness is rolled down from the top. Differences in status, in particular between Guangdong Province and the Hong Kong SAR, make integrated disaster risk governance and cross-border cooperation in the PRD more difficult than in China's other urban regions. This is the case not just because the Hong Kong SAR seeks to keep its prerogatives and resist integration guided from Guangzhou, but also because the Central Government may intervene directly to ensure security and stability, should even a disaster of contained proportions and limited geographical extent occur.

The Lamma ferry disaster, which happened on National Day, 1 October 2012, is a clear example of such direct intervention. While the collision between two ferries off Lamma Island was a tragic accident (the worst disaster at sea for Hong Kong since 1971) which resulted in the sinking of one of the ferries within minutes and the loss of 39 lives, the emergency was managed swiftly and professionally by frontline rescuers from Hong Kong uniformed services. The next morning HKSAR Chief Executive Leung Chun-ying paid a visit to survivors at Queen Mary Hospital and was joined by Li Gang, a deputy director of the central government's liaison office in Hong Kong. Although Li Gang had no role in the administration of the territory, he was not just stealing the limelight but also appearing to directly manage the disaster. On his request, four salvage ships were reportedly sent from Guangdong but never reached the area because the water was too shallow (Lau 2012). Observers were puzzled that such ships should be sent from the mainland, considering that the Hong Kong government was more than adequately equipped to deal with the contingency, as the collision occurred well within Hong Kong territorial waters and the local response by emergency services was swift.

As the Hong Kong polity has become more polarised in the past few years, the Central Government is increasingly concerned about any situation which could lead to loss of control in the city. Thus, the level of 'decentralised initiative' that

Guangdong Province, let alone any city in the PRD, would enjoy in case a disaster hits both sides of the border appears extremely limited.

Unlike on the mainland, Hong Kong authorities have not identified disaster risk reduction as a topic that deserves an integrated approach. Residents generally assess disaster risk as low (Chan et al. 2014). Regularly occurring events such as typhoons or rainstorms are not usually regarded as serious threats, considering the sophistication of weather forecasts and the effectiveness of frontline responders. At the same time, most residents would welcome additional information on disaster risk and preparedness (Newnham et al. 2015).

There is no disaster law in Hong Kong (International Federation of Red Cross and Red Crescent Societies 2012). In 2000, the Hong Kong SAR government issued a rather nimble Emergency Response Plan (ERP) which defines a disaster as an 'extreme emergency.' The Security Bureau is tasked with coordinating the responses of various government departments to sudden disasters, defined as 'exceeding the normal responses required of the public emergency services' (Government of the Hong Kong Special Administrative Region 2000, p. 3). While the organisational chart of the Emergency Monitoring and Support Centre (EMSC) may look complex, the arrangements are actually quite simple. Tier 1 emergencies are handled at the local level, with emergency services operating independently. The primary responders are the police force and the fire department. An incident is categorised as 'Tier 2' if it is 'likely to grow in terms of threats to life, property and security' or attract media interest. Tier 2 incidents are monitored by the Emergency Support Unit (ESU), which oversees the Security Bureau Duty Officers (SBDOs) (Government of the Hong Kong Special Administrative Region 2000, p. 7). The Emergency Monitoring and Support Centre (EMSC) monitors 'widespread threats to life, property and security', defined as 'Tier 3', across all government operations.

Two contingency plans instated by the HKSAR Government are of relevance to environmental disasters. The Contingency Plan for Natural Disasters (CPND), devised in the 1990s, was revised on several occasions and last updated in 2009 (Security Bureau 2009), while the Daya Bay Contingency Plan, issued in 1994, was last updated in 2012 (Government of the Hong Kong Special Administrative Region 2012).

The CPND specifies in great detail the responsibilities of a number of government services in response to various natural hazards on a three-level scale, while the Daya Bay Contingency Plan details measures to be taken in case of incident at the Daya Bay nuclear plant, located 50 km from Central Hong Kong in Guangdong Province. The HKSAR emergency response model is not as decentralised as the reading of official plans suggests. The territory is quite small, with effective transport and communication systems. Considering that even a localised disaster such as the Lamma ferry collision was declared a Level 3 emergency, command centres at responders' headquarters are immediately in charge.

8.4.1 Cross-Border Coordination

While Hong Kong disaster risk governance arrangements differ significantly from those on the mainland, cross-border coordination has improved over time. In particular, transboundary governance arrangements have been gradually institutionalised to respond to two types of environmental disasters which affect the region directly: tropical storms and epidemic outbreaks of infectious diseases.

Tropical storms are recurring meteorological events in Hong Kong and the Pearl River Delta. The refinement of weather forecasting models has led to a drastic fall in loss of lives over the past four decades. Hong Kong residents make informed choices on the basis of an elaborate warning system and online satellite maps of the tropical storm track, updated every 10 min. The Hong Kong Observatory is under pressure from the public to predict wind strength and flooding risk accurately, and is regularly criticised for overplaying, or underplaying an incoming storm (Lai 2015).

The Hong Kong Observatory has been exchanging real time data on tropical storms with its counterparts across the Pearl River Delta, first and foremost Macao and Shenzhen, since 1996. While the signalling systems are similar in Hong Kong and Macao, those on the Mainland differ. To make sure travellers across the border remain well informed in case of an approaching typhoon, the Guangdong Meteorological Bureau, Hong Kong Observatory, and Macao Meteorological and Geophysical Bureau have established a website (Greater Pearl River Delta Weather Website n.d.) that explains the different warning systems and shows the current status for cities in the region. One limitation is that Mainland data are only available in Chinese, making it more difficult for those who cannot read Chinese to access information or evaluate risks. At the same time, the Hong Kong Observatory has set up multiple communication channels, in particular through mobile phone networks. Nonetheless, disaster risk governance on both sides of the border is characterised by a top-down approach and fails to involve society in significant ways. Communities vulnerable to natural hazards are usually the primary responders across a range of disasters. Governance arrangements based on two-way communication between state and society have been shown to improve coordination and response.

The Hong Kong government realised, in hindsight, that it was underprepared for epidemic outbreaks of infectious diseases when SARS (Severe Acute Respiratory Syndrome) struck in 2003. A veil of secrecy shrouded the situation on the Mainland until a whistle-blower forced the Central Government to admit the severity of the situation and start sharing epidemiological information with the rest of the world through WHO. The episode was sometimes seen as paradigm-changing and heralding the end of the Westphalian system of epidemic surveillance and management (Fidler 2004). In the aftermath of the 2003 SARS outbreak, the exchange of epidemiological data across the border was improved. In 2004 the Department of Health set up the Centre for Health Protection with the mission to 'achieve effective prevention and control of diseases in Hong Kong in collaboration with local and international stakeholders' (Centre for Health Protection 2012). A tripartite group, the

Guangdong-Hong Kong-Macau Expert Group on the Prevention and Control of Infectious Diseases, was set up to improve the notification system for infectious diseases and facilitate mutual training and visits by field epidemiologists and other healthcare professionals (Department of Health 2004). Nonetheless, the Centre's 60-page strategic plan for 2010–2014 devotes only two paragraphs to regional cooperation, pledging to

> continue to foster close ties with the Mainland and Macao to enhance co-ordination and exchange of information and intelligence, to refine the notification system, to share experience and expertise, and to tackle the spread of infectious diseases among the three place [and] undertake reciprocal training activities. (Centre for Health Protection 2010, p. 52)

Doctors reported feeling confident that they are now better equipped to handle a SARS epidemic outbreak, should one happen again, not just because of better functioning cross-border communication channels, but also because Hong Kong hospitals have many more beds in isolation wards than prior to the SARS outbreak (Siu 2013). Private practitioners in Hong Kong are also better integrated in the epidemiological surveillance apparatus. Thus, in the midst of an avian influenza A (H7N9) epidemic outbreak, the CHP wrote a letter to private medical practitioners in Hong Kong urging them 'to pay special attention to those who presented with influenza-like illnesses and had history of visiting wet market with live poultry or contact with poultry in Guangdong Province and other affected areas within the incubation period' (Centre for Health Protection 2014).

The real-time sharing of meteorological information and cross-border notification of cases of known infectious diseases point to regular exchanges within informal professional networks, signalling improved transboundary disaster governance. At the same time, cross-border cooperation in disaster risk governance shows limited inclusiveness. Authorities on both sides of the border do not include NGOs specialised in disaster response in government contingency plans, although the Hong Kong SAR government regularly funds Hong Kong-based NGOs involved in disaster response on the mainland. While the media are also important actors in disaster risk governance, they are tightly controlled on the mainland, in particular in sensitive disaster situations.

Despite improvements over the past decades, disaster risk governance and cross-border coordination in the Greater Pearl River Delta generally appears to be reactive rather than forward-looking. The next section makes the case that more attention should be directed to future risk and threat assessments.

8.5 Assessing Future Risks

Hong Kong, Macao and cities across the Pearl River Delta have regularly improved disaster warning and response systems in the last decades but a 'lack of sufficient strategic planning' has been noted with regard to flood risk (Chan et al. 2012, p. 41). In a survey of 616 metropolitan areas, the Greater Pearl River Delta was ranked third highest among cities at risk from all natural hazards worldwide, just behind Tokyo and Manila. The GPRD was even found to be the metropolitan area most at risk of storms, storm surges and river floods (Swiss Re 2014). While the region is well-prepared to respond to routinely occurring hazards, cities across the delta may not be as well-equipped to tackle less usual events or hazards of an exceptional intensity. For instance, tsunami preparedness and response plans remain sketchy, as the likelihood of a significant tsunami is assessed as very low. Should such an event occur, cities across the delta would likely struggle to respond adequately, as was the case with the March 2011 earthquake off Japan's East coast and the ensuing tsunami. While a rational, statistical approach to such an unlikely event does not necessarily warrant costly investments that may appear as losses to current and future generations, such risks should nevertheless be taken into account in land use and future infrastructure planning.

At the same time, investments offering a better return may make sense as an insurance against a possible super-typhoon strike on the delta. Both Cyclone Nargis in 2008 in Myanmar and Typhoon Haiyan in 2013 in the Philippines triggered storm surges in excess of 5 m (Fritz et al. 2010). The impact of these storms was staggering, both in human lives and economic losses, but could be dwarfed by the impact of a storm of equivalent strength on the Pearl River Delta, because of its dense population and vastly higher economic weight than either the Visayas or the Ayeyarwady delta.

The Pearl River Delta houses millions of people, in low-lying areas or on reclaimed land, vulnerable to super-typhoon strikes. Much of the land on the Western side of the delta is lying below mean sea level, in particular in the Zhuhai, Panyu and Zhongshan prefectures, which were assessed as most at risk from storm surges in Guangdong Province's coastal districts (Li and Li 2011). Assessing the return period of extreme weather events simply on the basis of historical records may be misleading in a context of climate change, as extreme weather events become more frequent. Besides, the Pearl River Delta is strongly affected by subsidence, which is predicted to further affect the mean sea level baseline used for risk assessments, and was listed among the deltas in greater peril (Syvitski et al. 2009).

Anticipating the potential consequences of extreme weather events on the basis of historical records would be equally hazardous. Even if the magnitude and return period of extreme weather events remain unchanged, such events are likely to have more severe consequences in the future because of increases in the urban population of the Pearl River Delta, in particular in low-lying areas. In addition, the region's population is ageing rapidly. One of the consequences is the rapid increase in the number of disabled or chronically ill people. In Hong Kong alone, the number of people suffering from physical disability, more than half of them aged over 70, rose 60% from 2007 to 2013. Those suffering from chronic diseases represented 19.2% of the total population in 2013 (Census and Statistics Department 2015). Focusing on such large numbers of vulnerable people, whether in terms of evacuation, sheltering or medical assistance, will be a daunting challenge for emergency services across a range of possible hazards including heat waves, storm surges or nuclear incidents.

Besides, the expansion of critical infrastructure such as bridges and tunnels into areas at risk of natural hazards is likely to magnify economic losses where the flow of goods or circulation of workers is disrupted. The consequences of such compound disasters would no doubt reverberate throughout the global economy, as the PRD region is one of the manufacturing hubs for light industrial goods worldwide, with Hong Kong a key provider of financial and other professional services. The lack of strategic vision on both sides of the border calls for a wide-ranging, inclusive and cross-border consultation process focused on long-term risk assessment and reduction and, more generally, the governance arrangements required to handle disaster risk across the region in the coming decades.

8.6 Conclusion

Disaster risk governance arrangements in the Greater Pearl River Delta have improved significantly on both sides of, and across the Hong Kong Guangdong border over the past 20 years but have not fully kept pace with the region's meteoric development. Reactive adjustments rather than strategic planning have guided governance changes. Rapid urbanisation, demographic changes and anthropogenic climate change are the key drivers of long-term risk and vulnerability across the region.

Should a major disaster strike the Greater Pearl River Delta and, more specifically, the Pearl River Estuary Bay Area, local authorities across the region will almost certainly appeal to the central government. Given the political sensitivities of the Hong Kong case and the economic importance of the Greater Pearl River Bay Area for the country, the central government would have to respond directly and overrule a fragmented GPRD/PRE institutional set-up yet to be tested by a major disaster. Authorities at all levels would be under significant pressure from the public, while no existing mechanisms or channels currently involve society in disaster risk governance on either side of the border. At the same time, a major disaster would equally likely generate a strong involvement of the international community, in particular the United States and the European Union, considering their many interests in the territory and long-established concern for the well-being of Hong Kong residents.

With regard to the governance of 'usual' disasters, further research is needed on the institutional cross-border mechanisms devised by the Hong Kong Observatory, the Security Bureau or the Centre for Health Protection and their counterparts in Macao and on the Mainland. Further research is also needed on the informal professional networks which connect for instance meteorologists or epidemiologists across the delta with a view to assess their contribution to the cross-border governance of environmental disasters.

The high meteorological risk, fast rate of subsidence, extent of low-lying areas, huge population, runaway urbanisation, location on a major river delta and institutional fragmentation possibly make the Greater Pearl River Delta region an extreme case but by no means an isolated one. More empirical studies of comprehensive risk profiles and relevant governance arrangements in other sub-tropical, coastal megacities are needed to help prepare for, contain and mitigate the impact of transboundary environmental hazards.

A mere 'evidence-based' retrospective approach is insufficient when assessing and anticipating future disaster risk, especially with regard to megacities located in coastal, subtropical areas and compound disasters. In any case, the historical baseline for most megacities in developing countries is sketchy since '[l]ess than 1% of all disaster-related publications are about disasters in the developing world [while] 85% of disasters and 95% of disaster-related deaths occur in the developing world' (Roy et al. 2011).

References

Armitage, D., de Loë, R. C., Morris, M., Edwards, T. W., Gerlak, A. K., Hall, R. I., Huitema, D., Ison, R., Livingstone, D., MacDonald, G., Mirumachi, N., Plummer, R., & Wolfe, B. B. (2015). Science-policy processes for transboundary water governance. *Ambio, 44*(5), 353–366.

Barquet, K. (2015). "Yes to peace?" Environmental peacemaking and transboundary conservation in Central America. *Geoforum, 63*(1), 14–24.

Barquet, K., Lujala, P., & Rød, J. K. (2014). Transboundary conservation and militarized interstate disputes. *Political Geography, 42*(1), 1–11.

Böhmelt, T., Bernauer, T., Buhaug, H., Gleditsch, N. P., Tribaldos, T., & Wischnath, G. (2014). Demand, supply, and restraint: Determinants of domestic water conflict and cooperation. *Global Environmental Change, 29*, 337–348.

Boin, A., Rhinard, M., & Ekengren, M. (2014). Managing transboundary crises: The emergence of European Union capacity. *Journal of Contingencies and Crisis Management, 22*(3(S)), 131–142.

Caruson, K., & MacManus, S. (2008). Disaster vulnerabilities: How strong a push toward regionalism and intergovernmental cooperation? *American Review of Public Administration, 38*(3), 286–306.

Census and Statistics Department. (2015). *Hong Kong monthly digest of statistics: Persons with disabilities and chronic diseases in Hong Kong*. Retrieved March 7, 2016. http://www.statistics.gov.hk/pub/B71501FB2015XXXXB0100.pdf

Centre for Health Protection. (2010). *Centre for health protection strategic plan 2010–2014*. Hong Kong: Department of Health. Retrieved October 1, 2015. http://www.chp.gov.hk/files/pdf/chp_strategic_plan_2010-2014.pdf.

Centre for Health Protection. (2012). *Vision and mission*. Hong Kong: Department of Health. Viewed 7 Mar 2016. http://www.chp.gov.hk/en/vision/8/22.html.

Centre for Health Protection. (2014). *Letter to doctors,* Centre for Health Protection, Surveillance and Epidemiology Branch, dated 9 January 2014.

Chan, F. K. S., Mithell, G., Adekola, O., & McDonald, A. (2012). Flood risk in Asia's urban megadeltas: Drivers, impacts and response. *Environment and Urbanization ASIA, 3*(1), 41–61.

Chan, E. Y. Y., Lee, P. Y., Yue, J. S. K., & Liu, K. S. D. (2014). Socio-demographic predictors for urban community disaster health risk perception and household-based preparedness in Chinese urban city. Paper delivered at the 12th Asia-Pacific Conference on Disaster Medicine (APCDM), Tokyo.

Cheung, P. T. Y. (2012). The politics of regional cooperation in the greater Pearl River Delta. *Asia Pacific Viewpoint, 53*(1), 21–37.

China Internet Information Center. (n.d.). *The local administrative system*. Viewed 7 Mar 2016. http://www.china.org.cn/english/Political/28842.htm.

Chomitz, K. M., Buys, P., & Thomas, T. S. (2005). *Quantifying the rural-urban gradient in Latin America and the Caribbean, Development Research Group, Infrastructure and Environment Team, Policy Research Working Paper 3634*. Washington, DC: World Bank.

Dallimer, M., & Strange, N. (2015). Why socio-political borders and boundaries matter in conservation. *Trends in Ecology & Evolution, 30*(3), 132–139.

Department of Health. (2004). *Tripartite meeting on infectious diseases held in Hong Kong*, DOH 2004, Hong Kong. Viewed 7 Mar 2016. http://www.dh.gov.hk/english/press/2004/040806.html

Eisner, R. (2015). Managing the risk of compound disasters. In I. Davis (Ed.), *Disaster risk management in Asia and the Pacific*. London: Routledge.

Emergency Management Office, the People's Government of Guangdong Province. (n.d.). *About Guangdong emergency management*. Viewed 7 Mar 2016. http://www.gdemo.gov.cn/english/gdem/

Epstein, J. (2013). Natural disaster risk levels of the world's largest cities, *GreenAsh*. Viewed 29 Feb 2016. http://greenash.net.au/thoughts/2013/03/natural-disaster-risk-levels-of-the-worlds-largest-cities/

Farber, D. (2011). Navigating the intersection of environmental law and disaster law. *Berkeley Law Review, 2011*(6), 1783–1820.

Fidler, D. (2004). *SARS, governance and the globalization of disease*. New York: Palgrave.

Fritz, H., Blount, C., Thwin, S., Thu, M. K., & Chan, N. (2010). Cyclone Nargis storm surge flooding in Myanmar's Ayeyarwady River Delta. In Y. Charabi (Ed.), *Indian Ocean tropical cyclones and climate change*. New York: Springer.

Ghai, Y. (1997). *Hong Kong's new constitutional order: The resumption of Chinese sovereignty and the basic law* (2nd ed.). Hong Kong: Hong Kong University Press.

Government of the Hong Kong Special Administrative Region. (2000). *Emergency response system: The policy, principles and operation of the Government's emergency response system*. Hong Kong: Emergency Support Unit, Security Bureau, Hong Kong Government Secretariat.

Government of the Hong Kong Special Administration Region. (2012). *Daya Bay contingency plan*. Viewed 7 Mar 2016. http://www.dbcp.gov.hk/eng/dbcp/.

Greater Pearl River Delta Weather Website. (n.d.). *Weather warning*. Viewed 30 Apr 206. http://www.prdweather.net/pda_index.htm.

He, D., Wu, R., Feng, Y., Li, Y., Ding, C., Wang, W., & Yu, D. W. (2014). China's transboundary waters: New paradigms for water and ecological security through applied ecology. *Journal of Applied Ecology, 51*(5), 1159–1168.

Intergovernmental Panel on Climate Change. (2012). *Managing the risks of extreme events and disasters to advance climate change adaptation: Special report of the intergovernmental panel on climate change*. New York: Cambridge University Press. Retrieved 28 Sept 2015. https://www.ipcc.ch/pdf/special-reports/srex/SREX_Full_Report.pdf.

International Federation of Red Cross & Red Crescent Societies. (2012). *Law and regulation for the reduction of risk from natural disasters in Hong Kong: A desk survey.* Retrieved September 25, 2015. http://drr-law.org/resources/China-Hong-Kong-Desk-Survey.pdf

Kapucu, N. (2012). Disaster and emergency management in urban areas. *Cities, 29*, S41–S49.

Lagadec, P. (2006). Crisis management in the twenty-first century: 'unthinkable' events in 'inconceivable' contexts. In H. Rodriguez, E. L. Quarantelli, & R. Dynes (Eds.), *Handbook of disaster research*. New York: Springer.

Lai, Y. (2015). Observatory defends alert for typhoon that never was, *South China Morning Post*, 11 July. http://www.scmp.com/news/hong-kong/health-environment/article/1835654/hong-kong-observatory-defends-hoisting-t8-signal

Lau, S. (2012).Questions raised over mainland role in rescue and Li Gang's appearance, *South China Morning Post*, 3 October. Viewed 23 Sept 2015. http://www.scmp.com/news/hong-kong/article/1052611/questions-raised-over-mainland-role-rescue-and-li-gangs-appearance?page=all

Li, K., & Li, G. S. (2011). Vulnerability assessment of storm surges in the coastal area of Guangdong Province. *Natural Hazards and Earth System Sciences, 11*(7), 2003–2010.

Ma, X. (2012). The integration of city-region of the Pearl River Delta. *Asia Pacific Viewpoint, 53*(1), 97–104.

Mitchell, S. M., & Zawahri, N. A. (2015). The effectiveness of treaty design in addressing water disputes. *Journal of Peace Research, 52*(2), 187–200.

Newnham, E., Patrick, K., Balsari, S., & Leaning, J. (2015). *Community engagement in disaster planning and response: Recommendations for Hong Kong, policy brief.* Harvard: FXB Center for Health and Human Rights. Retrieved March 7, 2016. https://cdn2.sph.harvard.edu/wp-content/uploads/sites/5/2015/11/Community-Preparedness-Policy-Brief-10.28.15.pdf.

Nurhidayah, L., Alam, S., & Lipman, Z. (2015). The influence of international law upon ASEAN approaches in addressing transboundary haze pollution in Southeast Asia. *Contemporary Southeast Asia, 37*(2), 183–210.

Office of the China National Committee for Disaster Reduction. (2006). *Disaster reduction action plan of the People's Republic of China (2006–2015).* Retrieved October 4, 2015. http://www.preventionweb.net/files/30351_chinadisasterreductionactionplan200.pdf

Swiss Re. (2014). *Mind the risk: A global ranking of cities under threat from natural disasters.* Viewed 3 Oct 2015. http://www.swissre.com/rethinking/climate_and_natural_disaster_risk/Mind_the_risk.html

Roy, N., Thakkar, P., & Shah, H. (2011). Developing-world disaster research: Present evidence and future priorities. *Disaster Medicine and Public Health Preparedness, 5*(2), 112–116.

Security Bureau. (2009). *Contingency plan for natural disasters (including those arising from severe weather conditions).* Hong Kong: Emergency Support Unit, Security Bureau, Hong Kong Government Secretariat.

Selby, J., & Hoffmann, C. (2014). Beyond scarcity: Rethinking water, climate change and conflict in the Sudans. *Global Environmental Change, 29*, 360–370.

Siu, P. (2013). Doctors confident Hong Kong prepared if SARS strikes again, *South China Morning Post,* 25 February. http://www.scmp.com/news/hong-kong/article/1157697/doctors-confident-hong-kong-prepared-if-sars-strikes-again

Stacey, N., Karam, J., Jackson, M., Kennett, R., & Wagey, T. (2015). Knowledge exchange as a tool for transboundary and coastal management of the Arafura and Timor Seas. *Ocean and Coastal Management, 114*, 151–163.

Syvitski, J. P. M., Kettner, A. J., Overeem, I., Hutton, E. W. H., Hannon, M. T., Brakenridge, G. R., Day, J., Vörösmarty, C., Saito, Y., Giosan, L., & Nicholls, R. J. (2009). Sinking deltas due to human activities. *Nature Geoscience, 2*(10), 681–686.

The Economist. (2013). 'China's fuzzy vision for Greater Beijing', *The Economist,* 13 December. http://www.businessinsider.com/chinas-fuzzy-vision-for-greater-beijing-2013-12

Uchida, H., & Nelson, A. (2010). *Agglomeration index: Towards a new measure of urban concentration, background paper for world development report 2009: Reshaping economic geography*. Washington, DC: World Bank.

United Nations General Assembly. (2001). *Resolution 56/195 (A/RES/56/195) international strategy for disaster reduction.* Retrieved March 2, 2016. http://www.unisdr.org/files/resolutions/N0149261.pdf

United Nations International Strategy for Disaster Reduction. (2009). *UNISDR terminology on disaster risk reduction.* Retrieved 27 Sept 2015. http://www.unisdr.org/files/7817_UNISDRTerminologyEnglish.pdf

United Nations Office for Disaster Risk Reduction (UNISDR). (2007). *Hyogo framework for action 2005–2015: Building the resilience of nations and communities to disasters.* Viewed 29 Feb 2016. http://www.unisdr.org/we/inform/publications/1037

United States Congress. (1992). *United States-Hong Kong Policy Act*, (P.L. no. 102-383m 106 Stat. 1448).

Williams, G. (2011). *Study on disaster risk reduction, decentralization and political economy: The political economy of disaster risk reduction.* Analysis prepared as UNDP's contribution to the Global Assessment Report on Disaster risk Reduction. Viewed 2 Mar 2016. http://www.preventionweb.net/english/hyogo/gar/2011/en/bgdocs/Williams_2011.pdf

World Bank. (2015). *East Asia's changing urban landscape: Measuring a decade of spatial growth*. Washington, DC: World Bank.

World Conference on Disaster Risk Reduction. (2005). *Hyogo Framework for Action 2005–2015 (HFA),* United Nations Office for Disaster Risk Reduction (UNISDR). Viewed 7 Mar 2016. https://www.unisdr.org/we/coordinate/hfa

World Conference on Disaster Risk Reduction. (2015). *Sendai Framework for Disaster Risk Reduction 2015–2030*, United Nations Office for Disaster Risk Reduction (UNISDR). Viewed 17 Mar 2016. https://www.unisdr.org/we/coordinate/sendai-framework

Yang, C., & Li, S. (2013). Transformation of cross-boundary governance in the Greater Pearl River Delta, China: Contested geopolitics and emerging conflicts. *Habitat International, 40*, 25–34.

Yeung, Y., Lee, J., & Kee, G. (2009). China's special economic zones at 30. *Eurasian Geography and Economics, 50*(2), 222–240.

Yi, L., Ge, L., Zhao, D., Zhou, J., & Gao, Z. (2012). An analysis on disasters management system in China. *Natural Hazards, 60*, 295–309.

Part III
Cross-Border Disaster Collaborations

Chapter 9
The Resilience of Islands: Borders and Boundaries of Risk Reduction

Karl Kim and Konia Freitas

9.1 Introduction

He wa'a he moku, he moku he wa'a is an *'ōlelo no'eau* is a Hawaiian proverb that speaks of the canoe as an island and the island as a canoe. It underscores how our daily living is contingent upon the values we hold, the practices we engage in and our functional understanding of the finite resources that exist in the world around us. It also intimates the necessity of social networks and relationships for the transfer of knowledge between generations. Thus, to the Oceanic dweller, what is essential to surviving a long deep ocean voyage is equally critical to successful living on an island. Notions of space and time are understood differently on small islands. Isolation from other communities encourages awareness of risks, strengthens identity and promotes collective, collaborative action. Islands are exposed to many man-made and natural hazards and threats—while emergency response, humanitarian relief and disaster recovery are hampered by location and limited transportation assets. In addition to understanding the geographies of risk, the capacity to cope with and manage threats and hazards is necessary for survival and sustainability. While islands have developed systems for managing internal social and political affairs, the challenges associated with globalisation have created new opportunities and threats. Globalisation is a sword that cuts both ways. It has provided opportunities for growth and income, but it also created dependencies and has weakened internal processes. International trade and tourism have exposed island communities to external forces: with new threats such as climate change, sea level rise and environmental change, islands need more than ever before to develop robust

K. Kim (✉)
Department of Urban & Regional Planning, University of Hawaii at Mānoa,
Honolulu, HI, USA
e-mail: karlk@hawaii.edu

K. Freitas
Kamakakūokalani Center for Hawaiian Studies, University of Hawai'i at Mānoa,
Honolulu, HI, USA

M.A. Miller et al. (eds.), *Crossing Borders*, DOI 10.1007/978-981-10-6126-4_9

systems for emergency management, risk reduction, mitigation and adaptation. In addition to national systems for coordinating response and recovery, civilian-military interactions and exchanges with international organisations play out in the planning, exercises and management of disasters. In addition to the place-based culture and systems of governance, institutional and organisational requirements are drawn into the mix, requiring effective communications, coordination and harmonisation of procedures, policies and approaches to risk management. There is a need for both generally agreed-upon principles and standards as well as a capacity to learn and adapt to localised knowledge and systems. While disasters are seen as rare events, they actually reflect and embody many of the underlying characteristics, problems and capabilities of the affected communities. Based on experiences in Hawaii, Samoa, Indonesia and other island communities, the resilience of islands is investigated.

Cultural knowledge, traditional systems of community resource management, and efforts to sustain local knowledge, wisdom and practices—amidst growing disparities in wealth, power, and access to information and technology—suggest the need for greater awareness, improved training and capacity-building, and stronger commitments on the part of both indigenous and globalised communities. Trosper (2002) describes how indigenous institutions in the Pacific Northwest coastal communities developed not just principles of governance, but also systems of exchange based on reciprocity and generosity for 'at least two millennia before contact with people from the old world'. Strategies and opportunities for not just building resilience on island communities are needed, as is transferring the lessons to other communities. With the growing scale of disaster impacts, and increased demands for outside resources, disaster managers, planners, emergency response personnel, humanitarian relief workers and those involved in longer-term recovery there is a need to understand and manage the crossing of borders and boundaries of risk reduction.

This essay focuses on the lessons from islands which face hazards and threats and a growing need to build resilient systems for disaster preparedness, response and recovery. It argues that disaster risk reduction is best understood in terms of different physical borders as well as social, cultural and institutional boundaries which manifest themselves in not just laws and policies, but also practices and approaches to resource management. The understanding and collective management of these borders and boundaries determine risk, vulnerability, exposure and the capacity to cope with harmful events. Such an approach necessitates a trans-disciplinary orientation combining physical, natural and environmental sciences alongside social sciences and policy sciences. Inherent in this discussion is an appreciation for different types of uncertainty: scientific, as well as uncertainty associated with development outcomes and political decision-making, which are central to governance.

There are three different sections. The first focuses on the physicality of disaster risk on islands, the relationship between hazards and threats and space and time. Drawing upon the experiences of small islands, with limited land and resources, we argue that islanders are much more sensitive to disaster risk than continental land mass dwellers because of an awareness of dangers and limitations, the interrelationships between and built and natural forms and the need to recognise and contend with both internal and external forces and realities. The concluding section focuses on the tensions that exist between local and global systems: between exogenous market forces and international development, trade, and a worldwide system of communications, transport, commodity flows, and engagement. This is governed by political and institutional agreements, tariffs, trade rules, and formal as well as informal partnerships. This not only affects development, but also the resilience and sustainability of communities. We conclude by drawing upon the lessons learned across islands and identifying caveats and limitations as well as other emergent forces and developments impinging on disaster risk reduction and resilience.

9.2 The Physicality of Disaster Risk Reduction on Small Islands

Small islands are at a greater risk of experiencing natural disasters. A large proportion of the countries that have experienced serious natural disasters are small island developing states. Not only are they exposed to multiple hazards and threats, but their "'intrinsic vulnerability"' is a function of their small limited resource base, high external transport costs, high exposure with limited mitigation, warning, and recovery capabilities and their demographic and economic structure. (Pelling and Uitto 2001). Local changes such as increased urbanisation and continued loss of traditional culture have also exacerbated risks associated natural hazards. Increased modernisation and implementation of new systems and technologies have also increased the likelihood of cascading events and longer- term environmental damage.

Perhaps nowhere is the sense of isolation due to its insular conditions greater than in Hawai'i, one of the most remote places in the world. While the Hawaiian archipelago consists of more than 130 different islands, most of these are uninhabited and quite small in size. Located far away from continental land masses, Hawai'i and other small remote islands have a heightened sense of disaster risk. There are no adjacent large, land masses nor communities to provide mutual aid or quick response. It is not connected by rail lines, superhighways or transportation other than aircraft and shipping. In addition to the tyranny of distance, there is also an accompanying difference in time zones. With isolated island communities, there is a deep understanding of not just the distances, but also time: the time to travel from one part of the world to another; the challenge of navigation from here to there, but also the ability to find one's way back home. The notions of distance and time among island communities may be further conceptualised in relational and dynamic ways. Johnson (1981, 2000) observes how the Hawaiian term *au*, signifies a structure to time and space by encompassing the meaning of both space and time. For Hawaiians, ka wā mamua, meaning the time in front, or the past, and ka wā mahope, meaning the time past, or the future, provide critical yet comfortable time orientation. Thus, as Kame'eleihiwa (1992) observes, "'the Hawaiian stands firmly in the present, with his back to the future, and his eyes fixed upon the past, seeking historical answers to present day dilemmas'" (p. 22). While some islanders never leave home, there has always been the voyager, the navigator, the traveler: the ones who venture out and explore different lands and places and bring back new ideas, cultures, practices, and knowledge.

On small islands, there is keen awareness of space. Surrounded by water and threatened by coastal hazards, the location of floodplains as well as safer, higher areas for refuge are understood. Often, much development has concentrated in the lower- lying coastal areas, places in proximity to ports and harbours or because of the relative ease of roadway construction at lower, flatter locations. With volcanic islands, the multiple hazards associated with lava, gas, and other geologic threats can also complicate siting and development. Often times there are tradeoffs between a location which is safe for one type of hazard but is riskier for another. The tradeoffs between safety, access to resources and, availability of land were part of the traditional aspects of land use management, such as in Hawaii (see Fig. 9.1 by Luciano Minerbi 1988) where there were zones for resource use (fishing and forest products) and farming but also refuge areas, and safe zones, as well as a system of trails and culturally significant locations. Hawaiians devised a system of management that linked carrying capacity to resource areas through a method of localised zones (Connelly 2014). In order to provide for a large population through agriculture and aquaculture, meant that the land had to be well defined and structured; that order was produced through palena – bounded areas and resources that resulted with a series of land divisions (Beamer 2014, p. 34). An orderly system of land divisions brought about greater productivity, lessened conflicts and provided for the people who lived upon and worked the land. This was a sophisticated system of resource management. The system encouraged sustainable use of resources and protected the environment and inhabitants from different hazards and threats.

There is much evidence of the importance of the physical world on islands. Beamer (2014) notes that in the Hawaiian context, palena created places and spaces of attachment and access to not only finite resources but to the metaphysical. By a system of boundaries and borders people were connected to material and spiritual resources of places, which were then kept visually and cognitively and passed on by the inhabitants of the area generation to generation. Not only was the spirit world embodied into by landscape features, mountains and rocks and water, wind, and other natural forces, but native peoples understood the different boundaries and zones and areas suitable for resource management. It made sense to locate canoe building and fishing and marine product extraction close to the ocean and to locate agricultural activities in the fertile alluvial plains, where flooding would also serve as a source of nourishment. It involved not just a keen appreciation for the flows of streams but also the creation of lo'i (taro production), irrigation channels, and other infrastructure to manage vital water resources. This system also minimised flooding, reduced sediment run-off, promoted a harmonious relationship between the natural and built environments.

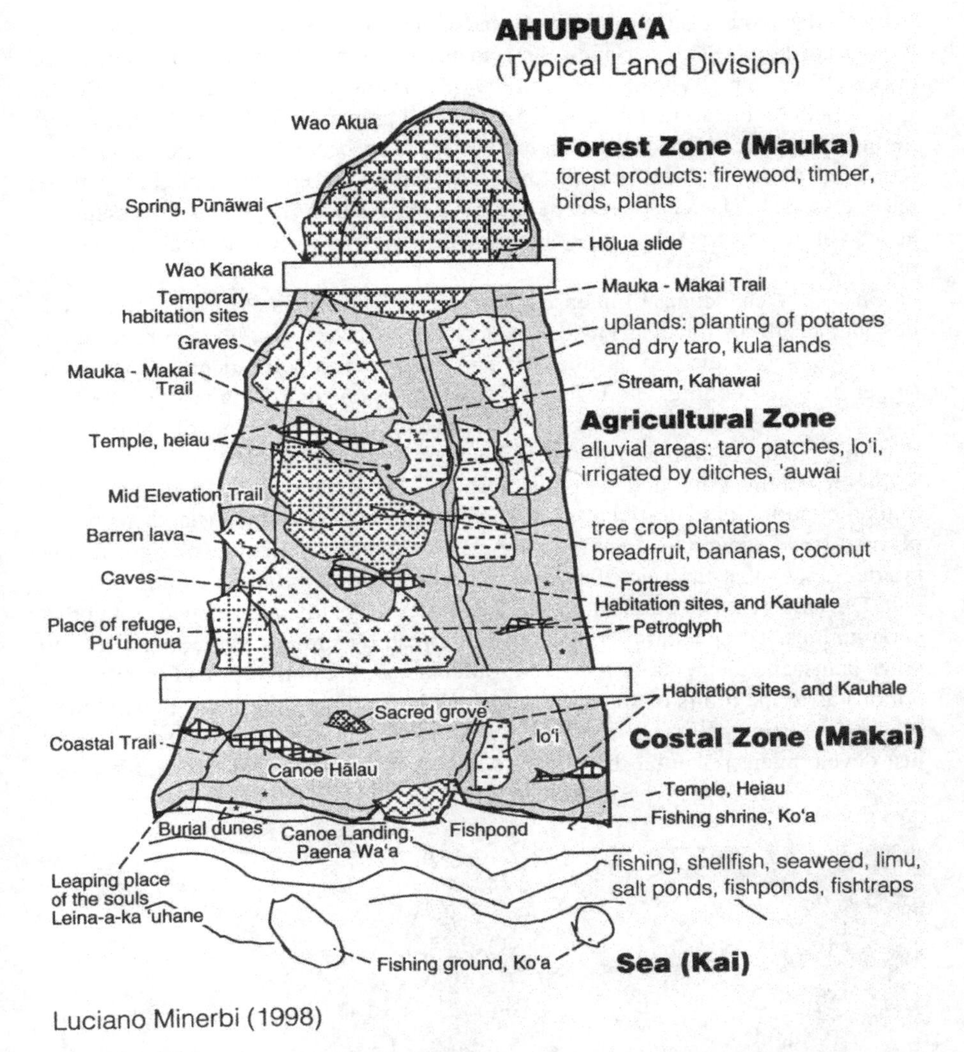

Fig. 9.1 Hawaiian traditional land use management system (Source: Minerbi 1988)

Throughout the Pacific, there is much evidence of the sensitivity to hazards and threats in terms of the location and siting of villages, the appropriate use of lands and resources, the style of construction and use of local materials, and building practices that support quick recovery from bad weather or hazardous events. The planting of crops such as taro or other root crops as well as breadfruit, banana, coconut and other resilient sources of food— combined with preservation techniques involving burying, fermentation, salt curing, and other strategies— have long helped Pacific island communities endure amidst various hazards and threats. Adaptation to threats and hazards has involved disassembling houses and structures and lashing them to the ground during hurricane and high wind events or filling canoes and boats with heavy stones to sink them in the harbors during coastal storm surge events. It also involved the development of warning systems and alerts, different ways of communicating threats and hazards and preparing for storms or hazardous events. One little- considered warning system is embedded in the names given to particular places and the oral histories that have evolved over time about these places. As such, place names can be sources of vital information about the nature of an area or demonstrate the range of natural hazards in a particular locale, district or region (King et al. 2007).

One of the challenges of intense coastal development is that much of the activities and resulting travel behaviour is located in hazardous, flood- prone areas. Figure 9.2 shows the trip destinations and estimated flooding depths from storm surge, riverine flooding, and tsunami threats. It reveals that a majority of the trips pass through areas of potential heavy flooding. Not only are destinations affected by flooding, but the overall transportation network is affected as well.

Figure 9.2 illustrates another point on small islands. There is often less space and limited opportunity to design alternative travel routes. Parts of the island can be cut off because of geography and topography. In the interior, there may be steep mountainous slopes that limit development and make roadway construction more difficult. Development has often extended deep into valleys that are isolated from other communities. These natural boundaries reinforced by streams, ravines, gullies and other landscape elements have served to demarcate communities and to reinforce notions as to the limits of space and time. With increased dependence on automobiles as the principal mode of travel, roadways become an essential aspect of not just development potential, but also resilience. While water and air transport are

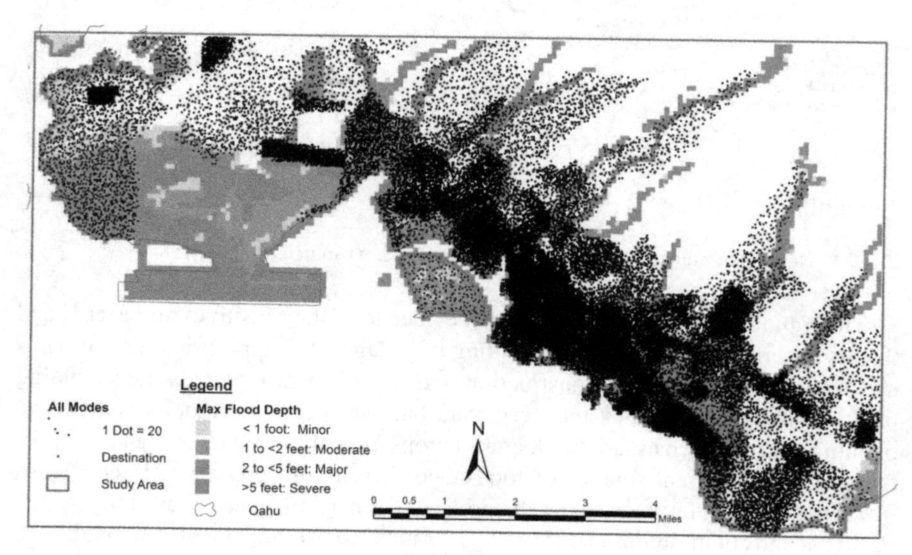

Fig. 9.2 Flood hazards and trip-making in urban Honolulu

important and relevant in many communities, roadways serve as the principal connectors between home, work, school, shopping and other activities.

Island communities that have had repeated experience with a certain hazard are generally equipped to manage those events. Cuba, for example, has had much experience with hurricanes and has developed a robust system of warning, evacuation, sheltering and recovering from this type of disaster. Similarly, after several devastating tropical cyclones in Guam, more attention to hazard mitigation and hardening of buildings and structures has occurred, making the community better able to withstand this type of disaster. With so many active volcanoes, Indonesian communities have developed robust systems for responding to and recovering from volcanic eruptions. Communities that have experienced a disaster better understand detection, alert and warning as well as well as the process of recovery. We found after repeated tsunami warnings, the transit agency in Hawaii was able to 'learn from the crisis' to institute changes following the tsunamis in Chile in 2010 and Japan in 2011 (Kim et al. 2013). The changes involved not just operating procedures, but also the coordination and management of information between central control and vehicle operators and the relationships between transit personnel, police, Red Cross and others involved in disaster management and response. While written procedures and after action reports as well as periodic review and updating of procedures following events is important, so too is the culture and professionalism that fosters a sense of duty and responsibility among the ranks. This involves not just the leadership and management of the transit agency, but also local values, culture and identity. Interviews with the agency revealed that Hawaii's isolation has promoted sense of family (*'ohana*) and commitment to serve others. While similar disasters in other communities have led to reported cases of absenteeism and job abandonment, in Hawaii, more transit workers showed up to provide assistance during the crisis. This is not just because of the close-knit community but also the recognition of the threat to family, friends and other members of the community. Proximity and familiarity is increased in island communities and there is evidence to suggest stronger social capital (bonding, bridging and linking) as well as trust, reciprocity and rules of engagement. We have studied this for activities such as waste management following disasters in Japan (Kawamoto and Kim in press).

9.3 The Social Life of Small Island Spaces

The subheading for this section is borrowed, in part, from William H. Whyte's (1980) documentary films about how people interact with 'small urban spaces' that show how factors such as environmental conditions, lighting and crowding affect the behaviour of individuals in urban areas. These methods of observation and the subsequent translation into design strategies could be adapted and applied to island settings. Similarly, on small islands, people interact with not just the natural and built environment. This is reflected in terms of how space is used and the pathways and movements of people and the various cultural, social, and human interactions with hazards and threats.

This is quite evident in communities such as Samoa where the social structure is based on extended families (*aiga*), headed by chief (*matai*). The *aiga* have a high chief (*ali'i*) as well as a 'talking chief' (*tulafale*). The *matai* from each *aiga* are organised into a village council (*fono*) headed by a 'paramount chief.' Moreover, much of the land is communally owned by the *aiga* and controlled by the *matai*. There are defined roles and responsibilities based on age and gender. For example, '*aumaga*' are young village males who provide security and support while '*aualuma*' are females who provide other important services. These networks provide a means of socialising members of the community and enhance the production of social capital. These roles and responsibilities strengthen not only the response and recovery from disasters, but also the management of resources and stewardship of communities during non-disaster times.

In Samoa, and in other small island communities, governance is characterised by formal/informal and modern/traditional dualities. The traditional village-based system exists alongside the territorial government which has executive (governor and lieutenant governor), legislative (senate and representatives) and judicial (high court and district court) functions. There are also village mayors, magistrates, and other service functions. This duality becomes all the more apparent because of the many external governments, donor agencies, and others involved in disaster response and recovery.

As with other disasters, the September 29, 2009 earthquake and tsunami which struck Samoa served to reveal both strengths and weakness in the systems of governance and collective risk management. On the one hand, the traditional roles and responsibilities and clear lines of authority, especially at the village level, enhanced response and recovery efforts. With community members already accustomed to sharing resources and supporting each other, the disaster served to demonstrate the value of social capital. Lives were saved. Conflicts were minimised. Communities were organised to act. Processes for sharing information, communications, deliberation and decision-making are critical during and immediately after a disaster. Yet the nature of a disaster such as the 2009 tsunami is that it overwhelms local capacity. It was an extraordinary event. It stretched the limits of local, village and territorial resources to manage the event.

Major disasters involve a host of different actors and stakeholders. In addition to the affected communities, many national and international agencies interact to support different aspects of response and recovery. While response functions typically involve search and rescue as well as initial damage assessment, recovery involves many different phases and longer-term commitments. Haas et al. (1977) identified different stages following disasters as communities progress through the emergency phase to restoration and early and longer-term reconstruction efforts. While these phases generally apply to all communities experiencing disasters, it is useful to understand that in small island spaces, the nature and pace of recovery and restoration may be limited because of physical conditions (described in the first section). Remote locations, limited transportation infrastructure, and higher costs associated with many of the recovery activities—including search and rescue, mass care, clearing rubble, storage and transport of waste, temporary housing, and post-disaster reconstruction—challenge local systems. There are often difficulties stemming from differences in culture and approaches to decision-making, especially as the range of actors and participants in disaster response and recovery widens.

Gavin Smith (2011) has analysed the range of different stakeholders and agents who typically come together following major disasters. There are different actors and intermediaries, governments, businesses, insurance, financial institutions and other organisations involved in many different activities related to recovery and restoration of damaged communities. Smith notes that the further away from the individuals and households actually impacted by the disaster, the less understanding of needs and the greater the resource rules. Some of the organisations are focused on disaster recovery services, while others may have other business, education, social or community functions that go beyond disaster recovery. Two types of challenges emerge in small island communities. First, there need to be clear pathways for integrating outsiders and the resources they bring into the local planning and decision-making processes. Second, while disaster recovery may be foremost, there are many other needs, programmes, services and functions that need to be addressed. The disaster event must be seen as part of a larger set of social and economic pre-existing conditions that affect not just the exposure and vulnerability to harmful events, but also the capacity to organise and respond effectively.

While small island spaces may be easier to comprehend because of pre-established systems, roles and responsibilities, they may also be more difficult environments to work in because of their uniqueness and differences from other communities. There is need to translate national and outside rules and procedures for resource allocation into understanding of local culture, protocol and ways of planning, deliberation and decision-making. See Robinson and Robinson (2004) for a description of Pacific ways of dialogue which include customary rules regarding '*talanoa*' (talk or discussion) common in Fiji, Samoa, and Tonga. This is especially the case over time, as the emphasis shifts from emergency and rescue operations (such as search and rescue) to more challenging tasks involving the planning and implementation of restoration and reconstruction activities.

The 2009 tsunami also affected Tonga and provides an opportunity to identify both common elements of small island resilience as well as some of the differences between island communities. Clark et al. (2011) describes the event and many of the impacts in Tonga. While both Tonga and Samoa share common characteristics as to their island cultures, Tonga is a constitutional monarchy with both a king and an appointed prime minister. Following the disaster, a team from the University of Hawaii and the Pacific Disaster Center worked on a plan to integrate disaster risk reduction into land use planning and development for the Kingdom of Tonga and Niuatoputapu which was most severely impacted by the 2009 tsunami. The project was challenging because of the attempt to integrate not just disaster risk reduction but also land use planning (which involved understanding a unique system of land ownership and tenure) and the goal of building back better, not just simply rebuilding structures in their same locations. The project involved mapping hazards zones, identifying safe new locations and then integrating livelihoods (fishing and farming) as well as other activities in the planning and design of new communities.

Several points emerge from this experience. First, recovery is among the most challenging and difficult aspects of disaster management. While disasters generally serve to expose and magnify weaknesses within the existing social, economic and political structures, recovery serves to draw attention to deficiencies within the system of land management and planning processes. Second, while tools and technology and decision support are essential ingredients in disaster management and recovery, there is a need for strong leadership and processes for information sharing, joint problem solving, deliberation and decision-making which integrates efforts across agencies and ministries as well as between local, provincial, national and international programmes in the public and private sectors and NGOs. In the immediate aftermath of a disaster, the social space of small islands may become quite congested; over time, there is a need to build capacity and the capability to plan, respond to and recover from disasters. This includes not just building disaster risk reduction capabilities, but also ensuring that such activities are mainstreamed into the routine business of government, the policies, practices and plans related to the development and management of communities. An assets-based approach (Freitag et al. 2014) in which communities inventory those qualities and conditions that are valued most might also considered. In this manner, the unique aspects of small islands might also be recognised and included in planning and recovery processes.

Islands, because of their isolation from other communities, often develop unique and diverse cultures and practices. Indigenous knowledge about tsunamis on Simeulue Island in Aceh, Indonesia contained in oral literature, poems and songs increased community awareness of the hazards and was used in developing a local warning system which saved many lives (Syafwina 2014). Mercer et al. (2009) have developed a useful framework and methods for identifying and integrating indigenous and scientific information for disaster risk reduction. They distinguish between local, traditional folk knowledge (accumulated through experience and community practices, and passed down through generations) and scientific knowledge which is 'understood to be with origins outside the indigenous communities'. Indigenous knowledge is local and scientific knowledge is global. The methods include the use of participatory techniques for engaging the community, identifying vulnerabilities, identifying indigenous and scientific strategies and then integrating knowledge to reduce risks and vulnerabilities.

9.4 Tensions Between Local and Global Approaches to Disaster Risk Reduction

Local traditional communities understand their vulnerabilities to hazards and disaster threats. Over time, especially with repeated exposure, these communities developed knowledge and understanding as to how best to cope with hazardous events. A disaster, however, overwhelms local capacity to manage, respond and recover from hazardous events. As such, a disaster is different from an emergency or routine event. It may require outside assistance from provincial, national and international governments and donor organisations. Disasters may start as local events, influenced by not just the geology and environmental conditions, but also by the local capacity to absorb, respond, provide relief and quickly recover and restore systems to pre-event levels of functioning. A disaster, however, brings extensive outside resources—materials, supplies, capital, and personnel—to assist in the relief and recovery process. A disaster declaration, therefore, replaces local systems with national or international resources. In the United States, the process is governed by the Stafford Act which guides federal emergency and disaster declarations and authorises the use of federal resources to support state and local governments. The principal agency for coordinating response and relief is the Federal Emergency Management Agency (FEMA), established under President Carter in 1979. The disaster declaration is triggered by a request for assistance from the state government. Once the declaration has been issued, the President can direct federal agencies, including the Department of Defense, to assist in relief and debris removal as well as provide federal financial aid for both relief and mitigation efforts (75% of eligible costs). Under the Stafford Act, the federal government can provide loans, technical support and coordination through the appointment of a federal coordinating officer (FCO) to support relief and recovery efforts.

Following the SARS outbreak in China in 2003, the Chinese Comprehensive Emergency Management System was created to improve coordination between national, provincial and local authorities. The Wenchuan earthquake in 2009 demonstrated the central government's role in response, relief and recovery. While the creation of the National Emergency Response Plan (2006) and the National Emergency Response Law (2007) clarified responsibilities, there is need for greater collaboration and coordination efforts—especially between the government and NGOs. The use of the military in China following the Wenchuan earthquake and rapid deployment of thousands of soldiers as well as '92 military trains, and about 110,000 military vehicles' and other military assets has been documented (Mulvenon 2010). The differences between the Chinese response to the earthquake disaster and the US response to Hurricane Katrina has also been describe. Among the most interesting programmes instituted in China was the National Counterpart Assistance Program where earthquake damaged communities in Sichuan were matched with cities and provinces not damaged by the earthquake to provide resources, technical support and recovery assistance. While the US and other countries have developed mutual aid agreements primarily for search and rescue and other response functions, there are fewer examples of coordinated support for disaster recovery.

Approaches to disaster response and relief typically follow a top-down approach, where the emphasis is on clear chain of command and coordination. The National Incident Management System (NIMS), which was established in 2003 in the United States by the Secretary of Homeland Security in response to the 9–11 attacks, provides a set of concepts, terms, and organisational structures for disaster response and recovery. It serves as a multi-agency coordination system which integrates the incident command system (ICS) with the overall National Response Plan and the 15 Emergency Support Functions (ESF) showing the roles and responsibilities of different federal and other organisations. Recently, under the National Disaster Recovery Framework, a similar approach was developed to articulate the Recovery Support Functions (RSF). It should be noted that recovery is different from response. It is more long-term in nature. As such, there are different phases. More agencies and stakeholders are involved in varying degrees. Although the agencies and stakeholders vary country to country, similar issues and concerns dominate the deliberations across countries as to how best to recover from disasters.

The first is whether to vest redevelopment and reconstruction activities within existing organisations or entities, or to create new special authorities such as the BRR (Rehabilitation and Reconstruction Agency for Aceh and Nias) in Indonesia or the Louisiana Recovery Authority (created by Executive Order of the Governor of Louisiana following Hurricane Katrina, eventually to be made permanent by the legislature). Similar efforts were enacted following 9–11 to rebuild lower Manhattan and to lead the efforts following Typhoon Haiyan in the Philippines.

A related concern involves the extent and nature of national and outside participation in the planning, design, and deliberation of reconstruction initiatives. In some countries, such as Indonesia and the Philippines, there has been great involvement of foreign NGOs and international donors who have implemented housing and community programmes focused on restoring livelihoods or rebuilding damaged health care facilities, schools and social services. Often these relief efforts are linked to poverty alleviation, development goals, and other humanitarian or social interests and concerns. The disaster becomes a vehicle for continuing efforts or programmes already underway. It may also serve as a focusing opportunity or as way to increase activity in a particular country.

Another concern involves the deployment of military assets in the recovery process. While most countries allow their militaries to be engaged in rescue and relief efforts once a national disaster declaration has been issued, countries vary in terms of the extent to which military assets can be used for clearing roadways, rebuilding infrastructure, and assisting in reconstruction efforts. There are both concerns regarding the appropriate use of the military as well as the idea that construction activities and funding for construction should flow to civilian organisations. At issue is the larger question and concern as to which private companies are engaged in and benefit from the reconstruction activity.

The need to audit and account for the spending and transfers of funds is a common concern with disasters. Corruption or mismanagement of funds is a frequent problem given the size, pace, and diversity of different needs and methods of providing aid. In addition to following standard accounting and reporting practices, there is also need to ensure flexibility and that programmes address actual needs and demands. These issues have arisen not just when large amounts of international funding are involved, but also with federal aid and transfers to state and local governments.

How are diverse stakeholders brought together and engaged in planning activities, and how do they work collectively to devise and implement solutions for hastening recovery and building back better? Olshansky and Johnson (2010) describe the processes and challenges to planners in their book *Clear as Mud: Planning for the Rebuilding of New Orleans*, highlighting the tensions between 'speed versus deliberation', opportunities to correct past wrongs, and needs for 'coordination, communications, leadership and planning'. Similar lessons from the recovery of the tsunami disaster in 2004, the earthquake and tsunami in Japan in 2011, typhoon Haiyan in the Philippines in 2013 and other disasters suggest that there are important common themes and opportunities to share knowledge and build capacity especially in terms of governance across different actors, stakeholders and organisations involved in disaster response and recovery.

These tensions between local and global agencies, processes and approaches to planning and decision-making are especially acute following disasters in small island communities. There is often resistance to national-level or outside forces. There are tensions between local and global ways of doing business, and replacing local supply chains and networks with international systems. The policy environment is also complicated by the presence of many existing trade alliances, security arrangements and other institutional arrangements. Some have been formed to support trade or security concerns, while other alliances are more focused on environment, energy or cultural exchange. Some of the partnerships are a function of proximity while others involve shared common country-level attributes, such as being a small island. Notably, overlapping and multi-level arrangements can function as national-level agreements. There are also subnational, city-to-city, professional and institutional arrangements, such as between universities, that also complicate these relationships.

There are two different perspectives on these relationships. Organisations such as the United Nations or large donor organisations involved in disaster response and recovery need to understand the larger geopolitical context for engagement and risk reduction activities. There is, obviously, a geopolitical context and many existing relationships, partnerships and programmes which affect the nature and extent of aid, both generally and with respect to disasters. The landscape is clearly quite crowded and there is a need to understand not just the boundaries and borders of existing programmes, but the various mechanisms for the transfer of knowledge, resources and support for disaster response and recovery. Also, from the perspective of individual countries, in terms of alliances, pre-existing levels of support and engagement and potential for disaster aid. It is much easier to work with and extend existing relationships and channels than to build new networks, hubs and field offices in order to implement aid for disaster response and recovery.

In addition to governments, there are NGOs and other alliances which are already engaged in international relations and development activities. Disasters stretch boundaries and often require the crossing of borders in order to provide international aid or assistance. At times, borders may need to be lifted in order to facilitate the flow of goods and services to distressed areas. While countries have sought to establish rules of engagement and authorities much of it is oriented towards response rather than longer-term recovery.

Because of globalisation and improvements in communications, Internet and transportation, there already is much cross-border traffic. Disaster aid needs to be seen as part of a worldwide system of commerce and trade. While disasters may provide an opportunity for businesses to establish new footholds, it is important to understand that such exchanges and interactions have been governed by bilateral as well as multilateral and regional agreements. Especially as time moves beyond the disaster-causing event, the need to normalise trade relations or get back to business as usual will increase. This in turn not only complicates the recovery process, but it also suggests that more time, engagement and dialogue are needed. The larger concerns related to trade, security, environment and cultural exchange will become increasingly relevant as situations improve.

While it is often tempting to focus on the latest earthquake or typhoon, the reality is that there are larger, looming threats that involve greater uncertainties and complexities. Climate change, sea level rise and corresponding changes in the environment will add further instability in terms of the prospects for sustained long-term recovery. Also, continued urbanisation and development in hazard zones will subject communities to continuing and increased threats. Man-made and technological hazards can also confound efforts to build resilience and reduce vulnerabilities. New threats such as cyber security or longer-term environmental damage resulting from cascading disasters require new technologies and more robust systems of governance.

9.5 Small Islands as Centres of Innovation for Collective Risk Management

In this essay, it has been argued that small island communities are not just more profoundly aware of disasters, but that in these communities, notions of borders and boundaries of risk reduction are stretched in at least three different ways. On small islands, the physicality of threats is more apparent than within large continents. We clearly see and visualise the boundaries, sometimes expanded by volcanic action or lost through coastal inundation by storms or tsunamis. Sometimes disasters like storms or tsunami originate in faraway places. They cross borders and physical boundaries, moving from one continent to another or traveling vast distances. There is a keen awareness of inside and outside forces and conditions before, during and after harmful events. There is a collective memory of disasters and harmful events.

The losses as well as the efforts by responders and those providing relief have been memorialised as part of the recovery process. By virtue of the small size of islands, land is far more precious and efforts to manage these and other resources (fresh water, agricultural and marine products and other locally produced goods and services) are more concentrated in time and space. Because of acute awareness of remote locations and high transportation costs, island communities also are more attuned to combining sustainability and resilience efforts. Beyond the physicality of small islands, systems for coping with and managing hazards and strengthening social capital have long been a part of the culture and identity of islanders. While globalisation and modernisation have threatened to replace local, indigenous knowledge and ways with 'scientific' exogenous ideas and practices, there is a clear and present danger to overlooking or de-emphasising the value of local knowledge. Some years ago, Lucy Lippard (1997) wrote a delightful book describing the 'lure of the local'. She argues that 'sense of place' is not just critical to understanding our identities, but also key to sustaining our future, and that 'local life is all about communicating across boundaries'. Perhaps nowhere is this more evident than on small islands where the notions of space and time are so keenly observed, understood and managed.

A second aspect of how disasters transcend boundaries relates to the multidisciplinary nature of understanding disasters and their impacts and the changes within society. In addition to different types of scientific knowledge about geophysical conditions, hydro-meteorological events, environmental and public health consequences of disasters, and the influences and forces which bear down upon a community, the social science is equally important, perhaps more so. Where people live and work, the types of buildings and activities they are engaged, and how they detect harmful events, alert others and take precautionary measures is related to the socio-economic, cultural and political conditions in the community. The extent to which mitigation efforts as well as training and capacity-building to reduce risks also says much about community. Those places that have had much experience with hazards are better prepared than those who have not experienced harm-causing events. Disasters must be seen within a social context. This involves not just knowing about hazards and threats but also understanding how these risks and vulnerabilities are addressed through land use planning, zoning, building codes and management of development. In some communities, these efforts are limited because of informal settlements or rapid urbanisation or the inability to update and manage property records. It is important to recognise that all disasters begin with local conditions, physical as well as social. Resilience is tied to other aspects of local governance and management of both risks and assets. Robust processes for risk and vulnerability assessment, determination of preparedness needs and requirements, mitigation and adaptation strategies and pre-event recovery planning can all help to build resilience at the local level.

The third aspect of border- and boundary-crossing with disasters involves the observation that a disaster overwhelms local capacity. As such, large-scale disasters necessitate cooperation across jurisdictional boundaries through mutual aid and by deployment of national and international resources (both governmental and non-governmental) and flows of goods, services and capital from undamaged to damaged regions. On small islands, it is easier to observe the presence of outsiders, which can overwhelm local conditions. Initially, there is a flood of relief workers, supplies, and other resources that may strain local transport and infrastructure services. Military and external assets may complicate coordination and communications. International donors, foreign governments and national-level resources converging on small island communities across different disciplines or clusters of activity require coordination and pre-event training. More attention to reconciling the differences between approaches such as the United Nations cluster approach and country-level approaches based on core capabilities or emergency support functions is needed, particularly as the frequency and intensity of international activity and cooperation in disasters increases. Sensitivity and awareness of local culture and practices can go a long way to enhancing effectiveness of response, relief, and recovery efforts. Understanding how people live and their networks and systems for communicating, sharing information, planning and managing collective resources is especially important in hastening the transition from relief to reconstruction and recovery. Small island communities provide an opportunity for observing, learning from and transforming disaster operations. More attention to best practices and focused training which aligns with generalised approaches or clusters of activity for mass care, health and environmental management—as well as specific capabilities such as planning, communications, damage assessment and other support functions required during the aftermath of disasters—is necessary and challenging across borders. There is a need for collaboration and information-sharing across different disciplines and boundaries associated with response, relief and recovery as well as different human, health, social and environmental needs.

The increased risk of small islands to disaster suggests two different types of opportunities. First, there is a need for more research and understanding as to border-crossing and movements across physical, social, intellectual and jurisdictional boundaries. Small islands provide an ideal setting for observation, analysis and experimentation with approaches for evaluating relief and recovery programmes, and also filtering and integrating local and external knowledge, science and practice.

More systematic research on the use of indigenous knowledge in island communities for disaster risk reduction is needed. Edwards (2013) uses the idea of 'third space' to give reference to a space in which Indigenous Peoples may finally offer their own interpretations of what they are observing and experiencing. In Hawai'i, we can take advantage of the accessible digital space that has made available vast repositories of Hawaiian-based information. Web-based resources include Hawaiian language newspapers; Ulukau, the Hawaiian electronic library; Pukewehewehe, the electronic Hawaiian dictionary; or Papakilo, a database collection of historic and culturally significant documents, places and events of Hawai'i; and the Kipuka database that links historic data sets to geographic locations using the latest mapping technologies. As Oceanic scholars continue to advance through the ranks of the academe they bring vitality, rigor and fidelity to indigenous methodologies such as 'ancestor lensing' or '*kupuna* optics' as a means to explore, question and rediscover how our *kupuna* or ancestors treated an activity, event or problem.

Discovering what is important about indigenous knowledge as a way to increase traction in the area of disaster research involves an understanding of the differences between terms such as indigenous knowledge, traditional knowledge and traditional ecological knowledge. It also involves determining the degree to which knowledge is transferable between communities, districts and regions and at what scales. It further involves being clear on their benefits, limitations and possibilities. For example, King et al. (2007) discuss the little-considered value of indigenous environmental knowledge in hazard identification, disaster management and prevention. They offer that Maori Environmental Knowledge, or *Matauranga taiao,* represents a 'cumulative body of knowledge, practices and belief that has evolved through adaptation' (p. 60). In this sense, then, knowledge is not just traditional, but inclusive of the 'totality of experiences of generations of Maori in New Zealand' (p. 60). Under this framework, they propose that information and knowledge involves observing and recording change in the physical environment, naming and classifying locations of risk and predicting environmental disturbances. Lastly, they consider the level of skill needed to interpret indigenous knowledge and offer a cautious approach to the limitations of oral traditions.

An indigenous research agenda may offer a means of establishing a 'baseline' that informs the nature and oral traditions that evolve around storm events, earthquakes, tsunami, landslides and flooding. Nunn and Reid (2016), for example, provide a method of dating stories belonging to Aboriginal groups from locations around Australia. They 'provide empirical corroboration of postglacial sea-level rise' for 21 locations and discuss the implications of the durability of these oral traditions (p. 11). Such a research agenda would consider not just the types of hazards and threats but also the extent to which intellectual as well as political boundaries are breached or present opportunities to be bridged. The concept of 'bouncing forward' examines a disaster recovery scenario that is inclusive of historically marginalised indigenous peoples. Ryks et al. (2014) reflect on how the recent developments among Maori serve as catalyst that enable greater participation in urban development including disaster planning. Experiences among Maori in post-earthquake Christchurch exposed the need for more formalised engagements with *iwi* and *hapu* to help with recovery processes. Engaging Maori (and other Indigenous Peoples for that matter) in disaster planning presents opportunities to address those factors that originally served to increase vulnerability. Sustainable urban development agendas can result with *iwi/hapu* positioning themselves to collect and record traditional knowledge of ancestral areas. This provides a national base of information on water, soil, land use and natural hazards. They document how the Treaty of Waitangi settlements have influenced urban-based *mana whanau* (traditional inhabitants) to increase employment opportunities in the areas of urban planning and resource management, retain collective control over their own development agenda concerning land and waterways that are returned as a matter of redress, and establish agreements with local government about the nature of decision-making regarding resource use. Some treaty settlement terms address cultural redress in ways that involve the restoration of environmental areas and/or involve the restoration of ancestral place names.

Understanding innovation and the diffusion of new ideas, practices and approaches to disaster risk reduction is essential. While there has been some progress in thinking about various localised adaptation strategies, more attention to understanding cultural, institutional and organisation barriers to change is needed. A central aspect of the research agenda must entail a more focused analysis of risk, risk assessment and risk management, across different hazards and vulnerabilities, inclusive of diverse disciplines and intellectual traditions. Mercer et al. (2007) have proposed a framework for combining indigenous and western knowledge with a focus on not just the hazards and risks, but also specific capabilities related to land use planning, building practices, agricultural systems, and social support systems. Collective risk management must embody both local and exogenous knowledge. A second, pressing need involves the translation of research findings on risk reduction strategies that integrate local and international knowledge and practice, and incorporate the development of training courses and learning opportunities so that communities can better understand how to collectively manage risks, hazards and threats and devise appropriate plans, programmes and actions for enhancing livelihoods, quality of life and resilience. Training becomes a mechanism for not just conducting risk and vulnerability assessment but also designing, building and evaluating communities. It also provides an opportunity to clarify values and consider measures and indicators of success. We need more active development of training programmes, exercises, drills and learning opportunities for those involved in risk reduction. Concerted effort to focus on island communities provides an opportunity to assess training needs, develop innovative new approaches and courses for building resilience, and measure outcomes and the return on investment on training and education. An investment in human capital and capacity-building will have long-term payoffs, particularly in terms of governance. Strengthening planning is about communicating across different boundaries, disciplines and communities. Engaging community members in the planning, design and implementation of resilient communities is a form of adaptive governance. Small island communities can serve as models or centres of innovation where the complex social, cultural, political and environmental forces can be better understood and hopefully managed.

References

Beamer, K. (2014). *No mākou ka mana: Liberating the nation*. Honolulu: Kamehameha Publishing.

Clark, K., Power, W., Nishimura, Y., Kautoke, R. A., Vaiomo'unga, R., Pongi, A., & Fifita, M. (2011). Characteristics of the 29th September 2009 South Pacific Tsunami as observed at Niuatoputapu Island, Tonga. *Earth Science Reviews, 107*(1–2), 52–65.

Connelly, S. (2014). Urbanism as island living. In A. Yamashiro & J. N. Goodyear-Kaopua (Eds.), *The value of Hawaii 2: Ancestral roots, oceanic visions* (2nd ed., pp. 88–99). Honolulu: University of Hawaii Press.

Edwards, S., & Hunia, R. (2013). *Dialogues of Matauranga Maori: Re-membering*. Te Awamutu: Te Wananga o Aotearoa.

Freitag, R., Abramson, D., Chalana, M., & Dixon, M. (2014). Whole community resilience: An assets-based approach to enhancing adaptive capacity before a disruption. *Journal of the American Planning Association, 80*(4), 324–335.

Haas, J., Bowden, M., & Kates, R. (1977). *Reconstruction following disaster*. Cambridge, MA: MIT Press.

Johnson, R. K. (1981). *Kumulipo, the Hawaiian hymn of creation*. Honolulu: Topgallant Publishing Company.

Johnson, R. K. (2000). *The Kumulipo mind: A global heritage in the Polynesian Creation Myth*. Honolulu: Publisher not identified.

Kame'eleihiwa, L. (1992). *Native land and foreign desires: How shall we live in harmony?* Honolulu: Bishop Museum Press.

Kawamoto, K., & Kim, K. (in press). Social capital and efficiency of earthquake waste management in Japan. *International Journal of Disaster Risk Reduction*.

Kim, K., Yamashita, E. Y., Ghimire, J., Burke, J., Morikawa, L., & Kobayashi, L. (2013). Learning from crisis: Transit evacuation in Honolulu, Hawaii after tsunami warnings. *Transportation Research Record, 2376*, 56–62.

King, D. N. T., Goff, J., & Skipper, A. (2007). Māori environmental knowledge and natural hazards in Aotearoa-New Zealand. *Journal of the Royal Society of New Zealand, 37*(2), 59–73.

Lippard, L. (1997). *The lure of the local: Sense of place in a multi-centered society*. New York: The New Press.

Mercer, J., Dominey-Howes, D., Kelman, I., & Lloyd, K. (2007). The potential for combining indigenous and western knowledge in reducing vulnerability to environmental hazards in small island developing states. *Environmental Hazards, 7*(4), 245–256.

Mercer, J., Kelman, I., Suchet-Pearson, S., & Lloyd, K. (2009). Integrating indigenous and scientific knowledge bases for disaster risk reduction in Papua New Guinea. *Geografiska Annaler Series B: Human Geography, 92*(2), 157–184.

Minerbi, L. (1988). *Traditional Hawaiian land use management system*. Honolulu: Department of Urban and Regional Planning, University of Hawaii.

Mulvenon, J. (2010). The Chinese military's earthquake response leadership team. *China Leadership Monitor, 25*, 1–8.

Nunn, P. D., & Reid, N. J. (2016). Aboriginal memories of inundation of the Australian coast dating from more than 7000 years ago. *Australian Geographer, 47*(1), 1–37.

Olshansky, R., & Johnson, L. (2010). *Clear as mud: Planning for the rebuilding of New Orleans*. Chicago: American Planning Association.

Pelling, M., & Uitto, J. (2001). Small island developing states: Natural disaster vulnerability and global change. *Environmental Hazards: Human and Policy Dimensions, 3*(2), 47–62.

Robinson, D., & Robinson, K. (2004). Pacific ways of talk—hui and talanoa. In I. Marin (Ed.), *Collective decision-making around the world: Essays on historical deliberative processes*. Dayton: Kettering Foundation Press.

Ryks, J., Howden-Chapman, P., Robson, B., Stuart, D., & Waa, A. (2014). Maori participation in urban development: Challenges and opportunities for indigenous people in Aotearoa, New Zealand. *Lincoln Planning Review, 6*(1–2), 4–17.

Smith, G. (2011). *Planning for post disaster recovery: A review of the United States disaster assistance framework* (2nd ed.). Washington, DC: Island Press.

Syafwina. (2014). Recognizing indigenous knowledge for disaster management: Smong, early warning system from Simeulue Island, Aceh. *Procedia Environmental Sciences, 20*, 573–582.

Trosper, R. (2002). Northwest coast indigenous institutions that supported resilience and sustainability. *Ecological Economics, 41*, 329–344.

Whyte, W. H. (1980). *The social life of small urban spaces*. New York: The Project for Public Spaces.

Chapter 10
The Empowerment of Local Community Groups as a New Innovation in Cross-Border Disaster Governance Frameworks

Yenny Rahmayati

10.1 Introduction: The 2004 Aceh Tsunami and Earthquake

Natural disasters will be the major form of transboundary disaster over the next few decades. Asia in particular remains a place prone to different sized natural disasters ranging from tsunamis to earthquakes, floods and forest fires, all of which have cross border impacts. Although there have been numerous studies about disaster governance, most of these discuss the subject from a policy level perspective, while giving less attention to the role of the community. Yet in many disasters, local communities are the first groups to engage in emergency efforts before the arrival of international assistance. Their role needs to be empowered through their inclusion in cross-border disaster governance frameworks so that they can drive policy innovations in building social resilience to future large-scale disasters.

Based on empirical evidence from the 2004 Aceh tsunami and undersea earthquake, this chapter explores the role of local community groups (local NGOs, local traditional groups, women's and youth groups) and how to empower them through their inclusion in the cross-border disaster governance frameworks. This includes recognising the ways in which these groups could fit into wider governance regimes in terms of their role and relative power to influence policy choices through their interactions and engagements with agencies that function across national contexts and jurisdictions.

Three projects carried out by local community groups are presented as case studies. The first was the 'Aceh Post-tsunami Cultural Heritage Project', conducted by the Aceh Heritage Community Foundation (AHC). The second project, entitled 'Strengthening Capacity of Community and Schools' aimed to develop psychosocial and creative development programs to prevent child labour in Nangroe Aceh

Y. Rahmayati (✉)
Centre for Design Innovation (CDI), Swinburne University of Technology,
Melbourne, VIC, Australia
e-mail: yrahmayati@swin.edu.au

M.A. Miller et al. (eds.), *Crossing Borders*, DOI 10.1007/978-981-10-6126-4_10

Darussalam' and was developed by Anak Bangsa Foundation (YAB). The third project, 'Strengthening Capacity of Community Learning Centres (PKBM)[1] in Nanggroe Aceh Darussalam to provide training and income generation options for older children', was initiated and implemented by Community Learning Centres/CLCs (PKBM). These case studies provide evidence about how the involvement of local community groups support disaster recovery efforts that foster the rehabilitation and reconstruction process in areas affected by natural disasters. This process eventually leads to the sustainability of the future development of the Aceh region. A recommendation for a model of cross-border disaster governance is provided at the end of the discussion.

On 26 December 2004, an earthquake measuring 9.1 on the Richter scale was followed by a massive tsunami, which struck Indonesia's westernmost Aceh province. That was one of the largest natural disasters in Aceh's history. The epicentre of this earthquake was located 250 km south-west of the region. Its rupture—a slippage of up to 10 m—resulted in the ocean floor being lifted and (permanently) dropped, pushing the entire water column up and down and generating a series of powerful waves. Tsunamis swept violently up to 6 km inland over the shorelines of Aceh and surrounding islands. The disaster affected the housing and settlement sector tremendously. Banda Aceh, the capital of Aceh Province, was one of the hardest hit spots, along with Aceh Jaya and Aceh Barat. In all affected regions, 126,741 lives were lost and, in the wake of the disaster, an additional 93,285 people declared missing. Some 500,000 survivors lost their homes leaving 20% of the Acehnese homeless (Bappenas 2005b), while as many as 750,000 people lost their livelihoods. Some Acehnese lost their land as the tsunami washed it away and partly even changed Aceh's coastline. Others found it hard to locate their lots in the ruined land, and many had no proof that the lost land was theirs (DTE 2005). In the private sector, which constituted 78% of the destruction wrought by the earthquake and tsunamis, up to 139,195 homes were destroyed or severely damaged, along with 73,869 ha of land with varying degrees of productivity. In the public sector, 669 government buildings, 517 health facilities, and hundreds of educational facilities were destroyed. By the time the reconstruction process was officially completed in 2009, 147,000 houses had been rebuilt for survivors (BRR 2009).

Two days after the disaster on 28 December 2008, the President of the Republic of Indonesia, Susilo Bambang Yudhoyono, declared the tsunami a national disaster. But due to the large-scale damages and casualties, the disaster not only attracted national concern, but also gained the world's attention. Billions of foreign aid dollars for reconstruction were offered, many international NGOs jumped into the fray, and Indonesian governmental agencies came to Aceh's rescue. In response to the disaster, 124 international NGOs, 430 national NGOs, dozens of donor and United Nations agencies, various government agencies (some military), and many others were involved in the emergency response, rehabilitation and reconstruction process in Aceh (BRR 2005).

---

[1] Community Learning Centres (PKBM) are local community groups at neighbourhood/village level, led by a *Keuchik* or local traditional leader (head of the village).

The local NGOs and local traditional groups that existed before the tsunami, especially those concerned with social, religious and human rights issues, were also involved in the disaster response. Aceh, which been isolated due to a long separatist conflict, suddenly opened to the world. Despite differences in terms of demographic, socioeconomic and political lines, local groups in Aceh worked together with other national and international agencies in the name of humanity, all of them joined in cross-border forces and flows in emergency response and rehabilitation (Miller and Bunnell 2013). The combined local, national and international aid resulted in large-scale post-disaster rehabilitation and reconstruction projects. That in turn led to enormous changes in urban space. In addition to new housing, much of the infrastructure had to be repaired. The economy had to recover, including people's means of subsistence and livelihood (DTE 2005). The rehabilitation and reconstruction phase was officially completed in 2009.

Banda Aceh, the capital of Aceh Province, was one of the hardest hit spots besides Aceh Jaya and Aceh Barat. However, among these three badly affected areas, Banda Aceh received more attention from the agencies and organisations. This because of the 29-year conflict between Aceh separatist movement and the Government of Indonesia has isolated Aceh and disrupted the urban-rural connection (Miller and Bunnell 2013). This study focuses on Banda Aceh and surrounding areas (part of Aceh Besar) as this is the concentration zone of tsunami emergency response, rehabilitation and reconstruction.

10.2 Local Community Groups in Aceh

Over the past two decades, 'non-governmental organisation' or NGO has become a familiar term not only for social actors but also academics (Martens 2002). Werker and Ahmed (2008:74) define NGO as "the subset of the broader non-profit sector that engage specifically in international development". The term has been used synonymously in the literature with several others, such as private voluntary organisation, private non-profit organisation and voluntary association/organisation (Tongsawate and Tips 1988). There are four types of NGOs based on the orientations: charitable, service, participatory and empowering while based on the level of operation, it was categorised into community based, citywide, national and international (Cousin, 1991). In Aceh, before the Indian Ocean Tsunami disaster hit the region in 2004, the existence of NGOs, either international, regional/national, or local, was less significant. Local traditional groups were more dominant and played an important role in the community: as stated in a 2005 report by the BRR (the Rehabilitation and Reconstruction Agency for Aceh and Nias),

> Aceh has a rich tradition of associations, ranging from faith-related and community-based organization (e.g. *Panglima Laut*, the association of fisherman, savings clubs, and funeral societies) to semi government structures, based on elected neighbourhood and community representatives (BRR 2005, p. 45).

These groups have been instrumental in maintaining the traditional values of the community.

The international, regional/national, and local NGOs in the Aceh region came into existence during the peak of the armed conflict between the Free Aceh Movement (GAM), who sought independence from the Government of Indonesia (GoI), and GoI forces—that is, from the 1990s until the tsunami occurred. During this period, most of the NGOs operating in the region focused on human rights issues while the rest were concerned about social culture issues and only a few focused on the development issues. According to the BRR:

> Aceh's recent history of conflict and international isolation means that civil society has evolved rather differently compared with other provinces. There are few local NGOs with large-scale operational capacity, but on the other hand there are many who are strong in advocacy and in protecting human rights, and there are strong associations, for example of fisher folk (BRR 2005, p. 182).

However, the 2004 tsunami catastrophe changed this situation. Immediately after the disaster, large numbers of organisations including UN bodies, international NGOs, regional/national NGOs and foreign government agencies suddenly established operations in Aceh. Many new NGOs established after the tsunami were located in urban centres, as there was more access to funding. As a result, the distribution of funds from the agencies and donors also concentrated in urban areas. Meanwhile, rural areas were isolated from aid not only because of accessibility but also due to safety reasons:

> Rural coastal Aceh has been particularly neglected; entire villages along the west coast had to be abandoned after the tsunami because humanitarian aid and reconstruction resources arrived too late, too irregularly, or not at all (Miller and Bunnell 2013).

The distribution of aid in rural areas relied more on the existing local traditional groups, such as *Panglima Laot* or a local community-based fishery management system.

Eventually there were around 490 organisations operating in Aceh and Nias, of which 291 were NGOs and donor agencies (BRR 2005). Although the international NGOs (INGOs), United Nations agencies (UN), regional/national NGOs, and government donors were dominant in number, this chapter argues that they could not make the best effort without support from local community. The chapter takes the position that the local community is a valuable resource, because they understand more about the location and situation than larger organisations. Their empowerment enhances the effectiveness and the outcomes of the recovery process as well as ensuring the sustainability of development in the future.

The 2004 tsunami disaster also brought a great change for the development of local NGOs in Aceh. New local NGOs concerned with the area of development began to appear, with areas of focus including education, health, livelihood, environment, social-cultural issues and community development. In addition, the existence of local community groups was strengthened by the rehabilitation and reconstruction policy which required every party involved in the reconstruction and rehabilitation effort to include local community groups in their programmes. This policy was part of the concept guided by the Master Plan of Rehabilitation and Reconstruction of Aceh and Nias.

> In the context of the rehabilitation and reconstruction of Aceh and Nias islands, the definition of participation must be materialized in a cooperation model and framework wherein the respective governments (central and regional) join hands with the community in reconstructing Aceh and Nias Island (Bappenas 2005b, p. 7–2).

Moreover, the government's rehabilitation and reconstruction policy recognised the importance of involving women and youth groups in the redevelopment programme in order to ensure the sustainability of programmes. Women are the backbone of the culture in Aceh, while youth are the future generation who will continue the development. Although youth is a group that still depends on adults and needs protection and support from them, especially in distressing and difficult situations like disaster (Fernandez and Shaw 2014), the youth are a significant population group in the demographic profile of the Aceh region and for this reason have significant influence on the development.

10.3 Opportunities and Challenges

Community organisations played a role in every phase of disaster response including the rescue, relief and rehabilitation phases—a role that was even most important than the roles of other related parties such as government and aid agencies (Nakagawa and Shaw 2004). The same was true in Aceh's post-tsunami recovery. They played a significant role because they are groups who have a good understanding of local situations including problems and obstacles that may appear. Therefore, their involvement becomes vital to ensure the effectiveness of the recovery programme. Lawther (2009) highlights that the best outcomes from the recovery efforts can be achieved through the promotion of high levels of community involvement (see also Pyles 2007). Secondly, they are active players in the development, not only passive recipients, so the benefit from the redevelopment will meet their needs. Thirdly, their participation will enable them to continue the development independently once the external parties have left. Community has many capacities that can be empowered. They can adapt successfully and function effectively in post-disaster situations (Norris et al. 2008). Given the nature of informal organisations, local community groups have a high level of independence as they are usually managed by volunteers and provide mutual aid and self-help (Pyles 2007). A similar idea was proposed by Lawther (2009), who argued that the extent to which communities can restore normalcy and reach sustainability depends on whether the recovery process is able to empower local resources.

The importance of the involvement of the local NGOs and local community groups has been addressed in the Master Plan of Rehabilitation and Reconstruction of Aceh:

> While many INGOs and donors have programs to help strengthen civil society's capacity, the most important roles civil society can play immediately are in ensuring citizens know and claim their entitlements and are aware of recovery programs intended to help them, helping communities voice their grievances, and tackling problems of corruption and abuse. They can also provide independent monitoring of recovery projects, ensure the needs of the most vulnerable groups are met and serve as an interface between citizens and all institutions involved in the recovery of their communities. Civil society can also play a valuable role in helping tsunami-affected communities understand the many challenges involved in delivering such a large and complex recovery program (Bappenas 2005a, b, p. 182).

Considering this important role, the involvement of local NGOs and local community groups therefore becomes a vital issue in ensuring the effectiveness of the post-disaster recovery programme. In post-tsunami Aceh, many funding donors and INGOs tried to engage the local community in collaborative activities in order to achieve better outcomes of the their recovery programmes. This was also part of the effort to strengthen the role and existence of the local community groups. As highlighted by the BRR, 'Civil society can and also should continue to strengthen the already strong foundations of community so essential to a stable working environment enabling all national and international contributors to deliver their products and services with confidence' (BRR 2005, p. 182).

However, the involvement of local community groups in Aceh's post-disaster recovery also had challenges. Community engagement is commonly believed to not be part of the solution, but part of problem (Pearce 2003). This notion has prevented community groups from fully participating in the aid distribution process. Meanwhile, the insufficient number of existing local NGOs also became a challenge. In addition, their capacity was weakened after the disaster due to the loss of resources, including financial resources and inadequate number of staff. After the tsunami, many local NGO staff were recruited by the INGOs, UN agencies and foreign institutions, leading to a lack of local human resources that became one of the challenges facing local NGOs (Bowden 1990).

Furthermore, there were some challenges in terms of collaboration of local community groups with external parties including agencies, donors, INGOs and government institutions. The first was that inadequate skills and facilities have prevented local groups from collaborating with other organisations. Skill factors cannot be underestimated in community engagement (Lawther 2009), both in technical terms but also in terms of non-technical skills such as managing conflict that potentially occurs in collaborative programmes. Quarantelli (1997) highlights that there are two possible conflicts that may occur: conflicts among the local groups, and between local and outside organisations and agencies. Meanwhile, Bowden (1990) argued that both NGOs and governments have a limited willingness and understanding to collaborate, which could lead to conflict. The second challenge was about communication: the local community groups received less support from the government because there was a lack of communication between them. Therefore, it is fundamental to build clear and consistent communication as well as building good relationships between agencies and community groups to handle issues efficiently in the shortest time (Lawther 2009). The third challenge is about the post-conflict and peace process between Aceh separatist group with the Government of Indonesia. The insecure political situation during the peace process after the tsunami has led to lack of support for the establishment of collaboration activities between local community groups and external parties. Many the international agencies were hesitating to work and distributing their aid to the conflict zones which were also affected by the tsunami. Security issues, safety and trust become the main obstacles in building mutual cooperation.

Furthermore, many existing local NGOs could not fully participate in the redevelopment process. Their expertise centred more on social and political issues, especially human rights issues, which was not in line with the post-disaster redevelopment programmes. They were not familiar with the redevelopment programme and this weakened their capacity to act and lead the initiatives. In many cases, the lack of technical expertise in certain issues has prevented the local groups from taking a lead in collaborative programmes (Lawther 2009).

10.4 Enhancing Local Community Groups' Capacities in the Post-disaster Redevelopment of Aceh

The enhancement of the role of the local NGOs and local community groups needs to be addressed in the disaster governance framework to ensure the success of post-disaster recovery and redevelopment in the future. As stated in the Master Plan of Rehabilitation and Reconstruction of Aceh and Nias (Bappenas 2005a, b, p. 7–2), 'The principle [sic] purpose of synergic cooperation between the government and the community (development actors) is to achieve greater development results by joint cooperation as compared to the results achieved by individual effort of each development actor'.

Recent trends show that disaster governance has acknowledged community participation as an integral aspect of the programme (Méheux et al. 2010). It is also believed to be a basic component for achieving resilience of the community (Norris et al. 2008). Such acknowledgement confirms that local community groups should take an active role and turn their position from the passive 'victim' who receives humanitarian aid into an active agent in the programmes (Davidson et al. 2007). Without the involvement of community leaders, social networks and hierarchies, meaningful and comprehensive community participation cannot be achieved (Daly and Rahmayati 2012). Participation of government institutions, NGOs and the general public becomes necessary in disaster governance (Newport and Jawahar 2003). This can be implemented through training programmes and opening access to other parties for collaboration, especially to the donors who can support redevelopment through financial sources. Regulska (1999) argues that as a newly established political actor on the local scene, NGOs have found themselves needing to affirm their identity and presence, establishing a variety of external relations with other local partners (local governments, other NGOs, citizens, or business communities), and building alliances to assist them in achieving their goals.

The enhancement of role of the local NGOs and local community groups in disaster situations can be achieved through the improvement of organisations' capacity, internally and externally. Internal capacity development can be carried out through the improvement of management, resources and capacity-building. Providing access to resources becomes important, as it can facilitate the process of problem solving that may occur in stressful life events (Dombo and Ahearn 2015). Meanwhile, the improvement of external capacity can be achieved by integrating the programmes of the local community groups into the government planning/actions. In post-tsunami Aceh, the integration of the local community groups into government and reconstruction programmes has been properly accommodated. It is proved by the inclusion of this issue in the Master Plan of Rehabilitation and Reconstruction of Aceh and Nias (Bappenas 2005a, b, p. 7–6) One of the efforts to reconstruct the Aceh community is through the policy of empowering the existing religious, customary and other social institutions in Aceh, especially for *Mukim*[2] and *Gampong*[3] institutions". These two institutions have a dominant role in community. This is confirmed by the Law No. 18 Year 2001 about Special Autonomy for Aceh Province.

Lu (2014) has argued that highlighting collaboration between various organisations in various communities is necessary for disaster governance. Partnership or collaboration is one of the schemes of an integration strategy. The coordination between the government and the local community groups is very crucial in this strategy. Each party can give support to others, while minimising the overlapping of programmes between them. Tongsawate and Tips (1988) argue that efficient coordination would allow the governmental organisations (GOs) and NGOs to learn from each other and to play positive and complementary roles, and thus contribute more to national development.

However, coordination is not easy to implement as conflicts sometimes occur, for example in the different perceptions and the lack of understanding between two parties. Sometimes the government or reconstruction authority does not have a clear strategy to support development programmes that involve the community. The unavailability of accurate supporting data means that the government programmes cannot achieve their goals. The Community Learning Centers (CLCs), especially CLCs who focus on the vocational training programmes for the community, are one example of this. They are very weak with a lack of training curricula, lack of funding, lack of teaching staff and limited facilities. Those problems could not have arisen if there was support from the government and inclusion in government programmes for improving curricula, provision of facilities or capacity-building

[2] *Mukim* is a term for a customary community unit, which has certain boundaries, customary apparatus and symbols, ownership titles and control of certain resources and infrastructure, possessing a locally-specific social order.

[3] *Gampong* is a legal community unit that constitutes the lowest administrative organisation under the direct supervision of the *mukim*, and is headed by the *Keuchik*.

training for the teaching staff. Nevertheless, these problems can be solved through development of good communication between GOs and NGOs, conducting deep discussions and building each other's trust. Some assistance from third parties, such as the donor agencies, as the facilitators is sometimes required. Financial support from the government becomes an additional important collaboration to enhance the success of redevelopment programmes.

10.5 Case Studies

As noted previously, three examples of projects carried out by local community groups in Aceh's post-tsunami recovery and redevelopment are highlighted for discussion in this chapter. The types of activities and collaboration carried out as well as the outcomes from each project were analysed. The finding shows that each project has different implications and demonstrates different strengths or lessons learned. The case studies presented are based upon the author's experience in Aceh's post-tsunami response between 2005 and 2009 while working as the Project Coordinator of the ILO-IPEC Aceh Programme and as the Executive Director of Aceh Heritage Community Foundation, a local non-profit community based organisation in Banda Aceh. They are also based upon extended observation several years. The data collected through series of fieldworks conducted in Banda Aceh from 2007 to 2012. During this period, site visits for direct observation and semi-structured interviews with the coordinators in-charge in the projects were conducted. The research was supported by reviewing project documentation and reports from public authorities, NGOs and aid agencies. The observations focused on the urban centre of Banda Aceh and the peripheral areas (western part of Aceh Besar). These areas were among the most devastated by the tsunami, and also the concentration zones of post-tsunami responses. The highlights issues were emphasised in the process of activities, how they collaborate and build synergy and the outcomes.

10.5.1 Aceh Heritage Community Foundation

The Aceh Post-tsunami Cultural Heritage Project by the Aceh Heritage Community Foundation (AHC), was an example how local NGOs worked within the regional network for Asia and the Pacific. After the 2004 tsunami, one danger in the process of reconstruction was that old, historically and culturally valuable buildings and complexes fell victim to the immediate need to restore and rebuild the urban fabric of towns and villages. Responding to this concern, and initiated by the Aceh Heritage Community Foundation (AHC), the 'Aceh Post-Tsunami Cultural Heritage Project' was carried out. The project aimed to assess the condition of Aceh's cultural heritage after the disaster and investigate the cultural rehabilitation process in the post-disaster situation. The project contributed to strategic actions for cultural heritage rehabilitation, as well as providing lessons that may prove useful in other similar situations all over the world.

The project consisted of several activities including an on-site Inventory of 'before' and 'after' the tsunami; research of old photographs and historical information; and setting up a database of recent and old photographs, sketches, measured drawings, maps, plans, descriptions and historical information. The project was carried out with the assistance of the Lestari Heritage Network based in Penang-Malaysia (formerly known as Asia and West Pacific Network for Urban Conservation or AWPNUC).[4] This network connected AHC to foreign/international donor agencies and other heritage organisations in places outside Aceh such as Japan, Switzerland, the Netherlands and Singapore. The main actors in this project were young people: most of the project participants were students from a local university. They came from different backgrounds such as architecture, economics, history, and engineering, supported by senior academics as advisors. The key partners and allies of this project were the Lestari Heritage Network (Malaysia), the Nara Machizukuri Center (Japan) and the Asia Research Institute (Singapore). Contributions and partnerships carried out through this project included providing expertise, collaborative research, an exchange programme, and fundraising activities.

This project produced real outcomes that have been used by parties involved in the post-disaster reconstruction process of Aceh. The main products were a comprehensive Inventory of Heritage Buildings and Sites after the Tsunami; research reports; and brochures, catalogues, posters and calendars. In addition, this project has increased community awareness, especially among younger people, of the importance of caring about their heritage, and has provided them with the skills and the capacity necessary for further safeguarding and conservation efforts. This project has also given an actual example of how cultural heritage can be an important tool in rebuilding cities and communities after disaster, a model that can be used and applied in other disaster-stricken areas all over the world in the future.

10.5.2 Yayasan Anak Bangsa/YAB (Anak Bangsa Foundation)

Strengthening Capacity of Community and Schools to develop psychosocial and creativity development programs to prevent child labours in in Nangroe Aceh Darussalam is a collaborative project between the Anak Bangsa Foundation (YAB) with the Education Office, Social Affairs Department and the Sharia Court. It was an example of how local community groups can work with national or local governments. The project was funded by the International Labour Organisation (ILO) for International Programs on the Elimination of Child Labour (IPEC) as part of their programme in assisting the community in Aceh after the earthquake and tsunami disaster. The project focused on education and child protection issues, and offered

---

[4] The network changed their name again to Asia Heritage Network (AHN) in 2012.

alternatives of intervention to respond to the condition of children in Aceh in the disaster recovery phase. The main objective was to prevent vulnerable children from entering into child labour. This was done through the provision of relevant services which assisted 'at-risk' children and their families.

The project used two strategic approaches, a community-based programme and a school-based programme. The community-based programme was implemented through the establishment of a children's creativity and recovery centre (CRCC) in the area where many vulnerable children lived, while the school-based programme was implemented (in collaboration with the District Education Offices) through the establishment of a mobile library and psychosocial services that served several schools and several temporary refugee barracks. This programme was delivered at three elementary schools in Aceh Besar District through the implementation of series of workshops which used media to highlight children's rights and the worst forms of child labour (WFCL) issues to the related stakeholders. In addition, YAB also worked closely with the Social Affairs Department to provide input to the Sharia Court to develop *Qanun* (Aceh provincial regulation) on child protection. The detailed collaboration was implemented through several activities such as joint research, a workshop and development of the draft of *Qanun* on child protection. Besides the Social Affairs Department and Sharia Court, YAB also collaborated with teachers in elementary and junior high schools in Aceh. The collaboration was conducted to reduce the number of illiterate school students and to mainstream student dropouts back to formal school.

This project succeeded in achieving its goal of preventing vulnerable children from becoming child labourers. According to the ILO internal report (2007), activities provided at the CRCC motivated children to continue their studies and produce creative works. Children's groups that were formed through the CRCC such as the art, drama, music, dance, and agriculture groups have encouraged children to develop self-confidence. These groups have become a positive model for children's groups in other locations. The CRCC also gave support to non-formal education by assisting children who had dropped out of school to return to school under the coordination of the Education Office. This project has been successful in preventing vulnerable children (boys and girls) from dropping out of school through literacy and numeracy assistance. When the project was completed, 319 boys and 302 girls in three elementary schools have access to a mobile library, psychosocial support, and creativity development programmes which are supported by teachers and community. Besides the children, other groups were indirect beneficiaries of this project including parents and families of the children, teachers, and communities.

10.5.3 Pusat Kegiatan Belajar Masyarakat/PKBM (Community Learning Centres/CLCs)

Strengthening Capacity of Community Learning Centers (PKBM) in Nanggroe Aceh Darussalam to provide training and income generation options for older children is a collaborative project between Community Learning Centres/CLCs (PKBM) and the Aceh Education Office. It was an example of how local traditional groups work with national/local government. The Community Learning Centres (CLCs), known locally as PKBM, is a community network existing across the Aceh region. The centres, run by volunteers from the local community, have a role in providing various non-formal education and life skills training within the local communities, and is a potential vehicle to deliver education and training services within the communities outside the formal education system. The PKBM programmes are mainly supported by the budget from the Local Education Office (*Diknas*); however, their capacity has been limited by lack of funds. The volunteers are not really capable of developing PKBM as a financially independent institution. This project was funded by the International Labour Organisation (ILO) as part of the above-mentioned YAB project. The project focused on community- and capacity-building for the youth of Aceh, seeking to respond to the medium- and longer-term needs of older children in Aceh (15–17 years old). The purpose of this project was to prevent vulnerable children from exploitation by the worst forms of child labour (WFCL) through the provision of education, life skills and training services. This project also assisted these children in finding decent work opportunities, and aimed to strengthen the capacity of the *Keuchik* (traditional village leaders) and PKBM volunteers to become more effective community service providers. The project was implemented through four approaches. First, surveys were undertaken to outline the on-going programmes of the PKBMs, calculate numbers of dropout school children, and identify marketing opportunities for products that these children could produce. The second approach was strengthening the capacity of the PKBMs, through collaboration with the Education Office to develop of training modules on management, motivation and leadership. The capacity-building also covered training on children's rights issues and how to work with children. The third approach was for the PKBMs to deliver some vocational training, covering various skills including automotive repairs, haircutting and make-up application, screen printing, traditional handicraft, sewing and embroidery, carpentry, and the production of mattresses and other products from cotton. The fourth approach was facilitating PKBMs to establish a KBU (small business group) to help participants develop a business plan. As the donor, the ILO also provided equipment to support this start-up business programme. The groups were monitored by the Education Office to ensure the businesses addressed market demand. Start Your Business (SYB) training was also provided under this project, to assist youth from 15 to 25 years old who wanted to start some form of micro-business.

This collaborative project has succeeded in preventing vulnerable children from becoming exploited by the worst forms of child labour (WFCL) through the provision of education, life skills and training services. According to the ILO internal report (2007), at the end of the project, 340 girls and 155 boys had been prevented from becoming victims of WFCL through the provision of skills training and activities developed by ten PKBMs. The capacity of the PKBMs also has improved, especially in managing and providing activities to children. Several groups of children who graduated from the training programmes have followed the SYB training and some have successfully established their own businesses. The internal capacity of PKBMs has also been improved following issues identified by tutors, PKBMs' staffs, village leaders, religious leaders, field workers, youth and women's groups who collaborated on the project. The partnership between PKBMs and the Education Office has strengthened the successes of this post-disaster recovery programme. The Community Learning Centres (PKBM), which are supported and managed by the community—including village leaders, *PKK* (women's group) and *Karang Taruna* (youth group)—were the key to sustainability. These stakeholders have understanding that the main function of PKBM is to support children's growth and development as important assets for the sustainability of the development programme in future.

10.6 Findings and Discussion

Before the 2004 tsunami, the existence of NGOs (international, regional/national or local) was less significant in Aceh, while the role of local traditional groups was dominant. During the armed conflict era, international and local NGOs started to emerge; however, they were mostly concerned with human rights issues. After the tsunami, a great number of organisations including INGOs, the UN and other aid agencies began to appear in Aceh to work on the emergency response, rehabilitation and reconstruction of the province. Although big agencies are large in number and played a major part in this process, the role of the local community groups (local NGOs, local traditional groups, women's and youth groups) was the most important. Post-tsunami Aceh also brought a great change to the development of local NGOs in Aceh. Many new local NGOs were established, with their focus covering a wider range of issues including education, health, livelihood, environment, housing and gender. The Master Plan of Rehabilitation and Reconstruction of Aceh encouraged this situation by applying policy that strengthened local community groups. This policy required any external actor involved in the post-disaster effort to build synergy partnerships with local community groups including local NGOs, local traditional groups, women's and youth groups. These groups have played a significant role in the post-disaster recovery of Aceh.

There are several points raised in this chapter that demonstrate why the local community groups are important in a disaster situation. First, the local community groups are the local actors who have deep knowledge and understanding of the local situation and conditions including the social cultural aspects as well the geographic location. Secondly, they are the recipients for the redevelopment and rehabilitation process, hence they know their needs well. Thirdly, their involvement ensures sustainability in the future when all the external agencies have left Aceh. However, a number of challenges have also been identified. The main challenge is the insufficient number of existing local NGOs. Another challenge is the lack of human resources and support from the government. In addition, inadequate capacity-building and facilities also prevent local community groups from building collaborations with other organisations. At the same time, many local NGOs that existed before the tsunami could not fully participate in the redevelopment process since their expertise was not in line with the post-disaster redevelopment programmes. This is another challenge that needs to be addressed.

The case studies discussed in this chapter showed the role of local community groups (local NGOs, traditional groups, women's and youth groups) in the post-disaster recovery efforts. Each case study demonstrates a different model in synergising roles and responsibilities. The first case showed how a local NGO empowered youth in their activities and utilised the regional network in their project. This local NGO also approached organisations from other countries that face similar disasters to obtain support regarding expertise and funding. The second case study demonstrated the beneficial partnership which was built between government institutions with participation from the youth of Aceh and the support of an international agency as the donor. The last case study showed how PKBM, pre-existing local traditional groups in each neighbourhood, developed a good synergy with several government institutions with the supporting funding from an international agency. The empowerment of the local traditional village structure (*Keuchik* or head of village) as well as the involvement of local women and youth groups demonstrates the significance of the partnership model.

Among the three, the first case study analysed in this chapter shows the most successful model of collaboration especially in terms of the independency, not only financially but also in the expertise which were built through the regional networking. The products resulting from their collaboration obviously can be (and has been) used by many parties including government institutions. In addition, this model is more resilient and sustainable as they are still able to continue their activity even after the rehabilitation and reconstruction phase was terminated in 2009. The second and third case studies are very dependent on donor funding and the role of local leader, a position which usually changes every 5 years. Therefore the sustainability of these projects is not assured, unless the programmes were duplicated or continued by the government once the donor agencies left. The external parties also have to support the establishment of self-reliance to discourage donor dependency in future (Cueto et al. 2015).

Despite the strengths and weaknesses of each case, the three case studies illustrate a good model of integration of local NGOs and the activities of local community groups in government planning and action. The collaborations and partnerships which have been built have also enhanced the management skills of the personnel involved. This is seen as one of the benchmarks of a successful model of empowerment of local community groups in post-disaster recovery and development.

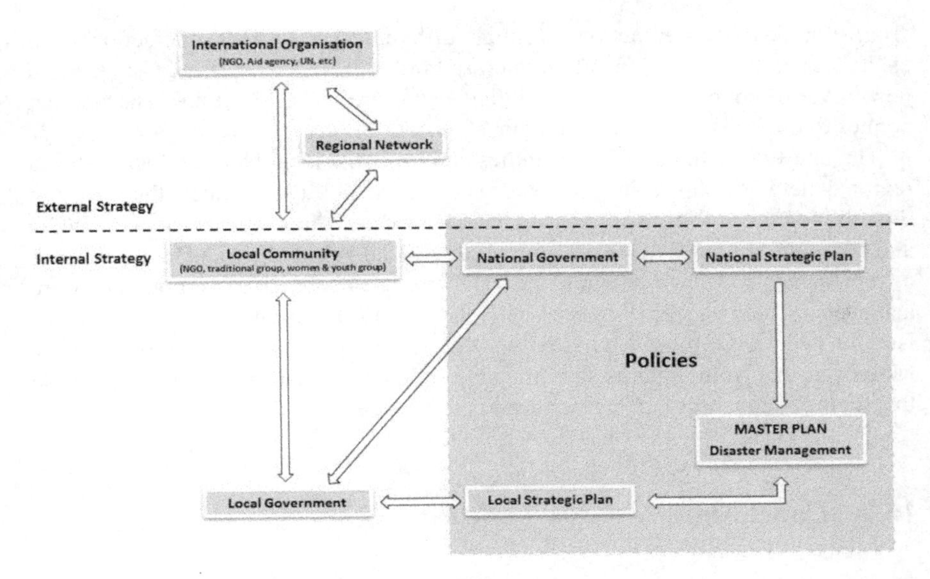

Fig. 10.1 Cross-border disaster governance framework

This chapter suggests a framework based on synergy models that have been demonstrated through case studies (see Fig 10.1). The first issue in developing the framework is the identification of the actors involved. They include the international NGOs, international agencies, regional networks, local community groups (local NGOs and local traditional, women's and youth groups) and government institutions (local and national). The second issue is the policy that supports the involvement and empowerment of local community groups in the master plan for post-disaster recovery and redevelopment. Figure 10.1 shows how the management works through two different approaches, the internal and external strategy. The external strategy is carried out through the development of direct collaboration or partnership between local community groups (local NGOs, local traditional groups, women's and youth groups) as the main actor of post-disaster recovery, with the international organisation (international NGOs, aid agencies, UN bodies, etc.) or through the regional network. The internal strategy is implemented through mutual work between community and government institutions at the national and local levels. This collaboration is supported by policy at both national and local levels—in this case, through the establishment of a National Strategic Plan by the central government and a Local Strategic Plan by the local government. These plans become the basis for the development of a Master Plan for Disaster Emergency and Recovery that applies to regular or sudden disasters.

The integration of local community groups into the government programmes has challenges, as it will depend on the governance regimes including political actors, political will and power, tension between local and central government. However, this framework gives some ideas of how to acknowledge the role of local community groups in post-disaster actions through different cross-border interactions and engagement; it is a new innovation in cross-border disaster governance as well as a new driver of policy invention in building social resilience for future disasters that applies to other urbanising world regions.

The analysis of the three case studies confirms that local NGOs which utilise a regional network shows independence and sustainability, although they are not directly empowered by the government. Meanwhile, the existence of local NGO and community groups that were empowered by the government cannot be assured, as represented in second and third case studies. Their successful programmes were achieved only where there was access to funding. Due to this weakness, these organisations need to be more supported with policy both at local and national levels. Moreover, their role requires strengthening in local and national strategic plans so that their effectiveness can be enhanced in the future.

10.7 Conclusion

Local community groups (local NGOs, local traditional groups, women's and youth groups) in Aceh that were less significant before the 2004 tsunami became important in the post-disaster period, especially during the rehabilitation and reconstruction phase of the Aceh region. Banda Aceh, the capital of Aceh Province, received more attention from the agencies and organisations in terms of the aid distribution. This condition has not only encouraged the increasing number of new NGOs or local community groups but also expanded the areas of focus for NGOs into education, health, livelihood, environment, social-cultural and community development. In contrast, rural areas which were also affected by the tsunami remain isolated from aid due to the long armed conflict. As a result, local community groups in rural areas were also less empowered. Cross-border relationships which have been built during the post-tsunami period have capacitated both government and local community groups for managing disasters. The collaboration has also built resilience to future disaster through the development of a structured agenda and policies.

However, to achieve the most valuable contribution from these local groups, some issues need to be addressed to strengthen their role in the redevelopment process. The role of the local community groups in post-disaster recovery can be enhanced in several ways. One important approach is the development of internal capacity through the improvement of management and resources. Another crucial issue is the improvement of external capacity by integrating the programmes of the local community groups (local NGOs, local traditional groups, women's and youth groups) into government planning and action through collaboration. The partnership between local community groups and government agencies will enhance the success of development programmes. In addition, the improvement of the local community groups as organisations that are supported by the community will ensure the sustainability of the development since their active participation makes them active agents of development, not just passive recipients. This condition enables them to continue the development in future independently, especially after the INGOs and other outsider parties have left Aceh. In conclusion, the role local community groups (local NGOs, local traditional groups, women's and youth groups) in the disaster recovery effort is essential to foster the redevelopment and to ensure the sustainability of the development in future.

References

Bappenas. (2005a). *Indonesia: Preliminary damage and loss assessment, The December 26, 2004 natural disaster*. Jakarta: Bappenas (Ministry of National Development Planning).

Bappenas. (2005b). *Master plan for the rehabilitation and reconstruction of the regions and communities of the province of Nanggroe Aceh Darussalam and the Islands of Nias, Province of North Sumatra, main book*. Jakarta: Bappenas (Ministry of National Development Planning).

Bowden, P. (1990). NGOs in Asia: Issues in development. *Public Administration and Development, 10*(2), 141–152.

BRR (2005). *Aceh and Nias one tear after the tsunami the recovery effort and way forward*, A Joint Report of The BRR and International Partners. Jakarta.

BRR. (2009). *Housing, roofing the pillars of hope*. Jakarta: The Executing Agency of Rehabilitation and Reconstruction for Aceh and Nias.

Cousins, William (1991). Non-governmental initiatives. In *ADB, The urban poor and basic infrastructure services in Asia and the Pacific*. Manila: Asian Development Bank.

Cueto, R. M., Fernández, M. Z., Moll, S., & Rivera, G. (2015). Community participation and strengthening in a reconstruction context after a natural disaster. *Journal of Prevention & Intervention in the Community, 43*(4), 291–303.

Daly, P., & Rahmayati, Y. (2012). Cultural heritage and community recovery in post-tsunami Aceh. In P. Daly, R. M. Feener, & A. Reid (Eds.), *From the ground up, perspective on post-tsunami and post-conflict Aceh* (pp. 57–78). Singapore: ISEAS Publishing.

Davidson, C., Johnson, C. A., Lizarralde, G., Sliwinski, A., & Dikmen, N. (2007). Truths and myths about community participation in post-disaster housing projects. *Habitat International, 31*(1), 100–115.

Dombo, E., & Ahearn, F. (2015). The aftermath of humanitarian crises: A model for addressing social work interventions with individuals, groups, and communities. *Illness, Crisis & Loss, 23*(1), 1–20.

DTE. (2005). Community-centred reconstruction needed. *Down to Earth, 64*, 5–10.

Fernandez, G., & Shaw, R. (2014). Youth participation in disaster risk reduction through science clubs in the Philippines. *Disasters, 39*(2), 279–294.

ILO (2007). *International programme on the elimination of child labour, Aceh post-tsunami response*, Internal Project Report. Jakarta.

Lawther, P. M. (2009). Community involvement in post-disaster re-construction: Case study of the British Red Cross Maldives recovery program. *International Journal of Strategic Property Management, 13*(2), 153–169.

Lu, Y. (2014). NGO collaboration in community post-disaster reconstruction: Field research following the 2008 Wenchuan earthquake in China. *Disasters, 39*(2), 258–278.

Martens, K. (2002). Mission impossible? Defining nongovernmental organizations. *Voluntas: International Journal of Voluntary and Nonprofit Organizations, 13*(3), 271–285.

Méheux, K., Dominey-Howes, D., & Lloyd, K. (2010). Operational challenges to community participation in post-disaster damage assessments: Observation from Fiji. *Disasters, 34*(4), 1102–1122.

Miller, M & Bunnell, T (2013). Urban-rural connections: Banda Aceh through conflict, tsunami, and decentralization. In Bunnell, T., Parthasarathy, D. & Thompson, E.C. (eds.), *Cleavage, connection and conflict in rural, urban and contemporary Asia* (pp. 83–98), Springer.

Nakagawa, Y., & Shaw, R. (2004). Social capital: A missing link to disaster recovery. *International Journal of Mass Emergencies and Disasters, 22*(1), 5–34.

Newport, J. K., & Jawahar, G. G. P. (2003). Community participation and public awareness in disaster mitigation. *Disasters, 12*(1), 33–36.

Norris, F., Stevens, S., Pfefferbaum, B., Wyche, K., & Pfefferbaum, R. (2008). Community resilience as a metaphor, theory, set of capacities, and strategy for disaster readiness. *American Journal of Community Psychology, 41*(1), 127–150.

Pearce, L. (2003). Disaster management and community planning, and public participation: How to achieve sustainable hazard mitigation. *Natural Hazards, 28*(2–3), 211–228.

Pyles, L. (2007). Community organizing for post-disaster social development, locating social work. *International Social Work, 50*(3), 321–333.

Quarantelli, E. L. (1997). Ten criteria for evaluating the management of community disasters. *Disasters, 21*(1), 39–56.

Regulska, J. (1999). NGOs and their vulnerabilities during the time of transition: The case of Poland. *Voluntas: International Journal of Voluntary and Nonprofit Organizations, 10*(1), 61–71.

Tongsawate, M., & Tips, W. E. J. (1988). Coordination between government and voluntary organizations (NGOs) in Thailand's rural development. *Public Administration and Development, 8*(4), 401–420.

Werker, E., & Ahmed, F. Z. (2008). What do nongovernmental organizations do? *Journal of Economic Perspectives, 22*(2), 73–92.

Chapter 11
Cities as Aid Agencies? Preliminary Prospects and Cautionary Signposts from Post-Disaster Interurban Cooperation in Asia

Kristoffer B. Berse

11.1 Introduction

Cities are no strangers to 'extending care across space' (Clarke 2011). From the rise of sister and twin cities in the rubble of World War II[1] to the rapid growth of 'municipal internationalism' in the last quarter of the twentieth century (Ewen and Hebbert 2007) to the continuing global expansion of 'translocal' development processes (Zoomers and Van Western 2011), cities have not shied away from forming alliances and extending support to one another whether in times of need or simply in the spirit of solidarity and cooperation, sometimes in spite of the geo-political borders and boundaries imposed by nation-states.[2]

Buoyed by rapid advancement in information and communication technologies (Castells 1996), changes in international development cooperation (Hafteck 2003), and the rise of transnational city networks (Keiner and Kim 2007; Niederhafner 2013)[3] and city mayors with global 'catalytic influence' (Acuto 2013), local governments have stepped up and bonded together to address various urban issues such as the growing threats of climate change (Blok and Tschötschel 2015; Funfgeld 2015;

[1] In the US, 'sister cities' was formally established in 1956 as part of the 'citizen diplomacy' initiative of then President Eisenhower to heal the wounds of war, political and otherwise, through cultural and community-to-community exchanges (Zelinsky 1991). In Europe, earlier 'town twinning' arrangements among French, German and English cities were preoccupied not only with rebuilding war-torn localities but also preventing the escalation of another intra-European conflict (see, e.g., Brown 1998; Campbell 1987; Dogliani 2002; Gaspari 2002; Vion 2002; Weyreter 2003).

[2] For examples of how cities have come to challenge nation-states in such thorny issues as the apartheid in South Africa, nuclear disarmament, human rights, and the Sandinista war in Nicaragua (see Fry et al. 1989; Hobbs 1994; Kübler and Piliutyte 2007; Shuman 1994).

[3] Keiner and Kim (2007) call them transnational city networks for sustainability; others refer to transnational municipal networks (TMNs).

K.B. Berse (✉)
National College of Public Administration Governance, University of the Philippines, Quezon City (NCR), Philippines
e-mail: kbberse@up.edu.ph

M.A. Miller et al. (eds.), *Crossing Borders*, DOI 10.1007/978-981-10-6126-4_11

Kurniawan et al. 2013). For Keiner and Kim (2007, p. 1392–1393), the transnational networking of cities has evolved to the point of 'taking over the role of international organizations and national governments … to expedite and advance the mission of sustainability.'

The opportunity for cities to work together has been accompanied by the necessity to adapt to the twin pressures of resolving deep-seated urban problems on one hand, and of protecting development gains from external stresses and shocks on the other (Revi et al. 2014). Among Asian cities, the need to cooperate is not only spurred by the urgent need to arrest the vagrancies of vagabond capital amidst intense economic competition (Douglass 2002), but also by the exigency to address the risks posed by natural hazards and the impacts of climate change, as compounded by poverty, environmental degradation and ill-planned urbanisation (Miller and Douglass 2016; Douglass 2010).

Between 2004 and 2013, Asia has borne the brunt of disasters compared to other continents in the world: it was hit by the most number of disasters, lost the most number of people, and suffered the largest economic damage (IFRC,2014). In terms of weather-related disasters, seven of the top 10 countries with the most number of affected people from 1995 to 2015 are in Asia, resulting in absolute economic losses of US$709 billion (CRED and UNISDR 2015).

Unfortunately, this is not likely to change in the near future as a recent global risk assessment revealed that 56 out of the top 100 cities most exposed to natural hazards are located in the Philippines, China, Japan and Bangladesh (Verisk Maplecroft 2015).

In light of the foregoing, this chapter's thesis revolves around the notion of cities as providers and receivers of post-disaster aid in the context of international city-to-city cooperation (C2C), an evolving form of public-public partnership involving cities from two or more countries. Following this introduction, it lays down the theoretical intersection of both academic and grey literatures on decentralised development cooperation and decentralising disaster governance. It then describes the processes by which selected Asian cities extend help to other cities in times of disasters under the umbrella of a regional city network. This is followed by an identification of the salient characteristics of decentralised disaster aid. The succeeding section analyzes the prospects and challenges of cities as aid agencies, as juxtaposed against the contemporary issues faced by traditional external service agencies (ESAs).[4] The paper ends with a recapitulation of key findings, and a discussion of its implications on the study and practice of cross-border disaster governance.

The study utilizes as unit of analysis the C2C experience of selected cities that belong to CityNet. CityNet is a membership-based association of local authorities in the Asia-Pacific region established in 1987, arguably the first and oldest proponent of institution-based C2C cooperation in Asia. Currently based in Incheon, South Korea, CityNet is one of the most active environment-related inter-city net-

[4] This includes the whole network of global development organisations involved in development work, namely United Nations agencies, multilateral and bilateral financing institutions such as the World Bank and Asian Development Bank, and international humanitarian NGOs.

works in the region (Ishinabe 2010) and prides itself on explicitly promoting the engagement of cities from the so-called South (i.e. developing countries) through its Technical Cooperation among Cities of Developing Countries (TCDC) programme. Data for this study were collected mainly from CityNet publications, reports, and databases and supplemented by interviews with officials from CityNet and selected member cities (e.g. Bangkok and Yokohama) between 2010 and 2011.

11.2 Decentralised Cooperation Amidst Decentralising Disaster Governance

Douglass (2002) observed that there are at least four types of intercity cooperation that may take place at both sub-national and international levels (Table 11.1). The most common type involves information and cultural exchange among partner cities and is usually carried out as part of friendship or sister city agreements. The second type involves co-sharing of resources in the provision of common infrastructure and services. At the international level, this is best exemplified by the International Association of World's Ports and Harbors (IAPH) which is composed of some 200 leading ports in 90 countries (IAPH 2016).

The third type of inter-city cooperation is aimed at fostering economic integration among cities in the region. An example of this is the so-called Sijori Triangle which involves the intentional linking of industries in Singapore, Johor Baru in Malaysia, and Riau in Indonesia to promote 'complementation in production' (Douglass 2002). The Busan-Fukuoka transborder economic partnership is another case, wherein the ties between local authorities have considerably expanded from

Table 11.1 Types and levels of intercity cooperation

Type of cooperation	Sub-national	International
1. Information and cultural exchange/friendship	Japan local government centre	Sister cities Korea Local Authorities Foundation for International Relations (KLAFIR) (Korea)
2. Infrastructure services sharing/ rationalization	Metropolitan management coordination bodies (Jabotabek. Bangkok)	International port city management agreements, Tumen river basin
3. Economic integration	Village and township enterprises with metropolitan enterprises (China)	Trans-border regions such as the Sijori growth triangle
4. Political association/ integration	Metropolitan governments Consolidated cities (Korea)	European Union

Source: Adapted from Douglass (2002)

its inception in 2009 to one that now includes the active involvement of quasi-public bodies such as economic associations and education institutions (Park 2012). The Beijing-Seoul-Tokyo (BESETO) corridor is another example of an attempt to integrate the capital cities in East Asia (Keum 2000).

The last type involves political integration in the form of a 'league of cities' with independent and legitimate decision-making powers. This is usually difficult to achieve. As noted by Douglass (2002), intercity networks in Asia are still at the level of information exchange, co-production of knowledge through research, and light diplomacy. Intercity cooperation could do much more to enhance the capacity of local governments in such areas as (a) information and technical expertise, (b) policy development and decision-making, (c) institutional strengthening and human resources development, (d) managing change and using external support, and (e) policy implementation (UN-HABITAT and WACLAC 2003).

Among the variegated manifestations of cities' growing role in international affairs is the worldwide proliferation of C2C arrangements.[5] Building on Hafteck (2003), it is defined here as an evolving form of decentralised development cooperation characterized by an international public-public partnership between local authorities in two or more countries, with or without the support of other subnational actors, where the primary goal is to enhance the capacities of one or both partners through the sharing of resources and, where appropriate, adaptation of best practices, based on the principles of trust and equality.

This means that in terms of objectives, C2C is distinguishable from traditional sister or friendship city relationships whose primary aim is to merely foster community, people-to-people or cultural exchanges (de Villiers 2009; Campbell 1987). In a C2C programme, the goal is to address a concrete urban management concern through the sharing of resources and transfer of best practices if appropriate. In linking structure, it is international in nature and cooperation is therefore not predicated on borderlands or cross-border regimes as described in other studies. The fundamental intersections and differences between C2C and other forms of international municipal linking are shown in Fig. 11.1 below.

C2C has previously been cited as viable mechanism for strengthening urban governance (Bontenbal 2009); improving urban services (Hewitt 1999b), housing (Hewitt 1998), urban planning and waste management (Tjandradewi et al. 2006); promoting environmental education (Tjandradewi 2011); and even bridging the gap between local institutions and citizens (Hewitt 2004; Bontenbal and van Lindert 2008). In the context of disasters, domestic cooperation among cities is not new. When Typhoon Haiyan (local name: Yolanda) struck the Philippines in November 2013, among those who responded were local governments from unaffected regions of the country. The aid came in the form of relief goods, search and rescue personnel, cash, and equipment, among others (Saligumba 2013; Tesiorna 2015; GMA 2013). Some city governments have even come to 'adopt' the affected municipali-

[5] C2C has been called by different names in the literature, including municipal international cooperation, international municipal cooperation, municipal internationalism, interurban partnerships, and local government to local government partnership, among others (Berse and Asami 2010).

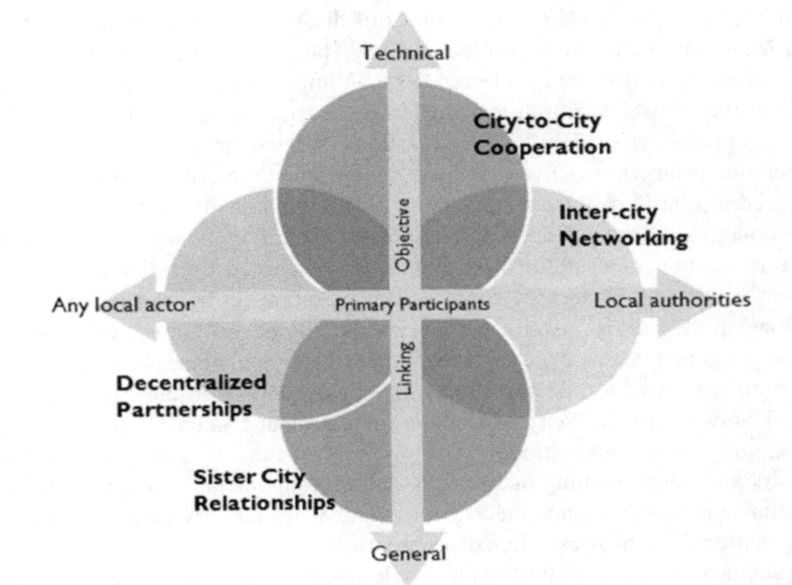

Fig. 11.1 International linking modalities involving sub-state and non-state actors (Source: Adapted from Berse 2013)

ties along the so-called Yolanda Corridor in the rehabilitation and recovery phase (Metro Manila Development Authority 2013).

The practice of post-disaster local cooperation is not surprising. Cities are in the front line of disaster response as many countries in the region go through varying levels of decentralisation (Miller and Bunnell 2012). In the Philippines, local governments are mandated to 'take the lead in preparing for, responding to, and recovering from the effects of any disaster' (PDRRM Act, Sec. 15).[6] The local Disaster Risk Reduction and Management units at all levels—provincial, city, municipal and *barangay*[7]—are vested with a plethora of responsibilities, among them the '[establishment of] linkage/network with other [local government units] for disaster risk reduction and emergency response purposes' (PDRRM Act, Sec. 12. c.20). However, the proactive approach to managing disaster risks as embodied in the law is relatively new to many cities and municipalities, with a good number of them having insufficient internal capacity to deal with disasters on their own (Commission on Audit 2014).

[6] The National Disaster Risk Reduction and Management Council through the Office of Civil Defense is only mandated by law to be in the frontline of disaster response when a disaster involves two or more regions.

[7] The *barangay* is the smallest political unit in the Philippines. At the provincial, city and municipal levels, the national law calls for the creation of Local Disaster Risk Reduction and Management Offices (LDRRMOs), while at the barangay level, a Barangay Disaster Risk Reduction and Management Committee is required.

On the other hand, aid-giving in times of disasters presents an opportunity for local leaders to widen their political capital. The media mileage that one can get from extending assistance is believed to be an important consideration in the decision whether to pay attention to an external disaster (Wisner and Gaillard 2009). This is supported by widely-held observations that domestic inter-local city-to-city cooperation is largely reactive (i.e. disaster-dependent) and ad hoc in nature, as they are based primarily on the personal and political relations of local leaders.

Internationally, the political motivations for extending aid are not that apparent. There is no immediate political gain that can be earned from helping a disaster-stricken city based in another country. The literature on twinning and sister city relationships highlight historical connections; shared economic, cultural, recreational, and ideological concerns; and similar or identical place names as important factors in the formation of partnerships (Zelinsky 1991). In the case of Japanese cities involved in C2C, reasons include environment-related factors (e.g. shared issues), degree of international orientation (e.g. the city's internationalization agenda), and other enabling factors (e.g. national support) (Nakamura et al. 2011). Unfortunately, these do not seem to prefigure why one city might help another, across national boundaries, when disaster strikes.

From the perspective of traditional development cooperation, the aid functions of external service agencies (ESAs)—United Nations organizations, multilateral and bilateral financial institutions, and international humanitarian NGOs—are quite well-established in their respective mandates. The same cannot be said of cities where managing disaster risks proactively, much less helping other cities at risk, is a relatively new responsibility that has yet to take root in local practice (Bang 2013; Bendimerad 2009).

Previous attempts to link international C2C cooperation and disaster governance have been very few. Sitinjak et al. (2014) explored how sister city relationships can be tapped to build the resilience of Jakarta and Bandung, banking on what they believed to be a prevailing 'culture of pragmatism' in Indonesia. Other studies have merely looked at the contributions of a city network's C2C programme in promoting resilience building (Tjandradewi and Berse 2011) and transferring best practices (Berse and Asami 2012). This chapter zeroes in specifically on the experience of selected cities in extending disaster aid to other cities in the context of a decentralizing Asia.

11.3 C2C for Aid: Selected Cases from CityNet

Between 1998 and 2009, CityNet has facilitated 66 cases of C2C among its members. Of these, five directly involved the city-to-city provision of aid in the aftermath of a disaster. As discussed below, these are (a) Bangkok and Yokohama (1997–2001); (b) Banda Aceh, Kuala Lumpur, Penang and Yokohama (2005–2008); (c) Galle, Moratuwa and Yokohama (2005–2007); (d) Islamabad, Muzaffarabad and

Yokohama (2006–2009); and (e) Khulna and Mumbai (2009). The evolution, nature and outcomes of these relationships are described in the succeeding sections.

11.3.1 Bangkok and Yokohama

The earliest CityNet-supported C2C tackling disaster risk took place between Yokohama and Bangkok in July 1997, following previous episodes of localized floods in the Thai capital. It initially involved a technical advisory service wherein the Japanese team from Yokohama's Sewage Works Bureau presented a simple yet innovative flood control method, which focuses on the development of reservoirs and 'retention ponds' for rainwater. The visiting delegation demonstrated to BMA officials how the system was developed and applied in Japan, particularly in the Tsurumi River Basin area around Yokohama, without necessarily disrupting the existing urban fabric.

For a week, Yokohama and Bangkok engineers worked together to establish the validity of the Japanese approach. This engagement eventually boosted the skills and confidence of the Thai engineers, who then went on to develop an application handbook and a computer program with Chulalongkorn University in order to assist architects and engineers in applying Yokohama's urban flood control system throughout Thailand (CityNet 1998). A team of Yokohama experts visited again in 1999 to provide further technical advice on flood prevention (CityNet 2001).

In February 2001, it was BMA's turn to send a delegation to Yokohama for a 6-day study visit. Bangkok wanted to enhance its understanding of the retarding and retention basin concept for flood prevention, and the use of natural methods for water treatment. The Bangkok team visited several sites including restaurants, schools, sports and recreational areas, and a construction site of a rainfall control pond, among others. BMA officials found the C2C experience very useful, particularly in terms of improving their own implementation of Monkey Cheek, a method of keeping rainwater for a while before allowing it be discharged gradually into the river, at a smaller scale—a challenge that they have grappled with ever since it was introduced in Thailand in 1996 (CityNet 2001).

11.3.2 Banda Aceh, Kuala Lumpur, Penang and Yokohama

Banda Aceh was one of the CityNet member cities affected by the 2004 Indian Ocean earthquake and tsunami, losing almost half of its officials in the disaster. Barely a month after the tsunami, Yokohama and YOKE (Yokohama Association for International Communications and Exchanges), both CityNet members, organized a donation drive to raise funds for affected areas in Sri Lanka and Indonesia (CityNet 2005a). In March 2005, Yokahama (through the CityNet Secretariat) delivered to

Banda Aceh a portion of the collected donation from Yokohama citizens, along with some equipment (e.g. a computer and facsimile machine) for the city hall.

Yokohama's efforts, however, did not stop at providing relief to the beleaguered Indonesian city.

From May to June 2005, it dispatched two urban planners to Banda Aceh—one from the Road Safety Facilities Division and another from the Urban Planning Division—to aid in rehabilitation and reconstruction efforts, along with a town planner and an architect from Penang, another city member from Malaysia.

Prior to this assignment, Yokohama and Penang had a long history of cooperation in such areas as urban design and solid waste management (Tjandradewi et al. 2006). This time around, they were joined by the Urban and Regional Development Institute (URDI) and the Municipal Association of Indonesia (AIM), both associate members of CityNet. Together with local experts from Banda Aceh, they chalked out initial plans for reconstruction, including the construction of a local fish market, a general town market (Pasar Setui), and a tsunami commemoration park (CityNet 2005b, d, e).

In October 2005, the Mayor of Yokahama visited Banda Aceh to 'make a tour of the tsunami-affected areas in Banda Aceh and review the progress made in reconstruction efforts' (CityNet 2005b). Of particular interest in this 3-day visit was a survey of the construction of Pasar Setui, which was being redesigned and rebuilt with partial support from funds pooled by Yokohama and other CityNet members (CityNet 2005c).

A second team comprised of two engineers from the Waterworks and Road and Highways Bureaus and one urban planner from Kuala Lumpur stayed in Banda Aceh from 15 January to 10 February 2006 (CityNet 2006a). This yielded tangible results, including a detailed plan for an integrated traffic management system featuring a mini-bus terminal, a proposal for the purification of natural water, and technical advice on housing reconstruction. The team also conducted a survey of the tourism potential of Banda Aceh, and drafted a proposal to realize this potential. The team of experts further submitted a proposal for a more detailed study to flesh out a long-term recovery strategy. Earlier, Yokohama Waterworks Bureau donated two sets of leak detectors and pack tests for iron and chlorine to Banda Aceh, to improve leak detection in the water distribution system (*ibid.*). Yokohama also invited Banda Aceh to participate in its annual training programme on waterworks management in September 2006 (CityNet 2006b).

Following these exchanges, the relationship between Yokohama and Banda Aceh evolved further with the signing of a Memorandum of Cooperation (MoC) between the two cities. Covering a 2-year period, January 2007–December 2008, the MoC was aimed at further strengthening the capacity of Banda Aceh in managing and operating its water supply—which was one of the city's critical concerns in the recovery process—and ultimately enhancing the cooperative relationship between Yokohama and Banda Aceh for the benefit of the entire communities in both cities. The cooperation assured Banda Aceh's continued participation in the training programme being administered by the Yokohama Waterworks Bureau (YWW) since 1998 (CityNet 2007b). As Japan's first city to have a modern water supply system, the programme was designed to transfer technical skills and knowledge in various aspects of waterworks operations, from the management of transmission pipes to maintenance of purification plants to leakage management, among others.

11.3.3 *Galle, Moratuwa and Yokohama*

In addition to support for Banda Aceh, cash donations from Yokohama citizens also reached affected cities in Sri Lanka. These were delivered through the CityNet Secretariat, which visited the municipalities of Colombo, Dehiwala Mt. Lavinia, Moratuwa and Negombo in March 2005. Most of the funds were meant for building community centres and public toilets, in collaboration with CityNet members SEVANATHA and HELP-O, two community-based organizations operating in Moratuwa and Galle, respectively (CityNet 2005b). Through CityNet, aid requirements were assessed and meetings with members, Mayors, community groups and potential partners were held to facilitate implementation of planned reconstruction projects (*ibid.*).

The community centre in Galle was completed by the end of 2006. It is a two-storey building which serves as a primary school for local children and vocational and computer training centre for the youth (CityNet 2006c). As stated earlier, HELP-O shared its experience in rebuilding the community with the help of funds from Yokahama and other CityNet members at a public forum held in Yokohama in February 2006 (CityNet 2006a).

Meanwhile, the Community Centre in Moratuwa, Sri Lanka, was completed a few months after. Moratuwa City and SEVANATHA, with financial aid from CityNet and Yokohama City, inaugurated it on 26 February 2007. Like in Galle, the centre is intended to give the community a safe venue for educational, cultural and economic activities (CityNet 2007a).

11.3.4 *Islamabad, Muzaffarabad and Yokohama*

While working with Banda Aceh in 2006, Yokohama also lent technical advisory services to Islamabad through the Capital Development Authority (CDA), in collaboration with the Pakistan Institute for Environment-Development Action Research (PIEDAR), one of CityNet's active associate members based in Pakistan. This was on the heels of a M7.6 earthquake that hit Pakistan on 8 October 2005, leaving between 74,000 and 86,000 people dead. Yokohama sent an architect and an urban planner in March 2006 to 'offer their advice and expertise to the local authorities.' They shared Yokohama's experiences in DRR and the Japanese Building Standards Law at a workshop organized by PIEDAR and CDA (CityNet 2006b). The two-man team also visited the heavily devastated towns of Muzaffarabad and Balakot to develop ideas for resilient reconstruction (CityNet 2006e).

Following this first mission, a site has been identified in the Sanwarian Village, Muzaffarabad, for building a six-classroom school and vocational training centre. Programmed as a CityNet project under the Disaster Cluster, the construction was to be implemented with CityNet managing and monitoring it offshore and PIEDAR closely coordinating with community members (CityNet 2006b). The project encountered some administrative difficulties, and it took 4 years (until December 2009) to finally complete the planned primary school in the area. Nevertheless, it has been estimated that the new classroom benefited 85 local children and their families (around 250 in total) in the village (CityNet 2009d).

In July 2006, Yokohama again sent two experts—one from the Safety Management Bureau and another from the Building and Repair Section of the Yokohama Waterworks Bureau—to visit Islamabad and the earthquake-affected Kot village in Muzaffarabad. They observed the capital's practices in building construction approval and inspection, earthquake resistance measures in the construction code, fire service preparedness, and disaster prevention capabilities. A workshop was also organised by PIEDAR and CDA for municipal officials and civil society groups. At the end of their visit, the experts presented a set of recommendations to Islamabad, including the setting up of a disaster information and communication centre, assignment of a 'Crisis Management General Administrator,' revamping of the fire service, promotion of community disaster prevention organisations, and disaster preparedness education in schools (CityNet 2006d).

A year later, on 19–30 August 2007, Yokohama sent a senior architect, who was also serving as CityNet's Assistant Secretary-General, to Islamabad to advise the CDA on its disaster prevention programme. The visit focused on the upgrading and extension of the 1960s Fire Headquarters Building, which was being equipped to effectively respond to disasters in the wake of the earthquake in October that year. The Yokohama expert also reviewed the plan and site for the school being built in Muzaffarabad by CityNet, through its NGO member, PIEDAR (CityNet 2007c).

11.3.5 Khulna and Mumbai

The cooperative relationship between Khulna (Bangladesh) and Mumbai (India) was a product of a devastating typhoon, Cyclone Aila, which wrought havoc along the Bay of Bengal on 25 May 2009. It triggered massive landslides, storm surge and flooding in 12 districts, especially Khulna and Shatkira, affecting about 4.8 million people and damaging more than half a million houses (Walton-Ellery 2009). Many of these communities were still recovering from the impacts of Cyclone Sidr, a much stronger typhoon that made landfall across 30 districts in Bangladesh in November 2007 (IFNET 2009).

In the aftermath of Aila, Mumbai and Dhaka through their respective CityNet Satellite Offices immediately worked together to come to the aid of Khulna. In close coordination with its counterpart in Dhaka, the CityNet Satellite Office in Mumbai, which was then directly under the direction of the Deputy Municipal Commissioner of Greater Mumbai, wasted no time in mobilising resources for the procurement of one of the identified needs at that time—water purification tablets. By 26 May, two million water purification tablets were airlifted to the beleaguered Bangladeshi city (CityNet 2009a).

The cooperation was short-lived and did not evolve into a formal and long-term cooperation between the two cities. There was no documentation on the beneficiaries and immediate outcome of the intervention either. Nevertheless, Mumbai's quick response and relief effort was later formally recognized by CityNet and Khulna city (CityNet 2009b, c).

11.4 Salient Characteristics of C2C-Based Post-Disaster Aid

The foregoing has discussed cases where cities have extended assistance to other cities in times of disasters, complementing traditional aid channels. The following section discusses the salient points that are present in all cases, particularly in terms of modality, duration, third-party involvement, role of city network, and geopolitical implications.

11.4.1 Beyond Relief

The five cases demonstrated the variety of activities undertaken by cities to help others that have been stricken by disaster. These range from the delivery of material aid for relief (e.g. cash and equipment) to providing technical assistance for disaster recovery (see Table 11.2). The partnerships, it must be noted, were not confined to the provision of traditional emergency relief, which can sometimes be messy and problematic (Oloruntoba 2005). In fact, in the case of Bangkok–Yokohama, the assistance did not involve relief at all.

There were also cases wherein capacity building became part of the arrangement as a way of continuing the transfer of knowledge and skills even when the relief phase is already over. These, however, were not planned from the beginning and developed only as a byproduct of the peer-to-peer exchanges, as in the case of Yokohama's extended partnership with Banda Aceh.

The cases were also able to bank on one of the key strengths of C2C cooperation, that is, peer-to-peer knowledge transfer (Hewitt 1999a) as a form of horizontal technical assistance. In-kind contributions through the dispatch of city experts before, during and after a disaster is built on peer-to-peer learning that is generally considered to be superior to the external consultant-driven interventions (Johnson and Wilson 2006) typical in many multilateral and bilateral aid agency arrangements. City officials and personnel have a better understanding of local conditions (especially if the participating cities share certain similarities), and interaction is believed to be more open and meaningful as there is no hierarchical relationship among the

Table 11.2 Summary of disaster-related C2C cases

Resource city	Beneficiary city	Other partners[a]	Activities	Years
Yokohama, Penang, Kuala Lumpur	Banda Aceh	AIM. URDI	Cash donation for market reconstruction (Pasar Setui), technical advice (traffic mgt, water purification), plan formulation, waterworks training	2005–2008
Yokohama	Moratuwa, Galle	SEVANATHA	Cash donation for construction of public toilets and community centers	2005–2007
Yokohama	Islamabad, Muzaffarabad	PIEDAR	Technical advice, school construction, knowledge sharing workshops	2006–2009
Mumbai	Khulna	CITYNET Satellite Offices in Mumbai and Dhaka	Donation of 2 million water purification tablets	2009
Yokohama	Bangkok	None	Training, technical advice, study visits, software development	1997,1999, 2001

[a]Aside from CITYNET Main HQ

participants. Nevertheless, it must be pointed out that knowledge transfers also have their own inherent challenges, among them the absorption rate or stickiness of the capacity building programme (Campbell 2009).

11.4.2 'Third-Party' Involvement

Moreover, the cases show that while most of the partnerships involved local authorities as primary actors, cooperation has been enriched by the participation of a third partner such as a non-government organisation (HELP-O, SEVANATHA, PIEDAR), a municipal association (AIM) or a research institute (URDI) (Fig. 11.2). Except for the Yokohama–Bangkok collaboration, where there was no external actor involved, these organisations provided management support to the partnership, ensuring that funds were expended prudently and activities undertaken as planned. They also provided technical inputs to the design and conduct of the programme. These partners

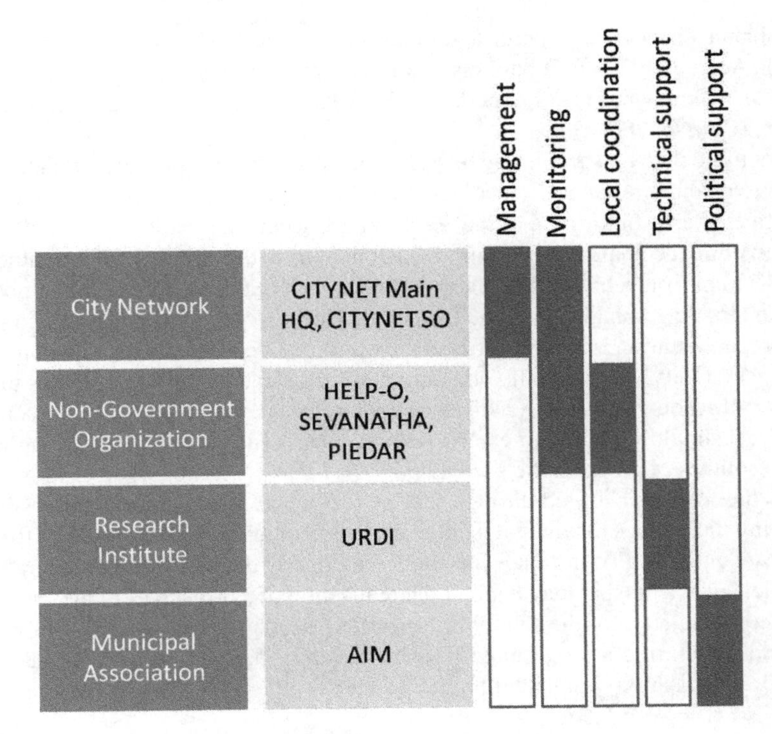

Fig. 11.2 Role of other actors in the delivery of post-disaster aid

were respected organizations in their areas of operation, known for their competence and credibility in variegated facets of disaster risk management.

Aside from their managerial and technical role, local partners were also critical in bridging the knowledge and cultural gaps among participating cities. This is particularly important in the provision of technical assistance, where knowledge of local conditions, social values and legal and institutional arrangements is fundamental. As reported by Yokohama experts dispatched to Banda Aceh:

> We had to understand the difference in the social set-up and infrastructure requirements from what we are used to in Japan. Once we got over that… we were able to deliver customized proposals and plans. (CityNet 2006a, p. 2)

11.4.3 Role of the City Network

Needless to say, CityNet played a pivotal role in gathering support for disaster-struck member cities. The city network usually issues a call for help to its wide membership immediately following a disaster. This was done through email, by putting up a dedicated webpage on its website, and even through the holding of fund-raising activities such as in the case of Yokohama. At a forum held a year after

Yokohama dispatched experts to Indonesia and Sri Lanka, representatives from Banda Aceh and HELP-O publicly acknowledged CityNet's part in facilitating the transfer of knowledge and expertise, which had benefited them in their recovery efforts (CityNet 2006a).

But more than just drumming up assistance, CityNet also provided a platform for cities to channel relief assistance without going or sending personnel right away to the affected city. In the case of the 2004 Indian Ocean disaster, CityNet took charge of delivering the financial and equipment donations to the affected cities in Indonesia and Sri Lanka on behalf of Yokohama. The CityNet Satellite Offices in Mumbai and Dhaka also acted on behalf of the local authorities in Mumbai and Dhaka to secure the procurement and delivery of needed water purification tablets to Khulna. The presence of CityNet regionally and nationally in selected countries ensures that the personnel resources of a city will not be unnecessarily drained for relief activities overseas, should it decide to extend help in times of disasters, thus leveraging the thin resources of cities on the giving end.

As the 'manager' of aid through C2C, the city network becomes responsible for ensuring that it goes to where it is intended, without taking the credit away from the donating cities. It also insulates the city from directly dealing with bureaucratic and technical problems that may hamper the realisation of a project, as in the case of the delayed completion of the school in Muzaffarabad, as the responsibility of managing and monitoring a programme falls in the lap of the city network and the local partner organisation designated for the project.

11.4.3.1 North-South and South-South Aid

CityNet takes pride in its South-South approach to C2C. In reality, however, such cases have been far and between and it is never truer than in the cases covered in this study where the provision of aid has been driven almost singlehandedly by Yokohama. The active participation of Yokohama may be attributed to the fact that it hosted the CityNet Secretariat at the time when the C2C interventions discussed above were implemented.[8] The Mayor of Yokohama him/herself was the ex-officio president of the network. Therefore, it can be said that the city itself was invested in the success of CityNet's flagship programmes such as the Technical Advisory Services (TAS) and TCDC, the association's main mechanisms for bilateral or even multilateral cooperation among members.

Yes, while this may be the case, it is also apparent that outside of Yokohama, other cities from developing countries have started to get involved in disaster-related aid as evidenced by the participation of Mumbai, Penang and Kuala Lumpur as resource cities. It can be argued, of course, that these cities are relatively better off than the recipient cities (i.e. Banda Aceh, Khulna), but it must be pointed out that in the history of C2C cases in CityNet, there have been cooperation programmes

[8] The Secretariat transferred to Incheon only in 2013.

involving cities of mutual socio-economic standing, albeit not necessarily in the area of disaster management (Berse 2013).

South-South cooperation or the presence of a similarly situated city in disaster-related C2C is important and necessary. For instance, the participation of experts from Malaysia, first from Penang and then from Kuala Lumpur, have been credited for ensuring that the team's proposals would be in line with the nuances and practices of a predominantly Muslim culture in Banda Aceh.

11.5 Cities as Aid Agencies: Prospects and Challenges

The five cases discussed here have illustrated the processes involved in the provision of post-disaster aid under the umbrella of network-facilitated C2C programmes. But how does this mechanism compare to traditional development aid? To shed some light on this matter, we refer to the four dimensions of 'best practice' in aid delivery as pointed out by Easterly and Pfutze (2008). According to them, an 'ideal' aid agency should have minimal overhead costs, less fragmented aid, effective delivery channels, and selective allocation in favor of deserving beneficiaries (i.e. less corrupt, more democratically free, and poor countries).

11.5.1 On Overhead Costs

Overhead costs appear to be minimal in the case of C2C-based aid. The administrative expenses of the 'resource' cities were mostly confined to the costs of sending local experts to 'beneficiary' cities (e.g. transportation and accommodation); the rest went directly to the substantive components of the programme or project. There are two reasons for this. First, the costs of administering the assistance were transferred to the city network as the 'manager' of the pooled funds. Other partners, whether on the giving or receiving end, were responsible for their own expenses. This arrangement leads to the second reason why overhead costs are minimised in a C2C setup: cities themselves are generally prohibited by law from paying for personnel other than their own.

To avoid potential legal and bureaucratic entanglements, official funds from local authorities had to be programmed as part of an official project with a partner organisation such as CityNet or directly with NGOs operating in the partner city. This was particularly true in the case of Yokohama's aid to the cities in Sri Lanka and Pakistan. Where this is not necessary, at least an official instrument had to be formally signed (i.e. MoC) between the two cities to provide the framework for continued engagement. In any event, the resource city still has to justify its own actions to its constituents, in the spirit of accountability and transparency. Absent this, external aid in times of disasters can easily become a thorny election issue that may hound the incumbent administration.

The biggest weakness of C2C-based assistance may not stem from overhead costs per se, as is typically the case for multilateral and UN organisations (Easterly and Williamson 2011), but in the overall availability of city funds for international cooperation. In Japan, it has been reported that local government budgets for C2C projects have been on gradual decline since 2002 (Yamazaki 2006). Even Yokohama has to raise donations from its own citizens through fund-raising activities; the city government could not just draw the money from its own coffers to donate to another city. The water supply equipment donated by Yokohama to Banda Aceh were also made possible with support from JICA.

It does not help that city networks like CityNet do not have a standby fund that can be instantly tapped to deliver aid in case a disaster strikes in one or more of its member cities. A quick response fund is not easy to establish given that CityNet is reliant primarily on financial and in-kind contributions from the host city, membership fees, and project funds from various partners. It has been reported that many of the C2C initiatives involving Japanese cities are usually subsidised by funding agencies, such as the Japan International Cooperation Agency (JICA) Grassroots Technical Cooperation Programmes, Council of Local Authorities for International Relations (CLAIR), Japan Fund for Global Environment (JFGE) and Global Environment Centre Foundation (GEC), among others (Maeda 2012). The development of the cooperative relationships was almost instantaneous, if not reactive.

11.5.2 On Delivery Channels

As can be gleaned from the cases, the cities themselves were involved not only as donors but also as aid agencies directly dispensing relief aid and providing technical assistance. This cuts the need for an intermediary aid agency in the traditional sense (e.g. INGOs) as the cities themselves are involved in the delivery of assistance, whether in the form of material aid or technical advice. Cities themselves are not lacking for expertise and experience in managing disasters before and after they happen, including dealing with the impacts of climate change (Blok and Tschötschel 2015).

The use of their own expertise has also helped cities avoid another problem associated with aid agencies: the overreliance on expensive external experts in technical assistance projects (IMF and World Bank 2006). This is where C2C arrangement stands out, as opposed to other delivery mechanisms. As pointed out earlier, the technical assistance in all cases was conducted utilizing local experts that were paid for by their respective organisations, either as part of a wider CityNet effort or as a component of an externally-funded project. The peer-to-peer learning process in C2C cooperation is also considered to be more promising than hierarchical and consultant-driven North-South programmes (Johnson and Wilson 2006).

Moreover, decentralised aid provides for a closer citizen-to-citizen connection, possibly bridging what human geographers call as the problem of 'care-at-a-distance' (Silk 2007)—generally, the absence of care out of ignorance—as it raises the awareness of citizens from resource cities about the plight of the 'distant strangers' (Corbridge 1993) in beneficiary cities. As demonstrated in the case of Yokohama, citizens were oriented to the situation of the affected cities in Sri Lanka and Indonesia through public advocacy campaigns, as a way of encouraging them to directly make donations to the city government instead of through the traditional humanitarian channels (e.g. international NGOs), where their contributions become miniscule in the overall scheme of things and are not easy to track. In turn, the money from Yokohama's citizens was pooled together and used to augment in-kind contributions from the city government to produce tangible outputs: community centres in Galle and Moratuwa, a primary school in Muzzafarabad, and a local market in Banda Aceh.

A major challenge in the delivery set-up of C2C-based aid is the issue of monitoring and evaluation. It is true that progress in Moratuwa, Galle and Banda Aceh was communicated to the public in the early days of the projects through a public forum and other activities organised by Yokohama and CityNet, and that NGO partners such as HELP-O and Sevanatha did have the responsibility to ensure the completion of the community centres in Sri Lanka as planned. However, beyond these, CityNet does not have a mechanism or strategy to monitor the status and measure the impacts of the assistance over the long run.

This is further complicated by the fact that in reality, CityNet, Yokohama and the other resource partners really have no leverage to demand accountability in case the C2C interventions get delayed, as in the case of the school construction in Muzzafarabad, or worse, discontinued to give way to other local priorities, as in the case of the demolition of the public market in Banda Aceh where most of Yokohama's financial aid went to. The local partners can only do so much, as they may be constricted by realities on the ground.

11.5.3 On the Fragmentation of Aid

Another issue that has beleaguered aid agencies is the fragmentation of aid or the extent to which aid is divided among many donors, countries, and sectors. As noted by Easterly and Pfutze (2008, p. 10), 'having multiple donors and multiple projects forfeits the gains of specialization and leads to higher-than-necessary overhead costs for both donors and recipients.' Problems attendant to fragmentation are not isolated, as reported in reports from the IMF and World Bank (2006) and the Commission for Africa (2005).

The multiplicity of donors, countries and sectors—which is characteristic of many aid agencies (Easterly and Pfutze 2008)—was minimized, if not avoided, in the studied cases. The contributions from the participating resource cities were pooled together under the umbrella of CityNet, thereby removing unnecessary redundancies in the coordination of both technical assistance and material aid. Even the one-time donation from Mumbai to Khulna was facilitated by CityNet's offices in Mumbai and Dhaka, and did not require additional resources to happen.

The types of 'relief' that cities provided were also more specific. Aside from the actual donations from its citizens, Yokohama has been consistent and realistic in offering its expertise in urban planning and water resource management in the context of post-disaster rehabilitation and recovery. The C2C cases also did away with food aid, which is generally considered to be less effective (IMF and World Bank 2006).

Moreover, the aid from Yokohama and Mumbai came with 'no strings attached.' It must be noted that one of the oft-criticized aspects of traditional development aid is the so-called 'tied' aid that requires the recipient country to exclusively purchase goods and services from the aid-granting country (Commission for Africa 2005). It must be noted that the quid pro quo approach in traditional development aid has given rise to the 'relief-and-reconstruction complex' (Bello 2006; Tzifakis and Huliaras 2015) that promotes 'disaster capitalism' (Klein 2007) at the expense of strengthening local resilience.

If anything, what weakens C2C-based aid is the lack of enough participants, either as aid agencies (i.e. involved in the provision of material aid or technical assistance) or simply as donors. Even for a well-established association like CityNet, the participation of members in C2C in general has been confined to a limited number of Northern and some Southern partners.

Within CityNet, the participation of small and intermediate cities, in particular, has been lacking (Tjandradewi and Berse 2011). Many of these cities have been constrained by financial means to participate in C2C, an unfortunate circumstance given that some of them may not be economically rich but have a lot to offer in terms of expertise when it comes to dealing with disasters. Within CityNet, for example, there are other cities from the Philippines, Vietnam and elsewhere that have also demonstrated competence in various aspects of disaster risk management, but remained untapped as resource cities in times of disasters.

11.5.4 On the Selection of Beneficiaries

The selection of beneficiaries is likewise less politicised compared to the case of traditional ESAs. The diversion of money to questionable regimes is avoided as the provision of aid is based on a call for help from fellow member cities that were not totally unknown to the city network, resource cities and supporting organizations. The involvement of Yokohama, Penang, Kuala Lumpur and Mumbai in offshore post-disaster assistance was triggered by the occurrence of actual disasters, while the relationship between Bangkok and Yokohama was prompted by perennial localized flooding episodes in Thailand's capital.

The role of the local partners has proved helpful in identifying needs and setting expectations. In Sri Lanka, HELP-O and Sevanatha acted as the bridge between the people and the local governments of Galle and Moratuwa, respectively. In Pakistan, PIEDAR closely coordinated with the community to ensure that the funds for the construction of the school in Muzafarrabad would not go to waste.

A need-based approach to beneficiary selection is good for as long as CityNet is dealing with only one or a few cities in need of assistance. A problem may arise when multiple cities get struck by a disaster at the same time. Without clear guidelines on how to prioritise help, the identification of cities on the receiving end then becomes a political process. In such a case, it would not be far-fetched to speculate that cities which do not have substantial influence within the network, for one reason or another, may be further marginalised, if not neglected.

11.6 Conclusions and Way Forward

Discourses on international disaster aid often focus on the role of UN agencies, multilateral and bilateral financing institutions, and international humanitarian NGOs—collectively called 'external service agencies' or ESAs (Atkinson 2001)—rarely looking at the contributions of cities themselves. This 'neglect' of course is not particularly confined to the praxis of disaster governance alone; cities after all, as one author puts it, are the 'invisible gorillas' of international studies (Acuto 2010).

The cases presented here demonstrate how international city-to-city cooperation can be a viable mechanism for 'globalising care' (Clarke 2011) through the institutionalised horizontal diffusion of aid in disaster-stricken areas. As it involves much fewer resources and more direct interaction between the source and recipient of assistance, it is able to minimize, if not avoid, some of the major issues that are thought to be hampering traditional aid channels. In this context, the role of city networks as an apolitical 'go-between' is critical as they pave the way for the participation of sub-State and non-State actors from both North and South.

The study does not by any means suggest that cities should supersede the role of multilateral, bilateral and UN aid agencies in international development cooperation. As can been from the small sample of cases from CityNet, C2C-based decentralised aid is not without its weaknesses. Foremost among these is the need to come up with a functional monitoring and evaluation mechanism to ensure accountability and sustainability and measure impact of aid interventions. The issues of limited resources and lack of policy guidelines for aid are equally important.

The academic literature linking C2C and disaster governance is fairly nascent. This study hoped to provide a first-cut analysis of the potential of cities as disaster aid agencies under the rubric of network-based intercity partnerships. Future areas for research are therefore wide open.

For one, the experience of CityNet cities in C2C-based aid has shown that post-disaster cooperation is possible even among localities that do not share a border. This brings to home the artificiality of geopolitical borders and boundaries in 'geographies of care' (Lawson 2007). How cities help each other across space and time in the aftermath of disasters would make for an interesting study.

A related trajectory of inquiry is the notion of 'borderless' disasters and 'borderless' disaster governance. Recent large and sudden-onset disasters, such as the 2015

Hindu Kush earthquake and 2004 Indian Ocean earthquake and tsunami, have affected multiple localities in different countries with disproportionate impacts on different societies. Given the wide-ranging implications of such events, it may be time to start talking as well about 'borderless' disaster governance in both practical and academic contexts. Possible interventions that could strengthen local resilience include redundant tsunami early warning systems and interurban solidarity funds as part of a sub-national cooperation regime. The post-disaster cooperation of cities over and above the foreign relations of nation-states is another area for possible research (e.g. Bangkok-Phnom Penh relations, Taipei's deployment to mainland China during the 2008 Sichuan earthquake).

From a policy mobilities perspective, it would be interesting to see how disasters as 'focusing events' (Birkland 2006) accelerate the establishment of ties among cities and eventually forge cross-border policies. How can we harness such ties to effect wider cooperation among sub-State and non-State actors or even foster new cross-border disaster governance regimes? What would be its potential role in 'adopting' climate refugees?

The practice of C2C and its particular application as a framework for cooperation in times of disasters is not new and certainly not confined only to transnational partnerships. Thus, it will be equally interesting to see how cities from one country try to help each other before, during and after a disaster. Recent events in the Philippines have already demonstrated the burgeoning of this type of relationship, and with the continuing challenges brought about by unabated 'maldevelopment' in the middle of a changing climate, we can expect a continued growth of this cooperative practice in the future.

References

Acuto, M. (2010). Global cities: Gorillas in our midst. *Alternatives, 35*, 425–448.

Acuto, M. (2013). City leadership in global governance. *Global Governance, 19*(3), 481–498.

Atkinson, A. (2001). International cooperation in pursuit of sustainable cities. *Development in Practice, 11*(2–3), 273–291.

Bang, H. N. (2013). Governance of disaster risk reduction in Cameroon: The need to empower local government. *Jàmbá: Journal of Disaster Risk Studies, 5*(2), 1–10.

Bello, W. (2006). The rise of the relief-and- reconstruction complex. *Journal of International Affairs, 59*(2), 281–296.

Bendimerad, F. (2009). *Open-file-report on the state-of-the-practice of urban disaster risk management*. Geneva: Earthquakes and Megacities Initiative (EMI).

Berse, K (2013). *An analysis of city-to-city cooperation in Asia: CityNet experience and the readiness of Philippine cities*. Unpublished doctoral dissertation. Tokyo: The University of Tokyo.

Berse, K., & Asami, Y. (2010). An idea in good currency? Terminologies and spatio-temporal trends in city-to-city cooperation research. In *Proceedings of the Conference of Asian City Planning, 8th Conference of Asian City Planning* (pp. 119–128). Tokyo: The University of Tokyo.

Berse, K., & Asami, Y. (2012). From Yokohama with love: Transferring best practices through international municipal cooperation. In Y. Nishimura & C. Dimmer (Eds.), *Planning for*

sustainable Asian cities, APSA 2011 Selected Papers (pp. 190–201). Tokyo: Asian Planning Schools Association.
Birkland, T. (2006). *Lessons of disaster: Policy change after catastrophic events*. Washington, DC: Georgetown University Press.
Blok, A., & Tschötschel, R. (2015). World port cities as cosmopolitan risk community: Mapping urban climate policy experiments in Europe and East Asia. *Environment and Planning C: Government & Policy, XX*, 1–20.
Bontenbal, M. (2009). Strengthening urban governance in the South through city-to-city cooperation: Towards an analytical framework. *Habitat International, 33*(2), 181–189.
Bontenbal, M., & van Lindert, P. (2008). Bridging local institutions and civil society in Latin America: Can city-to-city cooperation make a difference? *Environment and Urbanization, 20*(2), 465–481.
Brown, M. (1998). Towns that build bridges. *History Today, 48*(8), 3–6.
Campbell, E. S. (1987). The ideals and origins of the Franco-German sister cities movement, 1945–70. *History of European Ideas, 8*(1), 77–95.
Campbell, T. (2009). Learning cities: Knowledge, capacity and competitiveness. *Habitat International, 33*(2), 195–201.
Castells, M. (1996). *The rise of the network society*. Cambridge, MA: Blackwell.
CityNet. (1998). *Guidelines for transferring effective practices: A practical manual for South-South cooperation*. Yokohama: CityNet.
CityNet. (2001). Monkey cheek in city: Lessons from Yokohama. *CityVoice, 10*(22), 3.
CityNet. (2005a). Fighting the ripples: Rebuilding after the tsunami. *CityVoice, 14*(33), 3.
CityNet (2005b). *e-News*, (20), March–April.
CityNet (2005c) *e-News*, (21), May–June.
CityNet (2005d). *e-News*, (23), September–October.
CityNet (2006a). *e-News*, (25), January–February.
CityNet (2006b). *e-News*, (26), March–April.
CityNet (2006c). *e-News*, (30), November–December.
CityNet. (2006d). Experts visit Islamabad. *CityVoice, 15*(36), 2.
CityNet. (2006e). Untitled. *CityVoice, 15*(35), 2.
CityNet (2007a). *e-News*, (31), January–February.
CityNet (2007b). e-*News*, (32), March–April.
CityNet (2007c). *e-News*, (34), July–August.
CityNet (2009a). *e-News*, (45), May–June.
CityNet. (2009b). Mumbai: Extensive contribution to Khulna City, Bangladesh. *CityVoice, 18*(45), 2.
CityNet (2009c). *e-News*, (46), July–August.
CityNet (2009d). *e-News*, (48), November–December.
Clarke, N. (2011). Globalising care? Town twinning in Britain since 1945. *Geoforum, 42*(1), 115–125.
Commission for Africa (2005). *Our common interest*. Report of the Commission for Africa.
Commission on Audit (2014). *Assessment of disaster risk reduction and management at the local level*.
Corbridge, S. (1993). Marxisms, modernities and moralities: Development praxis and the claims of distant strangers. *Environment and Planning D: Society and Space, 11*(4), 449–472.
CRED & UNISDR (2015) *The human cost of weather-related disasters, 1995–2915*. Geneva.
de Villiers, J. C. (2009). Success factors and the city-to-city partnership management process – from strategy to alliance capability. *Habitat International, 33*, 149–156.
Dogliani, P. (2002). European municipalism in the first half of the twentieth century: The socialist network. *Contemporary European History, 11*(4), 573–596.
Douglass, M. (2002). From global intercity competition to cooperation for livable cities and economic resilience in Pacific Asia. *Environment and Urbanization, 14*(1), 53–68.

Douglass, M. (2010). Globalization, mega-projects and the environment: Urban form and water in Jakarta. *Environment and Urbanization ASIA, 1*(1), 45–65.

Easterly, W., & Pfutze, T. (2008). Where does the money go? Best and worst practices in foreign aid. *Journal of Economic Perspectives, 22*(2), 29–52.

Easterly, W., & Williamson, C. R. (2011). Rhetoric versus reality: The best and worst of aid agency practices. *World Development, 39*(11), 1930–1949.

Ewen, S., & Hebbert, M. (2007). European cities in a networked world during the long twentieth-century. *Environment and Planning C: Government & Policy, 25*(3), 327–340.

Fry, E. H., Radebaugh, L. H., & Soldatos, P. (1989). *The new international cities era: The global activities of North American municipal governments*. Provo: Kennedy Center for International Studies, Brigham Young University.

Funfgeld, H. (2015). Facilitating local climate change adaptation through transnational municipal networks. *Current Opinion in Environmental Sustainability, 12*, 67–73.

Gaspari, O. (2002). Cities against states? Hopes, dreams and shortcomings of the European municipal movement, 1900–1960. *Contemporary European History, 11*(4), 597–621.

GMA (2013). Cebu City offers to send team to help identify Yolanda dead in Tacloban. *GMA News*, 21 November, viewed 12 February 2016, http://www.gmanetwork.com/news/story/336398/news/regions/cebu-city-offers-to-send-team-to-help-identify-yolanda-dead-in-tacloban.

Hafteck, P. (2003). An introduction to decentralized cooperation: Definitions, origins and conceptual mapping. *Public Administration and Development, 23*(4), 333–345.

Hewitt, W. E. (1998). The role of international municipal cooperation in housing the developing world's urban poor: The Toronto-Sao Paulo example. *Habitat International, 22*(4), 411–427.

Hewitt, W. E. (1999a). Municipalities and the 'new' internationalism: Cautionary notes from Canada. *Cities, 16*(6), 435–444.

Hewitt, W. E. (1999b). Cities working together to improve urban services in developing areas: The Toronto-Sao Paulo example. *Studies in Comparative International Development, 34*(1), 27–44.

Hewitt, W. E. (2004). Improving citizen participation in local government in Latin America through international cooperation: A case study. *Development in Practice, 14*(5), 619–632.

Hobbs, H. (1994). *City hall goes abroad: The foreign policy of local politics*. Thousand Oaks: Sage Publications.

IAPH (2016). *Data base of IAPH member ports*. Viewed 28 February 2016, http://www.iaph-worldports.org/data-base-of-iaph-member-ports.

IFNET (2009). *Flood information: Impact of cyclone AILA*. Viewed 8 June 2016, http://www.internationalfloodnetwork.org/aila.htm.

IFRC. (2014). *World disasters report 2014*. Geneva: International Federation of Red Cross and Red Crescent Societies. viewed 29 April 2016, http://www.ifrc.org/Global/Documents/Secretariat/201410/WDR%202014.pdf.

IMF & World Bank. (2006). *Global monitoring report 2006: Strengthening mutual accountability: Aid, trade, and governance*. Washington, DC: World Bank.

Ishinabe, N. (2010). *Analysis of international city-to-city cooperation and intercity networks For Japanese National and Local Governments*. Kitakyushu: Institute for Global Environmental Strategies (IGES).

Johnson, H., & Wilson, G. (2006). North–South/South–North partnerships: Closing the 'mutuality gap'. *Public Administration and Development, 26*(1), 71–80.

Keiner, M., & Kim, A. (2007). Transnational city networks for sustainability. *European Planning Studies, 15*(10), 1369–1395.

Keum, H. (2000). Globalization and inter-city cooperation in Northeast Asia. *East Asia, 18*(2), 97–114.

Klein, N. (2007). Blanking the beach: The second tsunami. In *The shock doctrine*. New York: Metropolitan Books.

Kübler, D., & Piliutyte, J. (2007). Intergovernmental relations and international urban strategies: Constraints and opportunities in multilevel polities. *Environment and Planning C: Government & Policy, 25*(3), 357–373.

Kurniawan, T. A., de Oliveira, J. P., Premakumara, G. J., & Nagaishi, M. (2013). City-to-city level cooperation for generating urban co-benefits: The case of technological cooperation in the waste sector between Surabaya (Indonesia) and Kitakyushu (Japan). *Journal of Cleaner Production, 58*, 43–50.

Lawson, V. (2007). Geographies of care and responsibility. *Annals of the Association of American Geographers, 97*(1), 1–11.

Maeda, T. (2012). Networking cities for better environmental management: How networking functions can enhance local initiatives. In *Institute for Global Environmental Strategies (IGES)* (pp. 137–158). Greening Governance in Asia-Pacific, IGES White Paper IV. Hayama: IGES.

Metro Manila Development Authority (2013). MDA bares final list of Metro Manila typhoon 'Yolanda' adopt-a-town program. *MMDA*, 20 November, viewed 15 May 2016, http://www.mmda.gov.ph/index.php/11-news/24-mmda-bares-final-list-of-metro-manila-typhoon-yolanda-adopt-a-town-program.

Miller, M. A., & Bunnell, T. (2012). Asian cities in an era of decentralisation. *Space and Polity, 16*(1), 1–6.

Miller, M. A., & Douglass, M. (2016). Decentralising disaster governance in urbanising Asia. *Habitat International, 52*, 1–4.

Nakamura, H., Elder, M., & Mori, H. (2011). The surprising role of local governments in international environmental cooperation: The case of Japanese collaboration with developing countries. *Journal of Environment & Development, 20*(3), 219–250.

Niederhafner, S. (2013). Comparing functions of transnational city networks in Europe and Asia. *Asia Europe Journal, 11*(4), 377–396.

Oloruntoba, R. (2005). A wave of destruction and the waves of relief: Issues, challenges and strategies. *Disaster Prevention and Management: An International Journal, 14*(4), 506–521.

Park, S.H. (2012). Post-cold war trans-border networks in Northeast Asia: The Busan-Fukuoka network, *ARI Working Paper*. National University of Singapore – Asia Research Institute, 193, p 1214.

Revi, A., Satterthwaite, D. E., Aragón-Durand, F., Corfee-Morlot, J., Kiunsi, R. B. R., Pelling, M., Roberts, D. C., & Solecki, W. (2014). Urban areas. In C. B. Field, V. R. Barros, D. J. Dokken, K. J. Mach, M. D. Mastrandrea, T. E. Bilir, M. Chatterjee, K. L. Ebi, Y. O. Estrada, R. C. Genova, B. Girma, E. S. Kissel, A. N. Levy, S. MacCracken, P. R. Mastrandrea, & L. L. White (Eds.), *Climate change 2014: Impacts, adaptation, and vulnerability. Part A: Global and sectoral aspects, Contribution of Working Group II to the Fifth Assessment Report of the Intergovernmental Panel on Climate Change* (pp. 535–612). Cambridge, UK: Cambridge University Press.

Saligumba, J.R.L. (2013). Duterte rejoins rescue, medical team to Tacloban. *Davao Today*, 11 November, viewed 12 February 2016, http://davaotoday.com/main/headline/duterte-rejoins-rescue-medical-team-to-tacloban.

Shuman, M. (1994). *Towards a global village: International community development initiatives*. London: Pluto Press.

Silk, J. (2007). Caring at a distance: (Im)partiality, moral motivation and the ethics of representation—introduction. *Ethics, Place and Environment, 3*(3), 303–322.

Sitinjak, E., Sagala, S., & Rianawati, E. (2014). *Opportunity for sister city application to support resilience city, Resilience Development Initiative (RDI) Working Paper Series* (Vol. 8). Bandung: RDI.

Tesiorna, B.O. (2015). Leyte solon recognizes Davao City's help during Yolanda. *CNN Philippines*, August 25, viewed 14 June 2016, http://cnnphilippines.com/regional/2015/08/25/Leyte-solon-recognizes-Davao-Citys-help-during-Yolanda.html.

Tjandradewi, B.I. (2011). City-to-city cooperation for environmental education: Initiatives implemented by CityNet. In *Lessons learnt from regional intercity networking to promote sustainable cities in Asia* (pp. 49–77). Kitakyushu: IGES.

Tjandradewi, B. I., & Berse, K. (2011). Building local government resilience through city-to-city cooperation. In R. Shaw & A. Sharma (Eds.), *Climate and disaster resilience in cities* (Vol. 6, pp. 203–224). Bingley: Emerald Publishers.

Tjandradewi, B. I., Marcotullio, P. J., & Kidokoro, T. (2006). Evaluating city-to-city cooperation: A case study of the Penang and Yokohama experience. *Habitat International, 30*(3), 357–376.

Tzifakis, N., & Huliaras, A. (2015). The perils of outsourcing post-conflict reconstruction: Donor countries, international NGOs and private military and security companies. *Conflict, Security and Development, 15*(1), 51–73.

UN-Habitat, & WACLAC. (2003). *Partnership for local capacity development: Building on the experiences of city-to-city cooperation*. Nairobi: UN-Habitat.

Verisk Maplecroft (2015). *Natural hazards risk atlas 2015*. Viewed 29 April 2016, http://maplecroft.com/portfolio/new-analysis/2015/03/04/56-100-cities-most-exposed-natural-hazards-found-key-economies-philippines-japan-china-bangladesh-verisk-maplecroft/.

Vion, A. (2002). Europe from the bottom up: Town twinning in France during the Cold War. *Contemporary European History, 11*(4), 623–640.

Walton-Ellery, S. (2009). *A review of the cyclone Aila response 2009*. IFRC-led Emergency Shelter Coordination Group. Viewed 8 June 2016, http://goo.gl/vO355u.

Weyreter, M. (2003). Germany and the town twinning movement. *Contemporary Review, 281*(1644), 37–43.

Wisner, B., & Gaillard, J. C. (2009). An introduction to neglected disasters. *Jàmbá: Journal of Disaster Risk Studies, 2*(3), 151–158.

Yamazaki, K. (2006). Chihou kokusaika ni okeru kokusai kyouryoku no genjou to kadai (Current state and issues of international cooperation in regional internationalization). *Jichitai Kokusaika Foramu (Local Governmental International Forum), 199*, 1–4.

Zelinsky, W. (1991). The twinning of the world: Sister cities in geographic and historical perspective. *Annals of the Association of American Geographers, 81*(1), 1–31.

Zoomers, A., & van Western, G. (2011). Introduction: Translocal development corridors and development chains. *International Development Planning Review, 33*(1), 377–388.

Part IV
Cross-Border Disasters and Conflict Potential

Chapter 12
Governing Cross-Border Ecology, Hazards and Population Movement: Narratives and Counter-Narratives from India and Bangladesh

Sarfaraz Alam

12.1 Introduction

> While the rivers of Assam flow into East Pakistan [Bangladesh], the explosive population of East Pakistan [Bangladesh] has a tendency to flow upstream into Assam. (Pakyntein 1964, p. 134)

Nation-states as spatial entities are separated from each other by neatly drawn territorial boundaries. In most cases, political boundaries are drawn on ethnic/religious/linguistic bases regardless of ecological characteristics that geographically contiguous states often share. In other words, political boundaries of nation-states often ignore their bio-physical boundaries. Functionally, political boundaries help to either legitimise or delegitimise certain kinds of movements between states. Problems emerge when nation-states seek to control or modify flows of transboundary resources at the cost of the rights of the neighbouring states—for instance, by managing rivers using water diversion and storage structures. Inter-state conflicts also arise when activities in one state—deforestation, discharge of industrial pollutants and urban sewage into inter-state rivers, etc.—start causing problems in the neighbouring state in the form of compound disasters (e.g. flood, drought, salinisation, sedimentation, water pollution, etc.) leading to destruction of habitats and loss of livelihoods. In such a scenario, the poor and the marginalised sections of the society tend to abandon their homes in search of new livelihood opportunities and living spaces. In cases of scarce opportunities and lack of living space in their home state, such people tend to cross over into relatively prosperous neighbouring states. In this way, in response to compound disasters caused by activities in the upstream state, there may be a counter-movement of population from the lower riparian state to the upstream state. This fact clearly reinforces a direct cause-and-effect

S. Alam (✉)
Department of Geography, Institute of Science, Banaras Hindu University, Varanasi, UP, 221005, India
e-mail: sarfarazalam05@gmail.com

M.A. Miller et al. (eds.), *Crossing Borders*, DOI 10.1007/978-981-10-6126-4_12

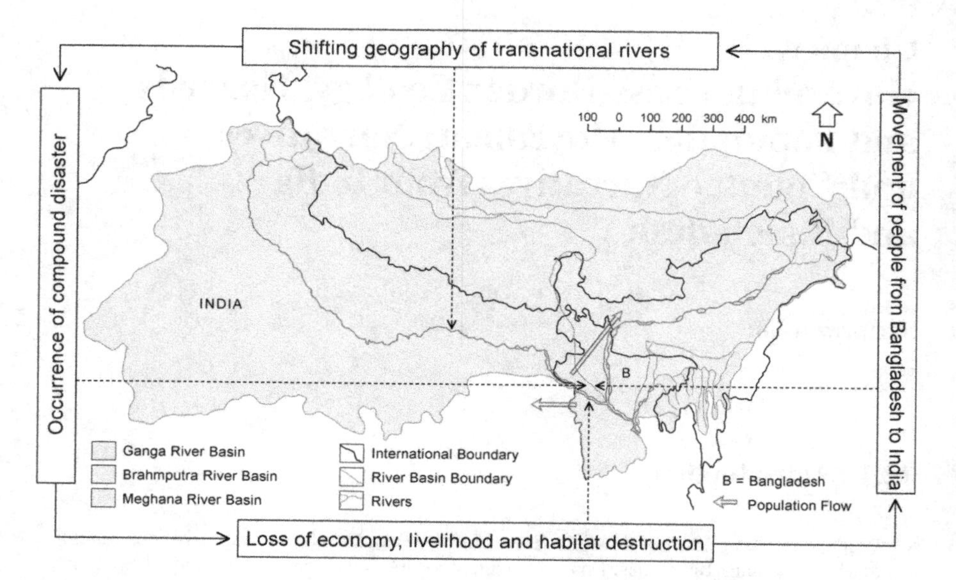

Fig. 12.1 Human-induced changes in transnational rivers and associated environmental disaster and population movement

relationship between the two kinds of cross-border movements. The relationship between cross-border flows of rivers, compound disasters and counter-movements of population can be explained in terms of the discordance between ecological and political boundaries of the neighbouring states (Fig. 12.1).

In this context, it is important to note that the poor and desperate people displaced by compound disasters are found to cross international boundary to clandestinely settle in anonymous urban environments to avoid detection. Within the urban setting, they settle in hazard-prone sites such as low-lying flood-prone areas, near dumping yards or in the river bed. The cross-border problems between India and Bangladesh continue to persist as both the countries have tend to approach at these essentially interrelated issues in isolation from each other and also from their respective narrow 'nationalistic' perspectives.

In this backdrop, the chapter seeks to address issues of cross-border ecologies, hazards and population flows along Indo-Bangladesh international boundary. It argues that an integrated approach to understanding cross-border issues between India and Bangladesh would be to consider the intrinsic connection between the existing incongruities of geographies between India and Bangladesh (i.e. between geography of the state and the geography of the ecosystem) as well as the linkages between anthropogenic shifts in the geography of transboundary rivers in India, occurrences of compound disasters in Bangladesh, and consequent population movements to India. The chapter analyses how the declining quality and quantity of transnational river waters in the upstream India, caused largely by human activities, produces adverse socio-economic impacts in the downstream state in the form of habitat destruction, economic decline and movement of population. It then goes on to critically assess responses of both the states and different interest groups therein to the shifting geography of cross-border rivers in India and counter-movement of population from Bangladesh. The chapter argues that it is the consideration of these cross-border issues from narrow 'nationalistic' perspectives, and their politicisation in domestic constituencies of India and Bangladesh, which have made these cross-border problems intractable. Further, they look at these cross-border issues independently of each other. Therefore, it concludes with a general suggestion that an effective management of cross-border issues would above all require an appreciation of interconnected causalities of cross-border issues involving a geographically vast and ecological unified space of India and Bangladesh.

The central arguments of this chapter are built on the existing scholarly literature on the past and contemporary conditions of cross-border ecologies, hazard and population flows between India and Bangladesh. It uses historical works pertaining to the process of peopling of the region in response to natural changes in river courses in the deltaic region of Bengal. It also utilises recent literature on cross-border ecologies, hazards and population flows to reveal how population movement in this region has been taking place in response to the adverse economic and ecological impact of human-induced changes in transboundary rivers.

12.2 The Political Geographical Context

A proper appreciation of interconnected causalities between bio-physical and socio-economic processes across Indo-Bangladesh border must be set in the political geographical context of Indian subcontinent. The geographical and historical realities of the Indian subcontinent present a distinct case in the context of discordance of political and ecological boundaries and cross-border ecology, hazards and population movement. The countries of the region have close geographical, historical, religious, economic, ethnic and linguistic affinities with each other. Geographically, all other states of this region either share land borders with India (Bangladesh, Bhutan, Nepal and Pakistan), or are separated from India by a narrow sea (Sri Lanka and Maldives). As a matter of fact, Pakistan (as West and East Pakistan) emerged as an independent nation following the partition of India in 1947. Later in 1972, Bangladesh broke away from Pakistan to emerge as an independent nation. Prior to the partition of India, and subsequent emergence of Bangladesh and Pakistan as separate politically organised sovereign territories, issues of utilisation of common resources such as rivers and inter-regional movement of population were considered normal affairs. However, the drawing and redrawing of political boundaries in this region raised questions about claims to common resources and the legitimacy of population movement between these states. In other words, in the aftermath of the creation of territorial boundaries between these states, new kinds of formulations and practices—often guided by a strong sense of territoriality and nationalist sentiments on the issues of cross-border management of ecology, hazards and migration—came to the fore.

In this context, it would also be pertinent to recognise that unlike the evolved democracies of Europe, the South Asian democracies cannot claim to be institutionally rooted and matured. As a result, political, religious, linguistic, regional and ethnic interests decisively influence the shaping of opinions on inter-state issues. Needless to say, the inability of these states to amicably resolve their longstanding inter-state issues and the existence of acrimonious relations between them ever since their independence, have created spaces for these kinds of interests. What is more, the lack of national consensus on many cross-border issues is frequently used by political parties and other interest groups within these countries to articulate their sub-national interests, thereby making their resolution difficult.

Against this backdrop, this chapter discusses the issues of cross-border ecology, hazards and migration between India and Bangladesh. Historically, this region is characterised by changing processes of interaction between ecological and political boundaries. The region has also been shaped by the shifting geographies of transnational rivers and consequent movement of population. Many similarities in biophysical and socio-economic characteristics are discernible between both the countries. Substantial parts of India and almost the entire geographical territory of Bangladesh are comprised of three mighty transnational rivers: the Ganga, the Brahmaputra and the Meghna (popularly known as GBM basin). The alluvial plains and deltas of Bangladesh are actually the lower section of the GBM basin. As a matter of fact, as many as 54 common rivers traverse the international boundary of the two countries and all of them enter Bangladesh from India. After entering into Bangladesh, these rivers are bifurcated into numerous distributaries before flowing into the Bay of Bengal. These rivers have been vitally important for the economic development of both the countries: the performances of agriculture, fisheries, agro-based industries and inland water transportation overwhelmingly depend on the health of these rivers, which have also been critically important for ecological security.

Among many transboundary issues between India and Bangladesh, the two most important are the migration of people from Bangladesh to India and the sharing of transboundary flows of resources (e.g. transnational rivers). Throughout history, migration in search of alternative habitats or alternative means of livelihood has been a standard strategy adopted by the people in response to ecological changes or environmental disasters in the GBM basin. The early movement of population was from mainland South Asia to the swampy and forested Delta of Bengal. According to Ranabir Samaddar (1999 p. 90), 'Deltaic Bengal has seen the agrarian frontier being continuously pushed eastward and the advance of agrarian civilization over forest.' However, since the early nineteenth century, people from the densely populated plains and deltas of Bangladesh (earlier known as East Bengal as a part of erstwhile British India, and East Pakistan as a part of Pakistan) started moving into the Assam valley. Interestingly, this movement of people did not cease even after the creation of Pakistan (after the partition of India in 1947) and consequent emergence of Bangladesh as a sovereign and independent political entity in 1971. The Indian

Government started viewing cross-border migration as a security concern for India. According to the Recommendations of the Group of Ministers Report on Reforming National Security System in India, this phenomenon has generated a host of destabilising political, social, economic, ethnic and communal tensions in many states and territories of the country (Government of India 2001, p. 13). Therefore, the Indian Government has adopted various strategies for tightly controlling its border—such as fencing, lighting and intensive patrolling of the border—as well as several legal measures for identifying, detecting and deporting of illegal immigrants from Bangladesh. The right wing political parties, a section of civil society groups of Assam and the press played an important role in securitising the issue of cross-border migration. However, the Bangladesh Government and political parties categorically deny the phenomenon of illegal immigration from their country to India. Moreover, they do not consider India's policy of a tightly-controlled border regime to be the gesture of a friendly nation.

The second important issue between India and Bangladesh is the sharing of transnational rivers and management of environmental hazards. As the economy and ecology of Bangladesh is critically dependent on these transnational rivers entering from India, the decline in the quality and quantity of their waters caused by activities on the Indian side produce many adverse effects in Bangladesh. Hence, Bangladesh as a nation is understandably concerned about continuing upstream water diversion and pollution and the adverse effect of these activities on its economy and ecology, primarily the destruction of habitat and loss of sources of livelihood. A perception has developed in Bangladesh that the country has been denied rightful claims over the common rivers by the upstream India—and, furthermore, that India releases excess water during rainy months and impounds cross-border rivers during dry seasons without care about the needs and concerns of Bangladesh. This has given rise to a sense of injustice and resentment in Bangladesh against India. The excessive politicisation of this issue by the political parties and press of Bangladesh has further aggravated this kind of perception.

It is noteworthy that the sharing of river waters and movements of people, which were not at all issues of consequence when the entire region was under the British Indian Government until 14/15 August 1947, became major causes of tensions between India and Bangladesh (then East Pakistan) after their emergence as politically organised sovereign territories. For example, a completely normal movement of people in response to ecological and economic changes from Bengal to Assam in the region became illegal migration after 1947. Similarly, all the rivers of this region became transnational rivers post 1947 partition. Thus, after the emergence of international boundary, cross-border, migration of people and the sharing of common rivers became matters of inter-state concerns.

These complex issues have become even more intricate to resolve due to the tendency of both the countries to look at them purely from their respective national perspectives. As a result, absolutely divergent views on cross-border issues prevail in India and Bangladesh. Due to this lack of willingness to consider these issues in holistic manner, a more nuanced understanding of cross-border issues is elusive. For example, there is not even a single major regional collaboration to manage the 54 common rivers. Another important reason why India and Bangladesh have failed to develop a holistic and sensible understanding of cross-border issues is their tendency to look at them in isolation from each other. For example, when they discuss the issue of sharing common rivers, their discussion remains confined to this topic only. They do not try to connect this issue with those of cross-border hazards and population displacement. In other words, both India and Bangladesh are not yet ready to accept the fact that cross-border issues of ecology, hazards, economy and migration are inextricable linked to each other.

12.3 Shifting Geography of Cross-Border Rivers

The location of a state in relation to its immediate neighbouring states and their relative geographical size and strength has important ramifications on control over cross-border resources. While India enjoys a central location, Bangladesh is peripherally located in the north-eastern part of Indian subcontinent. Bangladesh is surrounded from three sides by India. What is more, India is also geographically bigger, and economically and politically stronger than Bangladesh. The relative location and power asymmetry between India and Bangladesh have significant bearing on the sharing of cross-border resources and management of hazards and population movement.

Bangladesh and India share three large river basins, viz. the Ganges, the Brahmaputra and the Meghna (the GBM), which are also shared by with Nepal, Bhutan and China. The total drainage area of the GBM basins is around 1.75 million km^2: about 7% of catchment area is located in Bangladesh, 63% in India, 19% in China, 8% in Nepal, and 3% in Bhutan (Rasheed 2008 p. 38). The GBM river system drains about 87% of the total territory of Bangladesh.

Rivers are the most conspicuous geomorphic features of the plains and deltas of Bangladesh. as it is criss-crossed by over 220 rivers with a total length of more than 24,000 km, which cover about 7% of the national area (Chowdhury et al. 1996; quoted by Rasheed 2008 p. 37). Of these, 57 are transnational rivers. An important characteristic of these rivers is that they are exposed to frequent flood hazards with concomitant riverbank erosion caused by shifting of river channels (Baqee 1998 p. 1). The shifting of rivers has been so frequent that it prompted the famous South Asian geographer, Nafis Ahmad (1956 p. 388), to describe Bangladesh as 'a land of rivers wandering lazily over a level deltaic plain that is generally less than 50 feet above sea level'. The health of these rivers are utmost important from the point of view of the ecological and economic security of the country.

In the context of the management of cross-border resources and hazards, the location of Bangladesh in relation to all the cross-border rivers is quite significant. Bangladesh is a lower riparian country, as 54 rivers enter in its territory from India and 3 rivers enter from Myanmar. It is this unique setting of Bangladesh in the South Asian subcontinent that deprives the country of control over the headwaters of the cross-border rivers and exposes it to the potential hazards of flooding of these transnational rivers every year (Elahi 1991 p. 1). Against this backdrop, attention to the issues of management and sharing of 54 cross-border rivers between Bangladesh and India is in order.

As all the cross-border rivers flow from India to Bangladesh, any change in the geography of these rivers on the Indian side of the border invariably affects the downstream areas. For example, due to discharge of industrial effluent, municipal wastes and runoff from agricultural fields, water quality gradually deteriorates as the flow reaches downstream areas. The availability of water in these rivers also tends to decrease in the downstream particularly during the lean season. The adverse impacts of upstream pollution and diversion of waters of transnational rivers on the ecology, economy and habitat of the downstream areas in Bangladesh have been highlighted in many studies. For example, the reduction in flow of rivers due to large-scale upstream extraction and diversion of water in India has adversely affected the ecology and economy of Bangladesh, particularly of its north-west and south-west regions—including the Sunderbans, which is the world's largest mangrove forest (Ahmad 2004, p. 2; Haque 2010, p. 12). A major negative consequence of upstream water extraction is the increasing salinity in coastal areas of Bangladesh. Both inter-annual and long-term salinity levels in south-west Bangladesh are increasing. Among many other factors, this is also caused by the reduced flow into the Gorai river (and its gradual dying over time) due to abstraction of water from the Ganges river at the Farakka barrage in India (Brammer 2012, p. 82). These negative consequences are predicted to worsen in the coming years due to abstraction of river waters for upstream urban, industrial and agricultural needs. The reduction in freshwater discharge would cause further intrusion of salt farther and deeper into the mainland. The negative consequences of upstream water abstraction and diversion in India along transnational rivers (particularly the Ganges) have been highlighted by K.B. Sajjadur Rasheed (2008, p. 53), as follows:

- Water shortage in farmlands, soil moisture depletion and lowering of water table in irrigation wells.
- Low flow of water leading to steady habitat loss for fisheries.
- Accelerated channel siltation and extensive shoaling which alters the channel morphology of the rivers, with concomitant reduction in navigability.
- Incremental biodiversity loss in the mangrove forest of the Sunderbans (south-west coast of Bangladesh) as the delicate balance of the natural ecosystem alters due to fresh water flow reduction and salinity intrusion.

India has built diversion structures (a cross dam, rubber dam or a small barrage) near the borders of its territory on a number smaller rivers. Directly as a result of this many of these rivers are facing reduction of flow during the dry season. This has deprived the downstream people of precious water resources. For example, the building of diversion structures on Kodalia and Isamati rivers on Indian side of the border have rendered their downstream reaches completely devoid of water during the dry season (Faisal et al. 1999). Another example is the river Manu which enters Bangladesh from Indian state of Tripura. The Manu River Project was commissioned in 1986 by Bangladesh to facilitate irrigation of agricultural land in the

Moulvibazar district. *India also started constructing a barrage on this river to facilitate irrigation in the state of Tripura.* This created apprehension in Bangladesh that due to the upstream diversion of the river Manu in India, the availability of water on its side would come down significantly during the dry season, jeopardising the Manu River Project in Bangladesh (Financial Express 2000). Similarly, the Indian state of Tripura alleges that many irrigation, flood control and embankment projects along cross-border rivers such as Feni and Muhuri have been stopped following objections from Bangladesh thereby creating problems water for irrigation, river bank erosion and floods in the bordering districts of the state. Similarly, many small and medium projects on transboundary rivers (e.g. Manu, Khowai, Feni, Gumti and Muhuri) in Bangladesh territory bordering Tripura has been stopped due to objections from Indian side (ThaIndian News 2012).

The relative location of Bangladesh makes it vulnerable to many environmental hazards such as floods and cyclones. As a matter of fact, Bangladesh is considered as the most flood-affected country in the world, followed by India (Chowdhury 2002, p. 233). The source of many of its floods can be traced to the upstream territory of India. As Bangladesh is the geographical continuation of the Indian landmass, a natural hazard occurring in upstream India cannot be spatially confined to its political boundary. Thus, if there is a flood in any of the 54 transnational rivers flowing from India into Bangladesh, its impact is obviously experienced in Bangladesh as well.

Territories of Bangladesh bordering India (the north, north-east and north-west) are particularly vulnerable to flash floods from trans-border rivers. Flash floods in Bangladesh are caused by excessive runoff during exceptionally heavy rainfall occurring in neighbouring upland areas. The rivers that generally experience flash floods are Purnbhaba, Karatoya, Dharla, and Teesta in the north; Someswari, Kangsa, Juri, Surma. Kusiyara, Manu and Khowai in the north-east; and Gumti, Muhuri, Halda, Sangu and Matamuhuri in the south-east (Rasheed 2008, p. 74). The sudden onset of flash floods makes it difficult to employ sustainable flood preparedness schemes in such areas. In particular, cross-border rivers flowing into the Chittagong Hill Tract (CHT) of Bangladesh such as Feni, Karnaphuli, Sangu, Matamuhari and Bagkhali frequently experience flash floods. Similarly, many tributaries of Karnafuli rivers such as the Halda, Ichhamati, Chengi, Subalong, Rinkheong and Kasalong are prone to flash flooding. Heavy rainfall in the hills of the north-eastern states of India causes the water level in these rivers to rise by several metres within hours leading to fast and furious flows for several days (Rashid 2002, p. 31).

The problems of droughts and floods along the borders of the two countries are direct results of anthropogenic interventions in the natural flow of transnational rivers on both sides of the border. It is also found that borderland communities living on both sides of the Indo-Bangladesh border create structures on natural drainage for irrigation and flood control without taking into considerations their negative effects on the communities living on the other side of the border.

> Along the eastern border of Bangladesh, for instance, many Indian farming communities built cross dams to divert lean season flow from small hilly streams to their farmlands, leaving the downstream side of the river—in Bangladesh—high and dry. On the other hand, in

> the northeast and southwest of Bangladesh, embankments and polders have been built that block the natural drainage from the Indian side, leading to waterlogging. (Faisal 2002, p. 313)

The existence of an international boundary is a major hindrance to flood management along the Indo-Bangladesh border. The following example by Islam M. Faisal (2002, p. 315) clearly points out that a highly securitised border can create constraints in the management of floods.

> .... the Kushkhali-Boikari border embankment, which blocked natural drainage of some Indian land through *Dant Bhanga Beel* (a perennial wetland in Bangladesh later converted into a polder), was built some time back. India developed a drainage system parallel to the border to deal with the problem. During a devastating flood in 2000, however, the Kushkhali embankment was breached at a number of locations. After the flood, the Indian side opposed reconstruction of the embankment on the ground that no permanent structure can be built within the 'security zone' along the border. Now, with a breached embankment, the region faces the threat of frequent flooding. This is an example of how bitter political relations can create new problems where the original issue would otherwise have been taken care of long ago.

It is obvious from the above examples that the existence of an international boundary adversely affects communities in both countries.

12.4 Narratives and Counter-Narratives on the Sharing of the River Ganga

There is some kind of dispute between India and Bangladesh relating to all 54 of their shared rivers; however, it is the dispute over the sharing of the Ganga waters which has generated strong narratives and counter-narratives in India and Bangladesh. For many decades, this issue was the most important cause of acrimonious relations between the two countries. The Ganga was solely an Indian river before the partition of the country in 1947, becoming an international river with the partition of British India into India and Pakistan (consisting of West and East Pakistan). The relations between them began deteriorating when India started to divert increasing volumes of water from the Ganga river system for agricultural, industrial and domestic needs—and the negative effects in Bangladesh (then East Pakistan) were noticed. Due to upstream withdrawal and diversion of water from this river, water availability in Bangladesh during the dry season was reduced. The relations between the two countries over this issue further deteriorated in the 1960s, when India sanctioned the construction of the Farakka Barrage over the river Ganga. This project was envisaged to solve the problem of siltation in the river Hoogly/Bhagirathi, a distributary of the river Ganga. The Indian city of Kolkata is sited on this river. It is important to note that adequate quantity of water in this river was essential to maintain the functionality of the Calcutta (Kolkata) port. Despite the concerns of Pakistan, the Indian Government went ahead with the project and the Barrage was completed in 1970. Bangladesh emerged as an independent nation-state in 1971 and soon took up the matter with India. Two short-term agreements (1975 and 1977) were signed over the sharing of the Ganga waters, which were not viewed favorably by opposition political parties and other groups in the two countries. The then-Chief Minister of Indian state of West Bengal angrily reacted to the 1975 agreement as 'India's share of the water as driblets….in quantities that do not serve our purpose' (Franda, quoted in Gaan 2000, p. 148). On the other hand, the opposition political parties in Bangladesh described it as another pro-Indian act of the Bangladesh Government and accused it for overlooking the national interests. Maulana Bhasani, a popular political leader of Bangladesh, tried to arouse public outrage about alleged devastation and desertification caused by the reduced flow of the Ganga river by issuing threats to lead a march to demolish the Farakka Barrage. On 16 May 1976, he launched the 'Farakka Peace March' but due to intervention of the Bangladesh authorities it did not lead to any unsavoury incidents (Swain 1996, p. 132).

In the 1990s, due to a delay in the agreement on the sharing of the river Ganga, Bangladeshi intellectuals and scholars became very critical of India's decision to unilaterally withdraw waters. A Bangladeshi scholar alleged that Bangladesh was functioning as a 'drainage corridor' while neighbouring India made the most of the precious resources (Yakub 1994, p. 18). The unilateral withdrawal of water by India was interpreted by Bangladeshi strategic thinkers as a direct insult on the sovereignty of Bangladesh (Rahman and Islam 1992, p. 67). However, due to changes in the political regimes in New Delhi and Dhaka, a 30 year agreement over the sharing of the waters of Ganga was reached between India and Bangladesh in 1996.

A critical analysis of disputes over Ganga waters suggests that opposition parties and other groups in both countries tend to over-politicise the issue by making unsubstantiated claims about transboundary water sharing in order to create a political space for themselves or to increase their existing political base. In the process, they ended up creating public opinions which viewed any agreement between the two countries as conspiracy against national interests. As a result, even well-intentioned governments avoid entering into any mutually beneficial agreements. In this context, A. Richards and N. Singh (2000, p. 1915) have aptly described the emergence of typical unsubstantiated myths in Bangladesh in the case of the Farakka Barrage.

> … In Bangladesh, the Farakka Barrage has been widely portrayed in political and media discussions as a symbol of India's evil intent toward Bangladesh. Technical controversy about the 'flushing' process through which the barrage was expected to save the port of Calcutta and its industrial hinterland, as well as India's failure to recognize the downstream consequences of the project, left space for the assertion that the barrage was built *because of* its deleterious effects on Bangladesh (then East Pakistan). A second myth of Indian malice has also been widely repeated. This is the assertion that India can cause flooding in Bangladesh through the release of water stored behind the Farakka Barrage. Brief description of the barrage indicates that it is unable to store more than trivial quantities of water, far too little to have a significant effect on floods in Bangladesh.

In a similar vein, Ramaswamy R. Iyer (1999, p. 1512) has summarised the development of a rather over-simplified and unsubstantiated perception in Bangladesh on the issue of sharing of Ganga water.

> A simplified version of the Bangladeshi view of this dispute would be as follows: that there was a 'unilateral diversion' of the waters of the Ganga by India at Farakka to the detriment of Bangladesh; that the resulting reduction in flows had severe adverse effects on Bangladesh; and that this was a case of a larger and more powerful country disregarding the legitimate interests of a smaller and weaker neighbour, and callously inflicting grievous injury on it. That view of the dispute has been widely prevalent in Bangladesh, cutting across all kinds of divisions. A national sense of grievance grew and became a significant factor in electoral politics. In its extreme form the nationalistic position became a myth with India being cast in the role of a demon: whether Bangladesh was afflicted by drought or by floods, the responsibility was laid at India's door. 'Farakka' was blamed for all ills.

Some Indian scholars have regarded the prevalent rigid and exaggerated attitude of Bangladesh as a cause for delay in resolving cross-border issues. According to Ramaswamy R. Iyer (1999, p. 1513), the tendency in Bangladesh has always been to regard the undiminished continuance of historic flows as a birthright and not to recognise the needs of upstream populations. Due to this attitude, it refused even to consider other possibilities of meeting its water needs for fear of compromising its claims to Ganga waters. Iyer also points towards the emergence of a feeling of resentment and distrust of India as a big and powerful country. A similar negative reading of Bangladesh posture is present in India at the Government level.

> At the governmental level, a fairly common view was that Bangladesh was extremely rigid and unreasonable on this issue; that it had greatly overpitched its water needs and was claiming a disproportionate share of the waters in relation to the relevant criteria (contribution, cultivable area, etc.); that it tended to exaggerate the adverse effects of reduced flows; and that it had blown the dispute up into a big political issue in domestic politics, making inter-governmental negotiations difficult. (Iyer 1999, p. 1512)

Very characteristically, many of these myths and exaggerated beliefs have been deliberately created by the political classes in Bangladesh to address their respective domestic political constituencies. Beside this, the political class of Bangladesh is also very selective in taking positions on issues of pollution and hazards. While taking strong positions on issues of cross-border sources of pollution and hazards, they fail to even acknowledge contributing activities inside the territory of Bangladesh. For example, it is reported that industrial units such as textile factories, tanneries, pulp and paper mills, fertiliser plants and industrial chemical production facilities situated along the rivers of Bangladesh such as Buriganga, the Sitalakhya, the Balu, the Turag, the Bhairabh and the Karnaphuli are polluting them by releasing hazardous toxic water (United Nations Environment Programme 2001, p. 48).

The attitude on the Indian side also hinders the effective resolution of cross-border issues. Reflecting on the difficulties that emerged during the negotiation on the sharing of the river Ganga, Ramaswamy R. Iyer (1999, pp. 1512–13) has opined that India unconsciously tend to regard this river as essentially an Indian resource. Besides, in the national perspective of river planning in India, India left little room for a serious consideration of the needs of Bangladesh. Finally, there was inadequate appreciation of the ill effects suffered in Bangladesh because of the reduced flows in the Ganga.

It is fairly clear from above analysis that continuing disputes over cross-border issues on both sides of the border are due to India and Bangladesh approaching these issues purely from their national perspectives. Further, excessive politicisation of sharing transboundary waters on both sides of the border has created a perception of biases and lack of trust between India and Bangladesh. All these have resulted in the failure of India and Bangladesh to develop a more nuanced perspective on sharing and management of cross-border rivers.

12.5 Shifting Geography of Cross-Border Population Movement

The movement of people from the plain and deltaic regions of Bengal to Assam which started in the beginning of the twentieth century did not stop even after the emergence Bangladesh (then part of Pakistan) in 1947 as an independent sovereign country. Historically, it is the shifting geographies of rivers, water bodies and deltas which have been shaping patterns of human movement in the region in the past. The location and the physical geography of Bangladesh have had far-reaching effects on the trajectory of its history and life of its people. The natural barriers like innumerable rivers, marshes and swamps, forests and climate accounted for Bengal's isolation from the rest of India and shaped the way of life of its people (Ali 1974, pp. 69–76). Due to its peripheral location and geographical barriers, Bangladesh was a settlement frontier region of the South Asian landmass until the end of the nineteenth century. Historically, the plains and deltas of Bangladesh have experienced large-scale in-migration of peasants in response to the shifting geographies of its rivers and delta. Shahdeen Malik (2000, p. 12) describes this process as follow:

> During the last one thousand years the main rivers of the north-western and western Bengal delta decayed as their channels moved eastward. These shifting channels deposited silts in new areas making possible the transformation of dense forests into agricultural lands. This movement of geographic frontier made the region accessible and habitable, particularly the east-southern part of the Gangetic delta and settled agricultural communities moved eastward and southward.

Malik (2000, p. 12) goes on to describe the actual process of peopling of this region by the pioneer settlers:

> As late as the nineteenth century, the movement of agricultural communities in search of new arable lands; land reclamation in the deeper and previously inaccessible southern parts of the delta; clearing the dense forests and turning those into settled habitats; settling on islands in the southern coasts—all constituted a continuous movement of population within and from without the defined political and geographical boundaries.

This is clearly indicative of the fact that the continuing in-migration to the plains and deltas of Bengal (Bangladesh) was mainly in response to the changing ecological conditions of its river plains and delta. Ranabir Samaddar (1999, p. 90) describes this form of migration as chronologically deep ecological process. However, once the region was fully occupied and settled, the direction of migration changed completely in the beginning of the twentieth century. The emerging trend was characterised by the migration of landless labourers and marginal farmers from the overpopulated Bengal delta to the Assam Valley. This shift in the direction of population movement could be attributed to differences in government policies regarding land use, cropping patterns and land taxation in Bengal and Assam. The shift was also possibly due to wide gaps between the two regions with regard to the socio-economic conditions of people and the availability of land resources.

The prevailing socio-economic conditions in Bengal pushed peasants to look for alternative places to earn their livelihood, among which Assam was the nearest and most promising. The high fertility of the virgin wet soils and low rent prevailing at that time induced many farmers from Bengal to settle in this valley (Elahi and Sultana 1991, p. 20). The prevailing differences in socio-economic conditions of Bengal and Assam also encouraged the British Indian Government to adopt population transfer from Bengal to Assam as a matter of policy. This is evident from a report of a British administrator who had proposed transfer of population as a viable strategy to solve problems of under populated Assam and overpopulated Bengal: 'What is wanted for Assam are drafts of immigrants from the overcrowded, famine-stricken, districts of Bengal who might receive plots of waste land to break up on the most liberal terms' (cf. Dev and Lahiri 1981: 190).

The availability of large chunks of virgin land in Assam attracted the rice farmers of Bengal, as its wet and fertile soil provided ideal conditions for the cultivation of rice. The asymmetrical density of population between the two provinces also encouraged people to move. The rich Assam valley was sparsely populated (with 39 persons per km^2 in 1901) compared to densely populated Bengal Delta (with 195 persons per km^2). This movement was further reinforced in later periods by the rising density of population in Bengal (particularly East Bengal) which started to put pressure on agricultural lands there (Elahi and Sultana 1991, p. 21).

Once the stream of migration had begun, it was not easy to bring to a halt. Relatives and acquaintances of migrants were attracted to Assam. Interestingly, the stream of migration continued unabated even after creation of India and East Pakistan (as a part of Pakistan) as separate political entities. The economy of East Pakistan was most adversely affected by partition of Indian subcontinent. The most important agricultural raw material (i.e. jute) grown in East Bengal was separated from manufacturing centres located in and around Kolkata in India. As a result, there was a disruption in virtually every sector of the economy of East Pakistan. The creation of East Bengal (later East Pakistan and now Bangladesh) as a province of Pakistan had confined the majority population in the least developed areas, thus constraining its ecological and economic viability.

> Economic viability of East Pakistan was seriously questioned by some Muslim leaders even in those days … Mr. Syed Muazzam Husain, a former Minister of Bengal and an experienced administrator condemned the Mount Batten's scheme as a mutilated, truncated, and moth-eaten East Pakistan. (Farouk 1992, p. 5)

In subsequent decades, however, another triggering factor emerged as the main cause behind migration from Bangladesh to Assam. After independence, India started impounding and diverting the transboundary river waters to fulfil its developmental needs in the agricultural and allied activities as well as urban, transport and industrial sectors without thinking much about its repercussions on downstream Bangladesh economy. This resulted in shortage of water in Bangladesh for development of agriculture, fisheries and transportation.

As the economy of Bangladesh is dominated by renewable resource-based activities such as agriculture, agro-based industries, fishing, inland water transport and forestry, the decline in the quality and quantity of precious water resources has resulted in the degradation of the ecological foundations of people's livelihoods in the north-western and south-western parts of Bangladesh. The silting and pollution of rivers often lead to the economic dislocation of boatmen, fishermen, employees in water transport and businessmen who are dependent on rivers for their livelihood (Alam 2003, pp. 430–431). Faced with life and death situations, the poor and the marginalised people tend to escape from these areas and migrate to other parts of Bangladesh and even outside the country to India as a survival strategy. In other words, the contemporary phase of migration from Bangladesh to India can be attributed to a large extent to human-induced transformation in geography of transnational rivers and other water bodies in the plains and deltas of Bangladesh. The Bengal Delta, which was one of wettest places in the world, is now facing an unprecedented water crisis; the problem is not as much a lack of water as a lack of safe water (Van Schendel 2009, p. 243).

The interconnections between compound disasters, migration and urbanisation are clearly visible in both Bangladesh and India where urbanisation is both a cause as well as consequence of environmental hazards. As many urban centres in India are located along the banks of rivers, their high growth rate (as per 2011 Census, 31.16% of India's population live in urban in 2011 compared to 23.30% 1981) together with haphazardous expansion, have adversely affected their geomorphology, ecology and water quality. In this regard, a study on the impact of urbanisation on the hydrology of the Ganga basin in India points out that a large number of towns and big cities of India are located in the Ganga basin, which generate and discharge huge amount of sewerages and waste, the majority of which eventually reaches the river through the natural drainage system. The situation is worse in the case of smaller rivers. Most of the tributaries of Ganga and Yamuna have become the channels of transport of sewerage waste water and industrial effluents of the cities (Misra 2011).

In Bangladesh, rural to urban migration has led to a rapid increase in the share of urban population, from 15.8% of the total population in 1981 to 28.37% in 2011. Besides many socio-economic push factors, the rural-urban migration is caused by a combination of slow-onset environmental hazards such as siltation, water logging, erosion, deforestation, pollution and over-use and depletion of natural resource base, and sudden environmental hazards such as river bank erosion, floods and cyclones (Haider 1994, pp. 279–280). People displaced by compound disasters in rural areas tend to migrate and settle in risk-prone peripheral localities of urban centres. Such localities are located on either flood-prone low-lying lands or near hazardous dumping grounds or in socio-spatially marginal localities such as slums and squatter settlements.

In recent decades, due to rapid urbanisation and haphazard urban expansion, many large urban centres face problems of poor drainage and consequent stagnation of rainwater (Islam 1994, pp. 350–351); in addition, many townships such as Chandpur, Bhairav Bazaar, Sirajganj and Munshiganj face problems of severe bank erosion from river activities (Adnan 1994, p. 191). Curiously, the urban hazard management plans are formulated with a view to protect effluent localities at the neglect and expense of localities inhabited by the poor and marginalised people. For example, after the flood of 1988, which submerged about two third of Dhaka metropolitan area for 2 weeks, the Dhaka Flood Protection Project (DFPP) was initiated under which embankments were constructed around the city to protect it from water logging and floods. However, astoundingly, large chunks of human settlements were left outside the protected zone, the most notable of which was Char Kamrangi—a densely populated area containing slums inhabited by low-income households (Adnan 1994, p. 192).

12.6 Narratives and Counter-Narratives on Cross-Border Population Movement

From the Indian perspective, cross-border migration is the most pressing problem with Bangladesh. The continued movement of population from Bangladesh to India has generated a range of destabilising socio-political, economic, ethnic and communal tensions in India. This, in turn, has also embittered Indo-Bangladesh relations, causing tensions between the two countries (Alam 2003, p. 422). The responses of groups and political parties in India vary considerably. Liberal and left-leaning scholars and political parties take a relatively considerate approach toward migrants. They consider migration from Bangladesh to India to be a humanitarian problem which has its origin the economic and ecological problems and the policies of the British colonial power, and therefore advocate for a more compassionate approach to manage this problem. For many manufacturing and construction employers in India, Bangladeshi migrants provide a cheap and industrious labour force. The Muslim groups and political parties in Assam deny that there is any large-scale illegal migration in the state. They also decry the fact that genuine Muslim citizens of India are harassed by the state in the name of detecting and deporting illegal migrants.

There are also a number of anti-immigrant groups in India. The All Assam Students Union (ASSU) was created with an objective evict illegal Bangladeshi immigrants from Assam. They started anti-foreigner agitation which lasted for 6 years (1979–1985) and ended with the formation of a political party-Assam Gana Parishad (AGP). The AGP successfully contested the 1985 Assembly election and came to power in Assam. A Memorandum of Settlement was signed between the Central government and leaders of Assam Movement on August 15, 1985 with the objective of identifying and expulsion of illegal immigration from the state. Despite this, the issue of illegal immigration could not be resolved. Even now, the AGP demands the expulsion of illegal Bangladeshis from Assam without any discrimination between Hindu or Muslim immigrants. On the other hand, the Bhartiya Janata Party (BJP) and many other right wing groups tend to create a difference between refugees and illegal immigration along the religious lines. For them, Hindus coming from Bangladesh are refugees but Muslims coming from Bangladesh are illegal immigrants. In this context, it is important to note that the present Central Government led by the BJP has granted citizenship status to Hindus who have migrated from Bangladesh and Pakistan. Many anti-immigration groups in Assam have sharply reacted against this policy of the Central Government.

The right wing groups often tend to create a fear of an imminent threat of 'Islamisation of states' bordering Bangladesh. In their popular writings and speeches the ideologues of right wing political parties portray the phenomenon of migration along religious polarisation and jingoism. Some of their captions and statements have been documented by Indian scholar Priyankar Upadhyay (2006, p. 26):

- 'Demographic Aggression Against India: Muslim Avalanche from Bangladesh'
- 'Is India Going Islamic'
- 'The pushing of millions of Bangladeshi Islamic morons into India is fraught with gravest threats to our very existence'
- 'These millions of hungry Muslims are a shameless lot and resort to every conceivable stratagem to conceal their identity'
- 'The infiltration of Muslims from Bangladesh is cancerous'

These groups call for the identification and forceful deportation of illegal Bangladeshi immigrants to that country. It is important to note that the issue of illegal immigration in Assam and West Bengal is being raised by some political parties and social groups to build their political base in these states. Towards this they organise agitations and make efforts to build public opinion against Bangladeshi Muslim immigrants on purely religious and political grounds. Interestingly, the BJP, in alliance with AGP, has won the May 2016 Assembly election of Assam. The new government plans to seal the border with Bangladesh within 2 years to stop 'illegal immigration' from that country.

In Bangladesh, political parties cutting across the ideological spectrum vehemently deny any phenomenon of illegal migration from Bangladesh to India. Some academics concede the process of illegal migration from Bangladesh to India but they put things in perspective by linking it to environmental degradation and the geographically unfortunate location of Bangladesh. Any attempt to identify and subsequently deport the illegal Bangladeshi immigrants living in India to Bangladesh remains ineffective as Bangladesh refuse to accept them as their own citizens. For example, the Indian Government (led by the Indian National Congress) unsuccessfully tried to forcibly deport Bangladeshi Muslim migrants back to their own country from Delhi in 1992, but the Government of Bangladesh refused to accept these people (Gaan 2000, p. 165). The policy of 'push-back operation' has become defunct due to the hugely arduous tasks of their identification and subsequent refusal of Bangladesh to accept them.

In this way, similar to the issue of cross-border sharing and management of common resources, there is excessive politicisation and the tendency to look at the issue of cross-border movement of population from a purely national perspective in both India and Bangladesh. As a result, diametrically opposed perceptions have developed in both the countries which prevent an effective resolution of these issues.

12.7 Perspectives on Cross-Border Management

India and Bangladesh share their largest international boundary with each other. This boundary was drawn largely on the basis of religious demography. What is more, it was hastily drawn by the colonial power without taking into consideration the natural flow of river waters and the scope of natural disasters. The boundary-making process also did not take into account the historic movement of population in this region. As a result, despite the existence of an international boundary, one can find geographical continuities in cross-border ecology, hazard exposure and movement of population. This incongruity of political and ecological boundaries has given rise to at least three kinds of interlinked cross-border movements between Bangladesh and India: one-way flow of rivers from India to Bangladesh; two-ways flow of environmental hazards such as floods, droughts and cyclones; and one-way movement of population from Bangladesh to India. The management of these cross-border flows and movements poses major challenges for both countries. However, considering their political boundary as the sole criteria for managing cross-border issues, as policies in both countries on cross-border issues have done, has proved ineffective from the point of view of both the countries. As a matter of fact, some of these movements are so instant and natural that this boundary is not only meaningless but also unnecessary. Imtiaz Ahmed (2000, p. 60), a leading political scientist from Bangladesh, has discussed the impracticality of this boundary line in the lives of people in the event of major natural disasters, particularly floods. When floods of unprecedented magnitude affect both Bangladesh and India (West Bengal and Assam), the flow of environmental refugees (flood-induced migrants) happens both ways. Given the prompt availability of international relief on the Bangladesh side of the international boundary, some distressed people from the Malda and Murshidabad districts of West Bengal are found crossing over to Bangladesh. The case could also be the opposite if timely relief fails to reach the flood-affected border areas of Bangladesh.

For borderland communities, crossing the border is also a survival strategy in the face of natural and man-made adversities. People living in low-lying areas along the rivers often move up towards mainland during floods (Bhardwaj 2014, p. 27). The cross-border movement of population also tend to increase during times of political and communal disturbances. For instance, during 1950s and 1960s, incidences of the crossing of Bengali Muslims from Assam, West Bengal and Tripura into East Pakistan (Bangladesh) in the event of major communal riots have been reported by some scholars (Kamaluddin 1991, pp. 221–22). In recent times, Bangladeshi Hindus and ethnic minorities have tended to cross this boundary in the event of communal and ethnic disturbance in Bangladesh. These examples indicate that borderland people would incur far greater loss during the times of environmental hazards such as floods and cyclones if their movement across this international boundary is denied.

Infrastructural development projects in Bangladesh have also displaced large numbers of people. For example, the construction of the Jamuna Multi-purpose Bridge Project displaced over 100,000 persons. Some of them were rehabilitated but most disappeared from their homes (Salam 2002, p. 302). Earlier, the construction of the Kaptai Hydroelectric Project (1957–1962) on the river Karnaphuli in the Chittagong Hill Tract (CHT) had caused large-scale involuntary displacement of ethnic minorities. In the process of building a huge lake (reservoir) for the project, about 100,000 hill people lost their lands, 40% of their arable lands were inundated, and above all, 40,000 of them crossed the border and settled in India (Saha 2001, p. 74). Interestingly, most of the transboundary migrants returned to their homes after the situation stabilised.

These examples clearly indicate the futility of the Indo-Bangladesh boundary. Despite the existence of this boundary since 1947, its ineffectiveness is still so much embedded in the day-to-day practices of the border people that when they are displaced by environmental disasters or political disturbances they tend to cross it. For them, crossing the border represents escape from various types of insecurities. A tightly controlled border regime tends to create many hardships in the day-to-day lives of people living on both sides of this international boundary.

A strong linkage between bio-physical processes and socio-economic process with interconnected causalities is visible in the entire GBM basin. Every interference in the upstream of cross-border rivers through the diversion and storage of water or disposal of urban sewage and industrial toxic wastes is followed by adverse ecological impact downstream in the form of water scarcity, environmental degradation and hazards. Like in any other agro-based economy, a large number of people in Bangladesh depend on renewable resource-based activities such as agriculture, fishing, forestry and water transport. It logically follows that a decline in the quality or quantity of renewable resources such as land, water, forests and fisheries would lead to decline of economy and loss of sources of livelihood. A decline in agricultural productivity and production due to land and water degradation might threaten the livelihood security of poor peasants and agricultural labourers. The depletion of fisheries might lead to the economic dislocation of fishing communities, and declining quality and quantity of forest cover might cause economic dislocations of the forest communities. Similarly, the decline in the quality and quantity of water in rivers might adversely affect the communities depending on river transport and fishing. Finally, severe floods can destroy the habitats of the poor living in hazard-prone areas such as along the river banks, on river islands or in other low-lying areas. The decline of economy, loss of sources of living and destruction of habitats force the poor and the marginalised to leave their home lands in search of new livelihood opportunities and living spaces.

The linkage between the shifting geography of transnational rivers in upstream India and its impact on the economy, livelihood and habitat of downstream Bangladesh, and the consequent movement of people from Bangladesh to India has been aptly captured by Ashok Swain (1996, p. 189) in the following words:

> India and Bangladesh are in a long-standing dispute over the sharing of the waters of the River Ganges. Since 1975, India has been diverting most of the dry-season flow of the river to one of her internal rivers, before it reaches Bangladesh. At Farakka, this has affected agricultural and industrial production, disrupted domestic water supply, fishing and navigation, and changed the hydraulic character of the rivers and the ecology of the Delta in the down-stream areas. These trans-border human-inflicted environmental changes have resulted in the loss of the sources of living of a large population in the south-western part of Bangladesh and have necessitated their migration in the pursuit of survival. The absence of alternatives in the other parts of the country has left no other option for these Bangladeshis but to migrate into India.

In this context, it is also important to note that natural hazards such as floods and droughts have been occurring in this region since before the emergence of India and Bangladesh as independent states. However, these hazards were mainly caused by meteorological and fluvial processes and hence were perceived as natural phenomena. Moreover, both the territories during that period were under the single British rule and hence natural hazards were considered as internal matters. Similarly, the population movement from Bengal to Assam was an example of inter-regional migration. It is only after the emergence of India and Bangladesh (previously known as East Pakistan) as sovereign territorial units that perceptions about transnational flow of rivers and population movement have undergone significant changes. The changes in perception are also caused by the fact that in the contemporary times problems related to transboundary rivers such as floods, droughts, water quality deterioration, salinisation, and sedimentation in Bangladesh are now mainly caused by upstream human interventions in the form of extraction and diversion of transboundary rivers. The upstream disposal of urban and industrial wastes in these rivers deteriorates downstream quality of water. Therefore, not surprisingly, at present any act of unilateral intervention in the transboundary rivers on the part one country is perceived as a threat to the ecological and economic security on the other side of the boundary. Similarly, the migration of population from Bangladesh to India, which was an internal matter under the British Empire, is now a case of international migration and therefore, India considers these migrants as illegal and a threat to its internal security. Thus, the discordance between political and ecological boundaries between India and Bangladesh and the lack of meaningful collaboration on cross-border issues have created problems in both the countries in the form of compound disasters in Bangladesh and movement of population to India.

Like any other cross-border issues, issues of cross-border rivers and population movement are quite complex. There is a crisis of perception on cross-border issues in both countries. The politicisation of these issues adds to their complication and makes them intractable. A conflicting perception of these problems, shaped by narrow domestic political considerations, has developed in both the countries. Further, these cross-border issues are often used by various competing groups within territorial boundaries of both the countries to gain political mileage. Finally, there is also a tendency to look at cross-border problems from a religious angle. For example, the fact that some groups in India view Hindu migrants from Bangladesh as refugees and Muslim migrants from Bangladesh as illegal immigrants only complicates the issue of resolving trans-border movement of people.

The management of cross-border issues between India and Bangladesh is marked by crisis of reflection. There is a tendency on the part of Governments of India and Bangladesh to look at issues of cross-border rivers and population movement as independent of one another. In actual practice, they keep on highlighting those issues in which they have greater stakes. For example, India gives prominence to the issue of cross-border movement of population and Bangladesh to the issue of sharing of cross-border rivers. As a result, both sides do not take each other's positions on cross-border issues seriously. Interestingly, even scholarly treatment of cross-border issues is affected by national allegiance. For instance, one finds far too many popular and scholarly writings by Bangladeshis on the issue of cross-border rivers and by Indians on the issue of cross-border movement of population from Bangladesh to India. On the other hand, fewer articles are written by Indians on the issue of cross-border rivers and by Bangladeshis on issue of cross-border movement of population from Bangladesh to India.

12.8 The Way Forward

The partition of India and Bangladesh led to the partition of the same ecological and socio- economic space into two distinct politically organised territories. This process gave rise to many problems which were previously not thought of. Before 1947, India and Bangladesh were a single country as British India and hence there was no restrictions and limits on use of cross-border resources. Similarly, there was no restriction on population movement across the country. However, the situation has changed dramatically soon after the partition of India and emergence of Bangladesh as an independent state. In the new context, the unilateral utilisation of cross-border resources by one state without taking into considerations of rights of co-riparian states is a matter to be resolved within the framework of international law. It is considered unlawful if any activities of the upper riparian India create background for the occurrence of environmental hazards in the lower riparian Bangladesh and so is the clandestine movement of people from Bangladesh to India. However, as the discussion above clearly indicates, the management of these cross-border issues has been hampered due to conflict of perception and crisis of imagination in both the countries. Therefore, any policy for the management of cross-border issues between India and Bangladesh must be based on the fact of causal linkage between cross-border ecology, hazard and population movement as well as the value of inter-state cooperation in resolving cross-border issues. The solutions of environmental disasters which have cross-country also lie in cross-country initiatives (Chowdhury 2002 p. 235).

In view of the close linkage between cross-border ecology, environmental hazards and migration, it becomes imperative for both the countries to look at these issues from a holistic perspective. Considering the cross-border issues as causally and spatially interlinked processes in multiple ways may convince India and Bangladesh to look for their solution in an integrated way through inter-state cooperation. An integrated approach calls for the incorporation of linkages between fluvial-geomorphic behaviours of transboundary rivers and human adaptation to these changes, including their migration. In this context, Sadeq Khan (1997; quoted in Samaddar 1999, p. 90) has emphasised the value of an integrated approach for managing the problem of migration from Bangladesh to India. He suggests that this problem can be managed with the collaboration and cooperation of Bangladesh, the other regional countries and the developed countries of the world towards a territorial accommodation of her population growth through a combined strategy of water management, land reclamation, and integrated planning of growth centres and villages.

From practical as well as technical points of view, regional management of cross-border issues is not only desirable but also necessary. Bangladesh requires regular flow of its share of trans-border resources for its economic and ecological security. It also requires control of many hazards emanating from upstream. However, Bangladesh cannot plan the development its own water resources on long-term basis without support and cooperation from India.

In this context, Imtiaz Ahmed (1999, pp. 3–5) argues that the phenomenon of flood in Bangladesh is a regional issue, as the construction of numerous dams in both West Bengal and north-east India is to an extent responsible for increasing the severity of floods in Bangladesh. The geographical proximity between India and Bangladesh implies that border areas of both the countries may face the same environmental challenges, such as droughts, floods and river channel migration. Therefore, it is not prudent to plan for the utilisation of cross-border resources and manage hazards unilaterally. In contrast, India has so far not been able to effectively control the movement of population from Bangladesh to its territory by adopting various strategies inside its territory. As the cross-border population movement is casually connected to compound disasters in Bangladesh, inter-state cooperation for the development of common resources and management of environmental hazards would reduce the risks and vulnerabilities of the poor in Bangladesh.

This analysis suggests that looking at cross-border issues from a narrow nationalistic, political or religious perspective not only complicates the situation but also makes them intractable. Hence an important measure for resolving any of the cross-border issues between India and Bangladesh is the successful resolution of domestic political differences and consequent development of national consensus on these issues in both the countries. Alternatively, both the countries must attempt to detach domestic politics when negotiating cross-border issues.

12.9 Conclusion

The cross-border movements of rivers, hazards and population are linked together in interconnected causalities at manifold spatial scales. The upstream diversion and pollution of a cross-border river in one state can have far-reaching consequences in the lives, livelihoods and habitats of communities located at a great distance in another state. Affected people, unable to cope with these stresses, may migrate to the upstream state in search of life and livelihood. However, causalities involving various bio-physical and human processes are interlinked in complex ways operating in a vast geographical scale of continental size that involves more than one state. Therefore, the management of cross-border issues cannot be approached unilaterally.

Lack of awareness and understanding of such a complex linkage on the part of policymakers has given rise to many erroneous policies on cross-border issues. There is a tendency to look at various cross-border issues independent of one another (e.g. controlling illegal migration without taking into consideration the role of compound disasters caused by environmental degradation and hazards). There is also a tendency to look at cross-border issues from respective national perspectives (e.g. managing flood hazard in one state independently of the bordering country). In this policy paradigm, national territory and political boundaries assume far greater importance in the framework for the management of cross-border ecology, hazards and migration. This gives rise to strong populist narratives and counter-narratives on cross-border issues on both sides of the border which cause difficulties in their institutionalised management. Therefore, this chapter advocates for adopting a cross-border and interconnected causalities approach for managing cross-border issues, which would begin with the question: 'Can illegal migration from Bangladesh to India be controlled by cross-border collaboration for joint development of common resources and management of environmental hazards?'

References

Adnan, S. (1994). Floods, people, and the environment: Reflections on recent flood protection measures in Bangladesh. In A. A. Rahman, R. Haider, S. Haq, & E. G. Jensen (Eds.), *Environment and development in Bangladesh* (pp. 182–219). Dhaka: The University Press Limited.

Ahmad, N. (1956). The pattern of rural settlements in East Pakistan. *Geographical Review, 46*(3), 388–398.

Ahmad, Q. K. (2004). Introduction. In Q. K. Ahmad (Ed.), *Potential for sharing of common regional resources in the Eastern Himalayan Region* (pp. 1–8). Dhaka: Bangladesh Unnayan Parishad.

Ahmed, I. (1999). Planning against the unnatural disaster. In I. Ahmed (Ed.), *Living with floods: An exercise in alternatives* (pp. 3–5). Dhaka: The University Press Limited.

Ahmed, I. (2000). The plight of environmental refugees: Reinventing Bangladesh security. *South Asian Refugee Watch, 2*(2), 41–75. retrieved 22 April 2016, http://www.calternatives.

org/resource/pdf/The%20Plight%20of%20Environmental%20Refugees-Reinventing%20 Bangladesh%20Security.pdf.

Alam, S. (2003). Environmentally induced migration from Bangladesh to India. *Strategic Analysis, 27*(3), 422–438.

Ali, M. (1974). A note on the historical background of Bangladesh. *The Oriental Geographer, 28*(1), 69–76.

Baqee, A. (1998). *Peopling in the land of Allah Jaane: Power, peopling, and environment: The case of char-lands of Bangladesh*. Dhaka: The University Press Limited.

Bhardwaj, S. K. (2014). Building sustainable peace on Indo-Bangladesh border: The human security perspective. *Journal of International Relations, 5*(5), 17–36.

Brammer, H. (2012). *The physical geography of Bangladesh*. Dhaka: The University Press Limited.

Chowdhury, A. (2002). Disasters: Issues and responses. In P. Gain, S. Moral, P. Raj, & L. Sircar (Eds.), *Bangladesh environment: Facing the 21st century* (2nd ed., pp. 217–235). Dhaka: Society for Environment and Human Development.

Dev, B. J., & Lahiri, D. K. (1981). Assam in the days of Bhasani and league politics. *Journal of the Asiatic Society of Bangladesh, 25*, 189–235.

Elahi, K. M. (1991). Impacts of riverbank erosion and flood in Bangladesh: An introduction. In K. M. Elahi, K. S. Ahmed, & M. Mafizuddin (Eds.), *Riverbank erosion, flood and population displacement in Bangladesh, river bank erosion impact study* (pp. 1–12). Dhaka: Jahangirnagar University.

Elahi, K. M., & Sultana, S. (1991). Population redistribution and settlement change in South Asia: A historical evaluation. In L. A. Kosinski & K. M. Elahi (Eds.), *Population redistribution and development in South Asia* (pp. 15–36). Jaipur: Rawat Publications.

Faisal, I. M. (2002). Managing common waters in the Ganga-Brahmaputra-Meghna Region: Looking ahead. *SAIS Review, 22*(2), 309–327.

Faisal, I.M., Nishat, A., Tanzeema, S. (1999). 'Managing common water between India and Bangladesh', paper delivered at the International Conference on Cooperation in South Asia: Resolution of Inter-State Conflicts, Jawaharlal Nehru University, New Delhi.

Farouk, A. (1992). *Changes in the economy of Bangladesh*. Dhaka: The University Press Limited.

Financial Express. (2000). India builds barrage on Manu: Bangladesh may get no water in dry season, (Dhaka).

Gaan, N. (2000). *Environment and national security: The case of South Asia*. Delhi: South Asian Publishers Private Ltd.

Government of India. (2001). Reforming the National Security System, Recommendations of the Group of Ministers, Government of India, New Delhi.

Haider, R. (1994). Women, poverty and environment. In A. A. Rahman, R. Haider, S. Haq, & E. G. Jensen (Eds.), *Environment and development in Bangladesh* (pp. 276–304). Dhaka: The University Press Limited.

Haque, M.I. (2010). River and water issues: Perspective Bangladesh, Anushilan, Dhaka.

Islam, N. (1994). Urbanization and the urban environment in Bangladesh. In A. A. Rahman, R. Haider, S. Haq, & E. G. Jensen (Eds.), *Environment and development in Bangladesh* (pp. 332–358). Dhaka: The University Press Limited.

Iyer, R. R. (1999). Conflict-resolution: Three river treaties. *Economic and Political Weekly, 34*(24), 1509–1518.

Kamaluddin, A. F. M. (1991). Refugee problems in Bangladesh. In L. A. Kosinski & K. M. Elahi (Eds.), *Population redistribution and development in South Asia* (pp. 221–236). Jaipur: Rawat Publications.

Malik, S. (2000). Refugees and migrants of Bangladesh: Looking through a historical prism. In C. R. Abrar (Ed.), *On the margin: Refugees, migrants and minorities, refugee and migratory movements research unit* (pp. 11–40). Dhaka: University of Dhaka.

Misra, A. K. (2011). Impact of urbanization on the hydrology of Ganga Basin (India). *Water Resource Management, 25*(2), 705–719., viewed 21 February 2015. doi:10.1007/s11269-010-9722-9.

Pakyntein, E.H. (1964). India Census 1961 – Assam: General Report, Manager of Publications, Delhi.
Rahman, A. A., & Islam, K. (1992). *Environmental security: Concerns and issues in South Asia* (Vol. 1). Dhaka: Bangladesh Centre for Advanced Studies.
Rasheed, K. B. S. (2008). *Bangladesh: Resource and environmental profile*. Dhaka: AH Development Publishing House.
Rashid, H. E. (2002). River, water and wetlands. In P. Gain, S. Moral, P. Raj, & L. Sircar (Eds.), *Bangladesh environment: Facing the 21st century* (2nd ed., pp. 27–35). Dhaka: Society for Environment and Human Development.
Richards, A., & Singh, N. (2000). Impediments and innovation in international rivers: The waters of South Asia. *World Development, 28*(11), 1907–1925.
Saha, B. K. (2001). Changing pattern of agrarian structure in Bangladesh: 1984–1996. In A. Abdullah (Ed.), *Bangladesh economy 2000: Selected issues* (pp. 73–85). Dhaka: Bangladesh Institute of Development Studies.
Salam, F. M. A. (2002). Selected environmental issues – Jamuna multipurpose bridge project. In P. Gain, S. Moral, P. Raj, & L. Sircar (Eds.), *Bangladesh environment: Facing the 21st century* (2nd ed., pp. 302–303). Dhaka: Society for Environment and Human Development.
Samaddar, R. (1999). *The marginal nation: Transboundary migration from Bangladesh to West Bengal*. Dhaka: The University Press Limited.
Swain, A. (1996). The environmental trap: The Ganges River Diversion, Bangladeshi Migrants and Conflicts in India, Report no. 41, Department of Peace and Conflict Research, Uppsala University.
ThaIndian News. (2012). Bangladesh objections hold up Tripura Projects: Minister. Retrieved 6 Jul 2016. http://www.thaindian.com/newsportal/southasia/bangladeshobjectionsholduptripura-projectsminister_100635795.html.
United Nations Environment Programme. (2001). *Bangladesh: State of the environment 2001*. Thailand: UNEP.
Upadhyay, P. (2006). Securitization matrix in South Asia: Bangladeshi migrants as enemy alien. In M. C. Anthony, R. Emmers, & A. Acharya (Eds.), *Non-traditional security in Asia: Dilemmas in securitisation* (pp. 13–35). London: Ashgate.
Van Schendel, W. (2009). *A history of Bangladesh*. Cambridge: Cambridge University Press.
Yakub, N. N. (1994). Overview. In K. Haggart (Ed.), *Rivers of life* (pp. 1–30). Dhaka: Bangladesh Centre for Advanced Studies and PANOS Institute.

Chapter 13
China-Based Air Pollution and Epistemic Community Building in the Northeast Asian Region

Matthew A. Shapiro

13.1 Introduction

There is a 'common pool' resources problem in Northeast Asia with regard to air pollution, and potentially hundreds of millions of people in Northeast Asia are affected. This air pollution represents a violation of the 1979 UN Convention on Long-range Transboundary Air Pollution, as it originates primarily in Mongolia and northern China as yellow dust but "has adverse effects in the area under the jurisdiction of another State at such a distance that it is not generally possible to distinguish the contribution of individual emission sources or groups of sources" (Article 1(b)). Both natural and anthropogenic processes are at work, though, making it difficult to create practical policy prescriptions. Specifically, air pollution from Chinese industry, manufacturing, and transportation attaches to the yellow desert sand/dust that blows eastward out of northern China and Mongolia (presented in panel (b) of Fig. 13.1), settling eventually in the eastern region of China but also in Korea, Japan, and beyond.

Industrial practices in eastern China are responsible for up to 70% of the acid rain present in the Yangtze River Delta (Ge et al. 2015). Furthermore, pollution emanating from China in the form of particulate matter 10 μm or less in size (PM_{10}) amounts to 13–26% of contributing sources in Korea and Japan (Li et al. 2014) and 30% of contributing sources in Taiwan (Tsai et al. 2014), shown graphically in Fig. 13.2. Naturally occurring dust aerosols contribute 46.5%, 11.7%, and 11.0% of the PM_{10} concentrations in China, Korea, and Japan, respectively (Li et al. 2014). As of 2010, the highest concentrations of pollution 2.5 μm or less in size ($PM_{2.5}$) were in China (73 μg); in Korea, $PM_{2.5}$ concentrations were more than double those of countries in the European region (38 μg) (World Bank 2015). $PM_{2.5}$ pollution

M.A. Shapiro (✉)
Department of Political Science, Illinois Institute of Technology,
3301 S Dearborn, SH116, 60616 Chicago, IL, USA
e-mail: mshapir2@iit.edu

M.A. Miller et al. (eds.), *Crossing Borders*, DOI 10.1007/978-981-10-6126-4_13

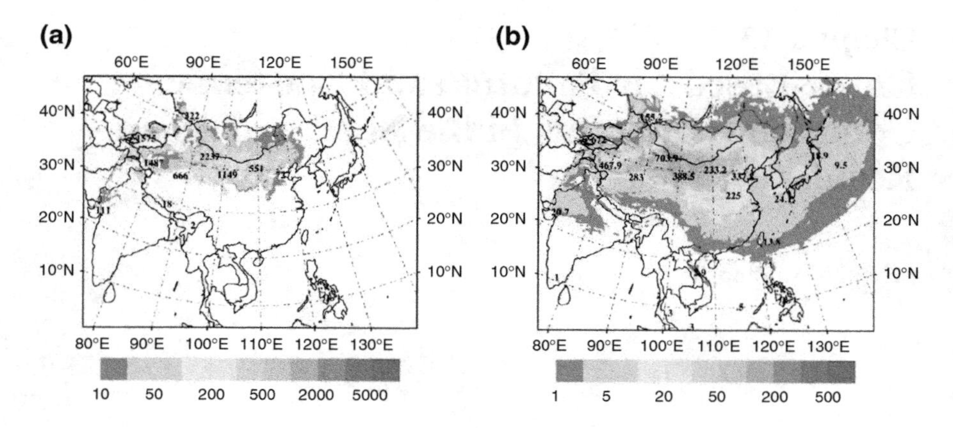

Fig. 13.1 Spatial distributions of annual emissions and deposition of *yellow* dust in 2010 (Note: Based on the ADAM2 model presented in Park et al. (2010). Source: EANET (2015) [copyright permissions pending]) (**a**) Total emission (ton/km^2) (**b**) Total deposition (ton/km^2)

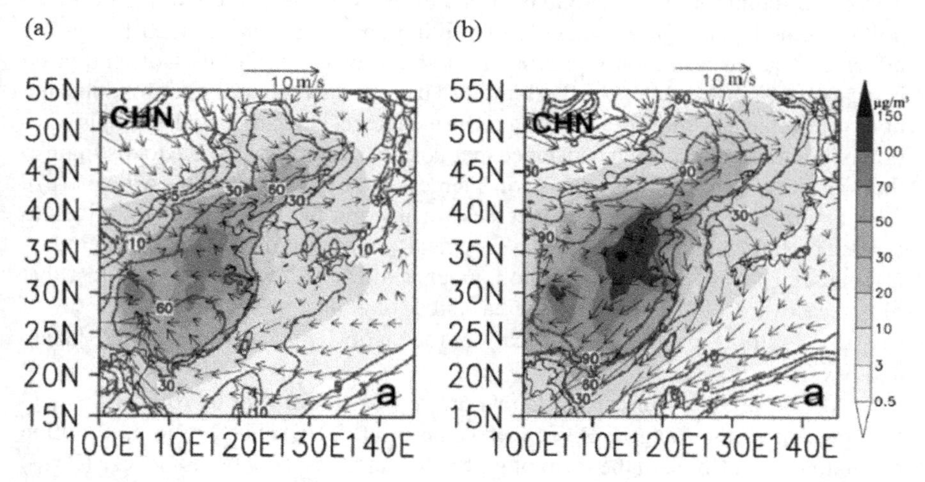

Fig. 13.2 Simulated contributions of anthropogenic aerosols from China to surface PM_{10} concentrations in 2010 (Note: Included are seasonal average wind vectors. Source: (Li et al. (2014) [copyright permissions pending]) (**a**) Spring (**b**) Fall

emanating from China has impacts on downwind areas (Han et al. 2015), and the health effects are extreme (see also Li et al. 2014; Sun et al. 2010). The construction sector in China is a particularly significant contributor to increases in $PM_{2.5}$ (Meng et al. 2015), with impacts measured as far away as the Pacific Northwest of the USA. (Fischer et al. 2009).[1] For Korea alone, the costs in 2002 were estimated to

[1] The effects of China-originating yellow dust and soot are felt even in the USA, where cities like Los Angeles receive one extra day of pollution per year from China's production of goods for export (Lin et al. 2014).

range from $3.9 to 7.3 million (Jeong 2008). Yet differences between immediate and delayed effects of yellow dust make it difficult to measure these costs, as shown in Ai and Polenske (2008).

The common pool resources problem under study here also reflects regional politics, which has been known to trump the ecological interdependence of Northeast Asia (Nam 2002). A prevailing view is that economic costs, a lack of regional agreement, and an absence of researchers to address the problem perpetuate the absence of coordination in Northeast Asia (Kim 2007). It is argued here that this position must be revised: environmental regionalism thrives, and the naturally and anthropogenically occurring air pollution arising from northern China is being addressed incrementally through increased urban-based and science policy-related efforts. The roots of Northeast Asian environmental regionalism go back 30 years, beginning with the efforts of the Asia Development Bank, the UN Environment Program, the state of environment reports prepared by the UN Environment and Social Commission for Asia and the Pacific (UNESCAP), the Northeast Asian Conference on Environmental Cooperation (NEAC) and the Environment Congress for Asia and the Pacific (ECO ASIA) (Shapiro 2012). Transboundary air pollution has been specifically addressed through the Tripartite Environment Ministries Meeting (TEMM), the Northeast Asia Sub-regional Program on Environmental Cooperation (NEASPEC), the Northeast Asian Training Center for Pollution Reduction in Coal-fired Power Plants and North East Asian Center of Environmental Data Training (NEACEDT) and the Acid Deposition Monitoring Network in East Asia (EANET).

There are two shared themes among regional organisations such as TEMM, NEASPEC, NEACEDT and EANET. First, they all focus on data collection and dissemination. Second, they all prioritise education and information provision to policymakers and the general public. Rather than engage in a comparative analysis of these different organisations (cf. Tsunekawa 2005), the focus here is on EANET, which fosters the sharing of knowledge and experience regarding air pollution-related data collection. More importantly, EANET is seemingly apolitical. Its members, despite being affiliated with their countries' respective environmental bureaucracies and science communities, are expressly focused on how to improve, standardise, and disseminate information to others about the accurate collection of data related to air pollution.[2]

The success of EANET and other similar organisations can be attributed to their focus on science, data accuracy, and data dissemination, all of which produce informal institutions such as intra-regional research collaborations. We must thus examine opportunities for these collaborations in the context of EANET and possibly other region-fostering institutions in Northeast Asia. Despite the political costs at the individual country level and the ecological costs at the regional level, science

[2] Even under these strict targets, EANET has not been free of politicisation. China was reluctant to join EANET initially, as its leaders correctly assumed that the group would exert pressure on China to cooperate with its neighbors to reduce pollution (Tsunekawa 2005). And Korea, in order to counter the hegemony of Japan in fostering EANET, founded the Joint Research Project on Long-rang Transboundary Air Pollutants in Northeast Asia (Yoshimatsu 2014).

and technology have continued to thrive across the region (Shapiro 2014a, b). More importantly, when researchers establish cross-national ties, the resulting environmental regime is at least partly determined by what Haas (1990) calls an 'epistemic community'—that is, a group of individuals who are politically empowered, knowledgeable, and motivated around a shared cause. The crucial aspect of this community is that its impacts are a function of its expertise. Its conclusions are thus rooted in scientific norms. Of course, this does not preclude scientists and researchers from being affected by domestic policies.

To better understand the nexus of EANET and the epistemic community of researchers in Northeast Asia, this paper attempts to answer the following questions: What is the nature of environmental coordination in Northeast Asia? Is there a role for polluted urban centres in affecting national priorities to address cross-border air pollution? What are some of the relevant developments thus far to address cross-national air pollution in Northeast Asia, including specific inventions/innovations developed to limit yellow dust-related effects? Finally, what future steps should be taken to ensure that any identified progress continues?

13.2 Understanding Environmental Coordination at the Regional Level

With or without the yellow dust-related pollution, there are two principal ways that one can describe prospects for environmental regionalism in Northeast Asia: politics predicts environmental regionalism, or environmental regionalism predicts politics (Lee 2001, 2002). The reality in Northeast Asia is that both situations are present, albeit to varying degrees. Regional environmental governance has been weak because non-state actors have been relegated to roles of less influence, there has been poor coordination around environmental initiatives, and there have been few clear outcomes from reducing environmental harms (Komori 2010). At the same time, seasonal fluctuations of the pollution arising from China has created a groundswell of public support for policy change, focused the attention of the scientific community on the problem, and focused the government's R&D budget on solutions for yellow dust-related pollution.

The creation of a regional environmental regime is extremely complex (Keohane and Victor 2011), and the Northeast Asian case in particular has been plagued with a number of confounding factors. For example, in China, top-down environmental mandates alone do not effect real reductions in environmental pollution (Kostka 2015; Lo 2015). Thus, increased decentralisation and the inclusion of non-state actors into the process must continue if the environmental components of the recent Five-Year Plan (No. 12, 2011–2015) are to be met (Kostka and Mol 2013; Mol 2009). The vehicle for such changes is the urban geography, as cities are typically the most environmentally challenged, politically motivated and technologically innovative, providing an ideal context through which improved sustainability efforts can be identified and employed (McHale et al. 2015). Public dissatisfaction has been documented particularly in Guangzhou (Zi et al. 2012), evidenced by the recent violent protests against the creation of an incinerator plant in Luoding (BBC News 2015).[3] Chinese citizens are even willing to offset economic gains with greater levels of environmental protection in urban areas ranging from Lanzhou (Zhao and Yang 2007) to Hangzhou (Chen and Shao 2007). Indeed, increased levels of air pollution over time have led to a relative shift in emphasis of China's urban centres away from economic growth and toward, among other things, environmental issues (Huang et al. 2016). With 50% of the population in China living in urban centres (719.4 million people) (World Bank 2015),[4] these concentrated concerns and demands for improvements in quality of life can no longer be ignored by the government (Zheng and Kahn 2013).

Other confounding factors in the development of a Northeast Asian environmental regime, such as improving the existing pollution measurement methods across the region and thus providing data upon which regional discussions can be based, are addressed directly through programs such as EANET. Other confounding factors, however, are exacerbated given historical tensions and concerns about hegemony. China, for example, has claimed that EANET challenges its national sovereignty (Tsunekawa 2005), and that its locally accumulated pollution and environment-related data does not have to be shared (Brettell 2007). Underlying these claims is the argument that dust storms carrying pollutants are natural despite evidence that desertification, the cause of the dust storms, is anthropogenic. Before 1992, China had even denied any effects of transboundary acid rain on its neighbors (Brettell and Kawashima 1998). Korea was also initially opposed to the institutional setup of EANET, protesting against the Japan-based headquarters of EANET's network centre. EANET has avoided these hazards primarily because it acknowledges at an institutional level both local and international interests, and its organisational structure reflects both domestic and regional concerns about the collection and dissemination of pollution data.

Regionally, we can expect that the decrease in the number of international players increases opportunities to address the common-pool resources problem. This is illustrated in the 'club' model, evident in East Asia (Kelley 2013), that describes how collective action problems such as climate change can be addressed with greater efficacy when international negotiations are limited to those countries that matter the most (Victor 2011). Ultimately, the club approach, through the outreach efforts of scientists and researchers, bolsters the effects of EANET and other similar institutions. This claim is rooted in existing research on international environmental regimes, such as Young's (1990) study of cross-national efforts to mitigate suboptimal outcomes with respect to environmental change, specifically ozone layer

[3] Chinese citizens also use social media to share images about air pollution in Beijing, including posts from celebrities (Gardiner 2014), prompting acknowledgement of the problem by public officials.

[4] In Korea, the urban population is even more concentrated: 82 % of the Korean population is urban-based, amounting to 41.2 million people (World Bank 2015).

depletion, global warming and biodiversity loss. In light of the club approach, Young's (1990) focus on non-state actors is invoked, 'epistemic communities' in particular: 'transnational networks of knowledge based communities that are both politically empowered through their claims to exercise authoritative knowledge and motivated by shared causal and principled beliefs' (Haas 1990, p. 349). The epistemic community that is particularly emphasised here is comprised of scientists and researchers that are able to resist political concerns while simultaneously informing policymakers. These scientists and researchers are not independent of the policy-making process, but can affect international cooperation at times.[5]

The transboundary air pollution and the relatively apolitical nature of EANET is assumed to result in scientists and engineers having a clear influence on overall regional attention to the problem of transboundary air pollution. What continues to confound the development of environmental regionalism in Northeast Asia is variance in how each state addresses the problem of yellow dust, particularly the extent to which China accepts its responsibility for the air pollution in its neighboring countries. There are a number of cross-state policies designed to address pollution in Northeast Asia (Shapiro 2012, 2014b), and perhaps EANET addresses this directly with its major goal of improving data collection and dissemination. We can attempt to measure at least the sharing of scientific information—inventions and innovations related to air pollution monitoring in particular—by identifying instances of international R&D collaboration within Northeast Asia. By doing so, though, we relegate well-established research that acknowledges the role of trade (Haggard 2013; Yoo and Kim 2015), finance (Sohn 2012), and public and private organisations (Abbott 2012; Abbott et al. 2013; Bulkeley et al. 2012).

International R&D collaboration actively contributes to a country's economic growth (Frantzen 2002; Kim 1999; Park 2004; Shapiro and Nugent 2012). Building on R&D-based endogenous growth theory (Aghion and Howitt 1992; Helpman 1993; Romer 1990), it has been shown that 'green' innovation benefits both the producing sectors' comparative advantage and their current output (Fankhauser et al. 2013). Yet there is also a disincentive for knowledge to be shared across countries if it results in economic losses. We know that China now plays a dominant role in global research output and networking (Wagner et al. 2015), and, with respect to yellow dust-related policies and programs, we also know that the economic benefits are likely to be greater overall than the environmental benefits, at least over time (Guo et al. 2008). This is likely because countries are conflicted about investing public R&D funds in pollution-reducing technologies, particularly air pollution, that have differing impacts between the origin and where it is deposited. How benefits are distributed across countries is also important. Greater benefits for more powerful countries that are disproportionately benefiting from green technology incentivise R&D investment, while concerns about non-reciprocity disincentivise

---

[5] A classic example of how this has occurred is the 1987 Montreal Protocol. Studies conducted in the pre-Montreal Protocol period showed that international controls on chlorofluorocarbons would help protect the ozone layer. This argument founded the efforts of a transnational epistemic community of atmospheric scientists to influence the positions of the UNEP and the USA (Haas 1990).

investment (Urpelainen 2011). We must thus assess whether research on yellow dust-related technology is occurring in Northeast Asia and whether it is being done collaboratively.

13.3 Methods

To understand the nature of the yellow dust-oriented epistemic community across Northeast Asia, descriptive and interview-based analyses are the primary vehicles. The analysis below begins with an examination of EANET itself, as there may have been changes in the orientation of this regional organisation that reflect both the increased threat of the yellow dust-related disaster as well as shifts in EANET's research focus. Given that EANET provides the air and water pollution data for scientists to analyse and model in order to make informed prescriptions to policy-makers, the connection between EANET's efforts can be correlated with the relevant scientific and technological output of these countries. To this end, the catalog of EANET's activities shall be triangulated with the patenting and publications record for air pollution-related work. These are compiled from several sources, primarily the US Patent and Trademark Office's (USPTO) database (USPTO 2015) and the Web of Science publications database (Thomson-Reuters 2015).

The following analysis is designed along the lines of a large body of research that uses patents as a proxy for innovation (see, for example, Griliches 1990; Hall et al. 2002; Schmookler 1966). Air pollution-related patents specific to either 'yellow dust' or 'yellow sand' or that reference $PM_{2.5}$ may fall within the classification of the USPTO Environmentally Sound Technologies (EST) Concordance (USPTO 2009), specifically patents addressing 'environmental purification, protection, or remediation—disaster in atmosphere.'[6] When these particular patents are not sufficient in number, data have been collected for "air pollution" patents as determined by a keyword search of patent descriptions. This also applies to USPTO patent data sourced from other entities, such as the OECD. With sufficient data—and if there are cross-patenting efforts within Northeast Asia—we can infer that an epistemic community is present that addresses the yellow dust issue (or at least air pollution). We know that China is now one of the world's leaders in terms of the generation of green R&D patents, where 'green' is based on patents falling under the EST Concordance. As well, China allocates a large amount of its public R&D budget to encourage collaboration across borders (Cainelli et al. 2012; Perkins and Neumayer 2008). Within Northeast Asia, however, Japan remains the overall greatest producer of green patents (Shapiro 2014a).

Complementary to yellow dust-related patents is publication output. Research on publications has shown, for example, that China's greatest overall international

[6] Included here are those patents dealing with electric or electrostatic field (e.g., electrostatic precipitation, etc. [class/subclass: 95/57+]) and liquid contacting (e.g., sorption, scrubbing, etc. [class/subclass: 95/149+]).

collaborator in terms of publication output is the USA (Wagner et al. 2014). If this is also confirmed with regard to yellow dust-related research, the epistemic community focusing on transboundary air pollution would have expanded all the way to the USA.[7] There are a number of ways in which one can tabulate a country's publication output on a particular subject: by journal or journal cluster or by topic/keyword. I have opted for a rather conservative measure of publication output by basing the keyword search of article abstracts on whether the term '$PM_{2.5}$' is included. As before, frequency of collaborations across the Northeast Asian region will be examined.

Where relevant, interview-based data will also be analysed. Interviews were conducted solely with stakeholders and experts in Korea; however, while the omission of Chinese, Japanese and Taiwanese experts is significant, the value of the Korean-based interviews should not be discounted. As shown in Shapiro and Gottschall (2011) and Shapiro (2012, 2014a), Korea has played a crucial role in engaging countries within the region as a 'middle power' (Kim 2014) and a regionally centred nation. Key actors in Korea were identified through the probing of members of the Presidential Commission on Sustainable Development, now called the Presidential Committee on Green Growth (see GNNCSDS n.d.). This commission arose from the Framework Act on Low Carbon, Green Growth, effective on April 14, 2010, to address climate change and energy issues and to target the growth of green industries. Employing a snowball sampling strategy, a sample of 19 individuals—eight affiliated with the commission and 11 recommended by members of the commission—was established, and interviews were conducted in the summer of 2014.[8] The survey instrument focused on the following topics: leadership roles within Northeast Asia, pollution's effects on environmental coordination efforts, political and economic forces affecting coordination, technology-oriented goals affecting coordination, and prospects for shared norms across the Northeast Asian countries in dealing with climate change via technology.

13.4 Results

A cataloging of all of EANET's activities over time, or at least since formalising its mission in January 2001, reveals a number of changes since it began its 'regular phase activities' (EANET 2010). Presented longitudinally in Fig. 13.3 and reflecting changes in the number of meetings/fellowships/etc. held, EANET

[7] This may not be inappropriate given the impact of yellow dust-related pollution on the west coast of the USA (Fischer et al. 2009).

[8] From the first group, specialists were interviewed from KEEI, the University of Science and Technology, Yonsei University, Seoul National University (2), Sejong University, Chung-Ang University and the KDI Graduate School. From those recommended by commission members, those interviewed are from the KDI Graduate School, STEPI (2), KISTEP, KETEP, GTCK, KEMCO, Jeju Technopark, KRIED and KEITI (interviewed together), Seoul National University and Dongguk University.

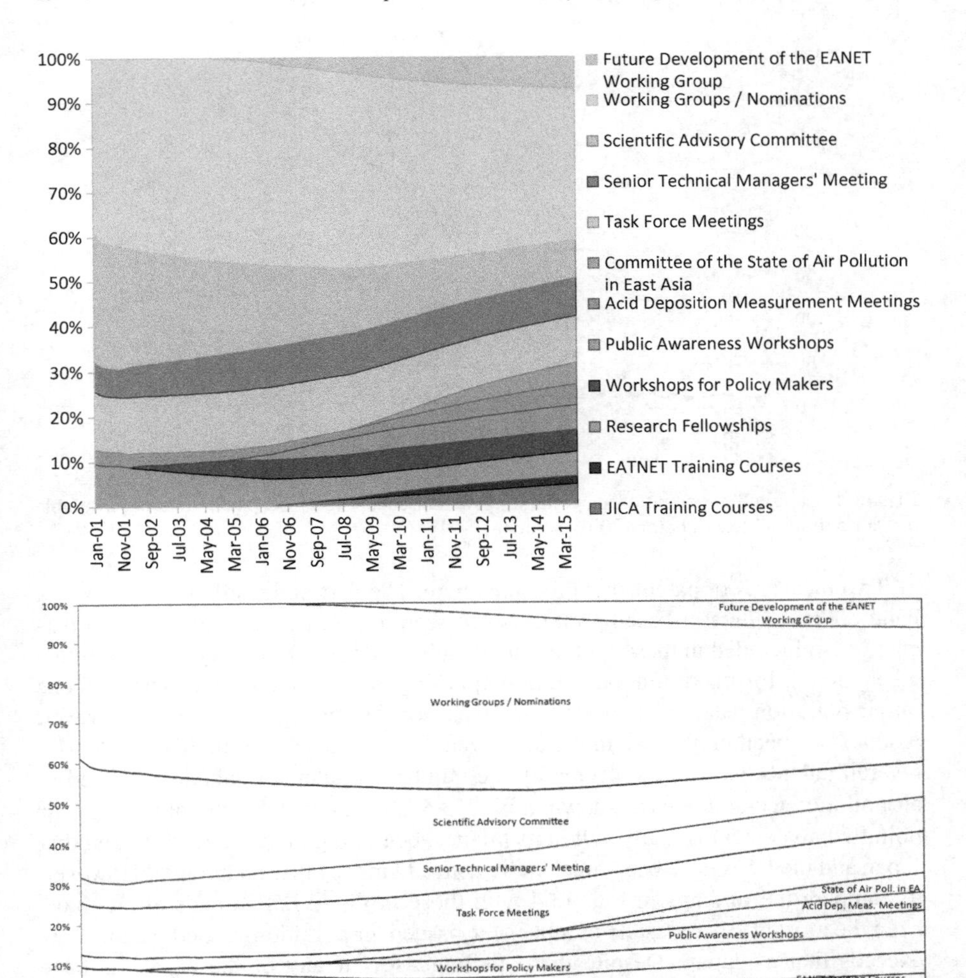

Fig. 13.3 Distribution of EANET's activities over time (Source: http://www.eanet.asia/schedule/index.html)

administrators have shifted the organisation's focus in several ways. The steady increase in research fellowships since 2006, the integration of workshops designed for policymakers, and the continued focus on public awareness is emblematic of its new approach to transboundary pollution. The committee meetings that address the state of air pollution in East Asia since 2012 also indicate that EANET is making a clear and deliberate connection between national reporting efforts and the region's overall environment.

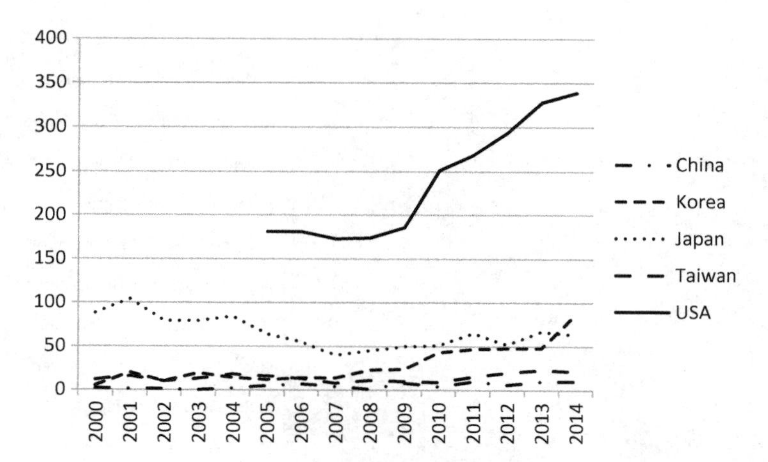

Fig. 13.4 Longitudinal count of air pollution patents issued (Note: US patent data not available for "air pollution" search before 2005. Source: USPTO (2015))

Two measures of patents are presented here. The first, reflected in Fig. 13.4, is a hand-collected dataset based on a keyword search of 'air pollution' in issued patents.[9] Also included in these findings are data for the USA, a country that represents a benchmark for maximum patenting output. Figure 13.4 shows that Japan's focus on air pollution patents has waned slightly since the turn of the century; however, Korea's has been on the rise in the last 5 years and especially from 2013 to 2014.[10] Overall patenting efforts, reflected by the number of patent applications related to air pollution technologies as shown in Fig. 13.5, indicate that the global focus on air pollution abatement is really still in its infancy, beginning in the late 1990s, and that Japan and the USA have been much more active than the other countries.[11] However, to compare the findings in Fig. 13.4 with those in Fig. 13.5, Korea's efforts have been the most effective, as the number of its issued air pollution-related patents now exceeds that of Japan. Despite these findings, few if any of these patents have occurred through Northeast Asian collaborations. With regard to air pollution-related innovations, the few collaborations that have occurred were primarily between Japan and the USA.

A search of articles indexed in the Web of Science (Thomson-Reuters 2015) for the keyword '$PM_{2.5}$' in the abstract reveals increased focus both overall and with specific regard to Northeast Asia. Shown in Fig. 13.6, publications focusing on $PM_{2.5}$ increased steadily from the late 1990s, dipped slightly in 2009 and 2010, and

[9] Note that 'air pollution' was the most appropriate proxy for yellow dust-related patenting output; the search for 'yellow dust' or 'transboundary pollution' yielded virtually no results in the USPTO's patent search engine.

[10] For ease of exposition, the focus of an individual country is representative of the focus of that country's research community.

[11] Declines in the most recent years are not evidence of declining patenting activity but, rather, represent the lag time required for patents to move from 'application' to 'issued' status.

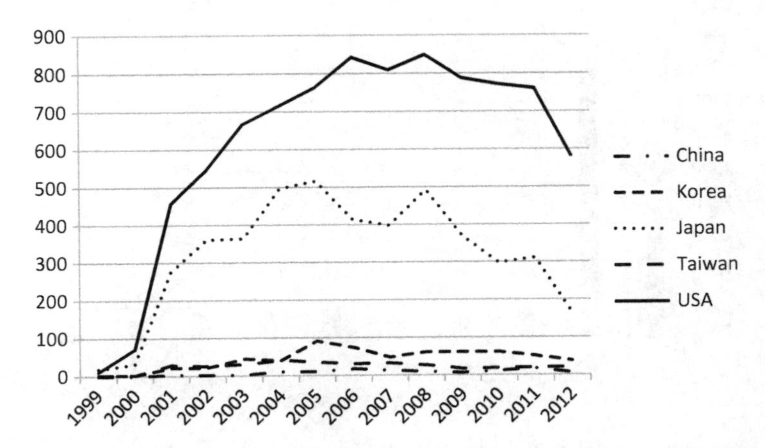

Fig. 13.5 Longitudinal count of air pollution abatement patent applications (Note: Based on USPTO data sourced from OECD.stat. Source: OECD.stat (http://stats.oecd.org/))

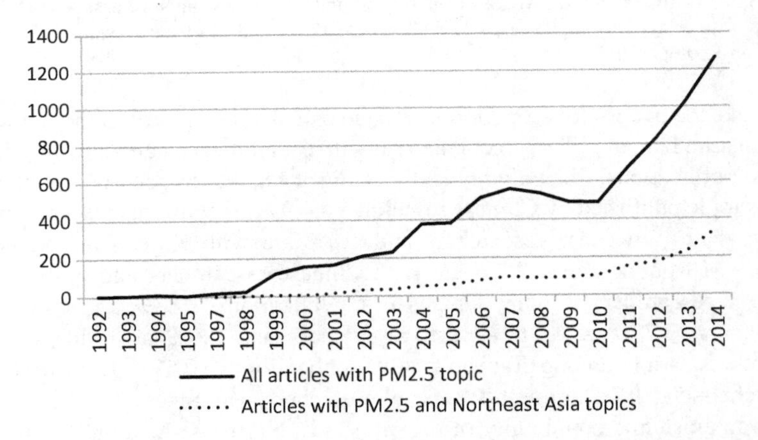

Fig. 13.6 Longitudinal count of $PM_{2.5}$ articles, overall and with focus on Northeast Asia (Note: "All articles with $PM_{2.5}$ topic" subsumes "Articles with $PM_{2.5}$ and Northeast Asia topics". Source: Thomson-Reuters (2015))

then have increased since then. Highlighted among these are publications that also have topics mentioning any of the Northeast Asian states, i.e. China, Korea, Japan, and Taiwan. Publications focusing specifically on the Northeast Asian region in whole or part have been increasing exponentially since 2010. Shown in Fig. 13.7, those articles covering the issue of $PM_{2.5}$ with regard to Northeast Asia are primarily produced in the Northeast Asian states, although researchers in the USA are the second-most frequent producers of research on this subject. China's efforts to address $PM_{2.5}$ have nearly paralleled the amount of published research produced by the USA.

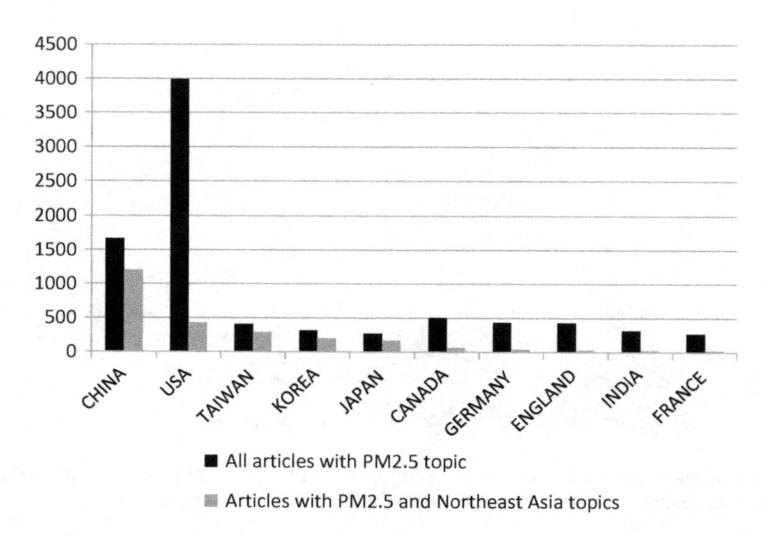

Fig. 13.7 All-time count of $PM_{2.5}$ articles, overall and with focus on Northeast Asia (Note: "All articles with $PM_{2.5}$ topic" subsumes "Articles with $PM_{2.5}$ and Northeast Asia topics". Source: Thomson-Reuters (2015))

Unlike the pattern for air pollution-related patents, researchers in the states under examination here are likely to collaborate with each other to generate air pollution-related publications. Shown graphically in Fig. 13.8, among the 1668 publications on this topic published by Chinese researchers, 4.9% are with Japanese researchers, 2.6 are with Taiwanese researchers and 1.6% are with Korean researchers. Of Korea's 318 publications, 8.2% are with Chinese researchers and 6.6% are with Japanese researchers. Among Japan's 271 publications, 30.3% are with Chinese researchers, 7.7% are with Korean researchers and 2.6% are with Taiwanese researchers. And, among Taiwan's 408 publications, 10.5% are with Chinese researchers and 1.7% are with Japanese researchers. The function of China in fostering an epistemic community of researchers in Northeast Asia on the subject of $PM_{2.5}$ is phenomenal, particularly in light of Japan's dominance in green-related research overall (Shapiro 2014a). While the USA is the prevalent partner for each of these countries save Japan, we do not know the extent to which USA-based publication partnerships are a function of USA-based foreigners collaborating with researchers back home. However, this should become less a concern given research identifying China as the dominant producer in global research output and networking (Wagner et al. 2015).

This publication-based data is corroborated with interview analysis, which revealed a common theme among most interviewees: specifically, that the Northeast Asian states are collectively focused not only on the yellow-dust problem but on prospects for increased nuclear power plants in China. Talks at the regional level tend to focus on the fact that the prevailing winds blow west-to-east and, thus, airborne pollution generated in China extends far beyond the country's political

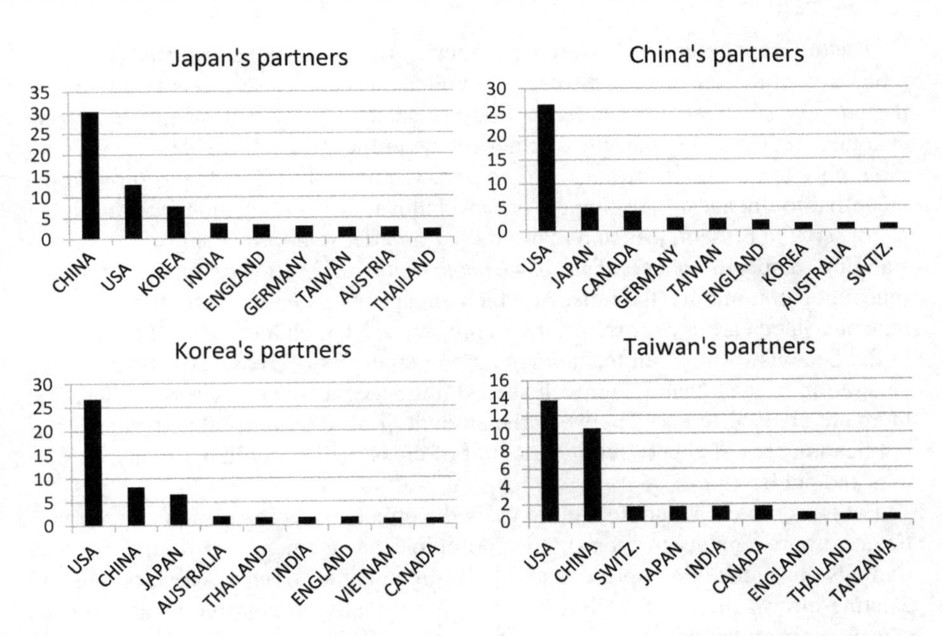

Fig. 13.8 Top partnering countries, percentage of collaborative publications with $PM_{2.5}$ topic (Source: Thomson-Reuters (2015))

borders. Given these winds, nuclear disasters are a point of concern for China's neighbors, as any failure in nuclear plants sited on China's coast will have impacts similar to those of transboundary yellow dust. Interviewees highlighted that the hazards from a power plant failure will be even more devastating than the impact of transboundary pollution in the form of $PM_{2.5}$.

13.5 Conclusion

Transboundary air pollution from China represents an environmental disaster in the clearest sense. When the facts of the pollution are debated, challenged, or disputed, EANET and other regional organisations like it provide an avenue out of the quagmire in which regional discussions may be trapped. EANET does not merely assign blame for transboundary pollution but highlights data collection and dissemination. Separating the politics and historical animosity from the environmental disaster is an ideal option to address the transboundary air pollution originating in China. Further, many of the changes that have occurred in patent and publication output overlap with the institutional changes at EANET. One cannot yet claim causality, but there are clear correlations between EANET's second-generation approach to transboundary pollution and the increase in attention to the issue by the research community. This is evidence of an epistemic community of researchers, and it is growing larger with each passing year.

The different outcomes between the patents and publications analysed above are a function of the science infrastructure. Publication-based research is generated for the purpose of expanding the larger body of scientific research, while patents are designed to limit the sharing of intellectual property and provide a temporary monopoly to the patent owner. The temporary monopoly is needed to provide the incentive to engage in the pursuit of new innovations. Publications, on the other hand, serve to present, test, and revise theory in order to improve our overall understanding of specific phenomena. If we separate the scientific community from the innovation community, the latter of which engages in patenting activities, the findings presented earlier confirm what was proposed in Urpelainen (2011): differences in the benefits from green technology among countries decrease investment in the generation of such technologies. It should not necessarily be a surprise, thus, that there are great differences between the amount of air pollution-related patents and publications but also between the amount of cross-national collaborations via patents and publications.

Future research would do well to track the effects of these epistemic communities on improvements in cross-border pollution management and mitigation. The findings presented here represent solely an alignment between research output and existing efforts such as EANET, but it will eventually be possible to identify the causal connections between the two if research efforts continue unabated. Future research must also account for parallel organisations' efforts to create an epistemic community around the issue of transboundary air pollution. It is worth noting that NEASPEC's meetings in 2013, 2014 and 2015 seemed to subsume the goals of EANET. Indeed, EANET's data were used as part of NEASPEC's modeling of the source-receptor relationship with regard to transboundary air pollution. NEASPEC also attempts to integrate EANET, the Scientific Research Institute for Atmospheric Air Protection (SRI Atmosphere), the Joint Research Project on Long-Range Transboundary Air Pollutants in Northeast Asia, and a host of universities and (domestic) government research institutes. Further examination of these cross-organisation dynamics will provide an even better understanding of how transboundary air pollution is being addressed.

Future research should also focus attention on the general public. EANET focuses on acid deposition monitoring, data curation and analysis, quality control of monitoring methods across countries, training of monitoring methods, and research on acid deposition and air pollution. There is also a clear focus now on disseminating these findings to the layman (EANET 2011), a practice which will ultimately challenge claims that EANET is apolitical. Thus far, the general public has been largely ignored, but it is nonetheless critical that we understand precisely how EANET's activities are connected to public perceptions of transboundary pollution. These perceptions can translate into policy action if there is a groundswell of support, and this in fact seemed to happen after the US Embassy in Beijing installed an air quality monitor on its roof. The accompanying Twitter updates of air quality based on EPA standards have led some to speculate that, since Twitter posts from this monitoring station became the preferred information source for many Beijing residents, it provided the impetus for Beijing to ultimately update its $PM_{2.5}$ monitoring infrastructure (Roberts 2015).[12] Whatever the case, a complete assessment of public opinion will provide the necessary bridge between EANET's recently revised orientation and policy outcomes.

Epistemic communities are key to solving common pool issues. Their foundation in scientific pursuits and their overarching expertise about a shared cause enables them to stand slightly above the political fray, as analyses of long-term international R&D networks reveal that they are not analogous to political networks (Wagner et al. 2015). On this point, Northeast Asia shows considerable promise in dealing with transboundary air pollution, as Chinese urban planners readily adopt best practices from abroad (Wu et al. 2014). This will be crucial, as air pollution levels are dropping in the coastal urban centres while increasing inland (Zheng et al. 2014). In this way, the current environmental challenges and responses in China—and thus the rest of the region—provide the foundation for sustainable development of inland/westward urban areas.

References

Abbott, K. W. (2012). The transnational regime complex for climate change. *Environment and Planning C: Government and Policy, 30*(4), 571–590.

Abbott, K.W., Green, J.F., Keohane, R.O. (2013). Organisational ecology and organisational strategies in world politics. The Harvard Project on Climate Agreements, Discussion Paper 13–57.

Aghion, P., & Howitt, P. (1992). A model of growth through creative destruction. *Econometrica, 60*, 323–351.

Ai, N., & Polenske, K. R. (2008). Socioeconomic impact analysis of yellow-dust storms: An approach and case study for Beijing. *Economic Systems Research, 20*(2), 187–203.

BBC News. (2015). China incinerator plan cancelled after protests. BBC News. http://www.bbc.com/news/world-Asia-China-32229589.

Brettell, A. (2007). Security, energy, and the environment: The atmospheric link. In I.-T. Hyun & M. A. Schreurs (Eds.), *The environmental dimension of Asian security: Conflict and cooperation over energy, resources, and pollution.* Washington, DC: United States Institute of Peace.

Brettell, A., & Kawashima, Y. (1998). Sino-Japanese relations on acid rain. In M. A. Schreurs & D. Pirages (Eds.), *Ecological security in Northeast Asia*. Seoul: Yonsei University Press.

Bulkeley, H., Andonova, L., Backstrand, K., Betsill, M. M., Compagnon, D., Duffy, R., Kolk, A., Hoffman, M., Levy, D., Newell, P., Milledge, T., Paterson, M., Pattberg, P., & VanDeveer, S. (2012). Governing climate change transnationally: Assessing the evidence from a database of sixty initiatives. *Environment and Planning C: Government and Policy, 30*(4), 591–612.

Cainelli, G., Mazzanti, M., & Montresor, S. (2012). Environmental innovations, local networks and internationalization. *Industry and Innovation, 19*(8), 697–734.

Chen, C., & Shao, L. (2007). Report on the environmental satisfaction in Xiacheng District of Hangzhou City [in Chinese]. *Zhejiang Statistics, 14*(10), 46–68.

EANET. (2010). About acid deposition monitoring network in East Asia. Viewed 29 Apr 2016, http://www.eanet.asia/eanet/brief.html.

EANET. (2011). Acid deposition monitoring network in East Asia. Retrieved 29 Apr 2016, http://www.eanet.asia/product/EANET_Brochure.pdf.

[12] Air pollution-related information has in fact been distorted by the Chinese government (Ravetti et al. 2014), although such practices seem to have ended (at least in Beijing) in 2012 (Stoerk 2015).

EANET. (2015). Review on the state of air pollution in East Asia. Retrieved 29 Apr 2016, http://www.eanet.asia/product/RSAP/RSAP.pdf.

Fankhauser, S., Bowen, A., Calel, R., Dechezleprêtre, A., Grover, D., Rydge, J., & Sato, M. (2013). Who will win the green race? In search of environmental competitiveness and innovation. *Global Environmental Change, 23*(5), 902–913.

Fischer, E. V., Hsu, N. C., Jaffe, D. A., Jeong, M.-J., & Gong, S. L. (2009). A decade of dust: Asian dust and springtime aerosol load in the U.S. Pacific Northwest. *Geophysical Research Letters, 36*, 1–5.

Frantzen, D. (2002). Intersectoral and international R&D knowledge spillovers and total factor productivity. *Scottish Journal of Political Economy, 49*(3), 280–303.

Gardiner, B. (2014). Air of revolution: How activists and social media scrutinize city pollution. The Guardian, 31 Jan. http://www.theguardian.com/cities/2014/jan/31/air-activists-social-media-pollution-city.

Ge, B.-Z., Liu, Y., Chen, H.-S., Pan, X.-L., & Wang, Z.-F. (2015). Spatial source contributions identification of acid rain over the Yangtze River delta using a variety of methods. *Atmospheric and Oceanic Science Letters, 8*(6), 397–402.

GNNCSDS [Global Network of National Councils for Sustainable Development and Similar Bodies]. (n.d.). Korea: Presidential Committee on GreenGrowth (PCGG). Viewed 29 Apr 2016, http://www.ncsds.org/index.php/sustainable-development-councils/86-country-profiles/profiles/155-korea.

Griliches, Z. (1990). Patent statistics as economic indicators: A survey. *Journal of Economic Literature, 28*(4), 1661–1707.

Guo, Z., Ai, N., & Polenske, K. R. (2008). Evaluating environmental and economic benefits. *The International Journal of Sustainable Development and World Ecology, 15*(5), 457–470.

Haas, P. M. (1990). Obtaining international environmental protection through epistemic consensus. *Millennium: Journal of International Studies, 19*(3), 347–363.

Haggard, S. (2013). The organisational architecture of the Asia-Pacific: Insights from the new institutionalism. In M. Kahler & A. MacIntyre (Eds.), *Integrating regions: Asia in comparative context*. Stanford: Stanford University Press.

Hall, B. H., Jaffe, A. B., & Trajtenberg, M. (2002). The NBER patent citations data file: Lessons, insights and methodological tools. In A. B. Jaffe & M. Trajtenberg (Eds.), *Patents, citations and innovations*. Cambridge, MA: MIT Press.

Han, L., Cheng, S., Zhuang, G., Ning, H., Wang, H., Wei, W., & Zhao, X. (2015). The changes and long-range transport of PM2.5 in Beijing in the past decade. *Atmospheric Environment, 110*, 186–195.

Helpman, E. (1993). Innovation, imitation, and intellectual property rights. *Econometrica, 61*(6), 1247–1280.

Huang, L., Yan, L., & Wu, J. (2016). Assessing urban sustainability of Chinese megacities: 35 years after the economic reform and open-door policy. *Landscape and Urban Planning, 145*, 57–70.

Jeong, D.-Y. (2008). Socio-economic costs from yellow dust damages in South Korea. *Korean Social Science Journal, 35*(2), 1–29.

Kelley, J. G. (2013). The potential for organisational membership rules to enhance regional cooperation? In M. Kahler & A. MacIntyre (Eds.), *Integrating regions: Asia in comparative context*. Stanford: Stanford University Press.

Keohane, R. O., & Victor, D. G. (2011). The regime complex for climate change. *Perspectives on Politics, 9*(1), 7–22.

Kim, L. (1999). *Learning and innovation in economic development*. Cheltenham: Edward Elgar Publishing.

Kim, I. (2007). Environmental cooperation of Northeast Asia: Transboundary air pollution. *International Relations of the Asia-Pacific, 7*(3), 439–462.

Kim, S. (2014) South Korea's climate change diplomacy: Analysis based on the perspective of "middle power diplomacy", EAI MPDI Working Paper, no. 5. pp. 1–39.

Komori, Y. (2010). Evaluating regional environmental governance in Northeast Asia. *Asian Affairs, 37*(1), 1–25.

Kostka, G. (2015). Command without control: The case of China's environmental target system. *Regulation & Governance, Early View, 10*(1), 58–74.

Kostka, G., & Mol, A. P. J. (2013). Implementation and participation in China's local environmental politics: Challenges and innovations. *Journal of Environmental Policy & Planning, 15*(1), 3–16.

Lee, S. (2001). Environmental regime-building in Northeast Asia: A catalyst for sustainable regional cooperation. *Journal of East Asian Studies, 1*(2), 31–61.

Lee, S. (2002). Building environmental regimes in Northeast Asia: Progress, limitations, and policy options. In P. G. Harris (Ed.), *International environmental cooperation: Politics and diplomacy in Pacific Asia*. Boulder: University of Colorado Press.

Li, J., Yang, W., Wang, Z., Chen, H., Hu, B., Li, J., Sun, Y., & Huang, Y. (2014). A modeling study of source-receptor relationships in atmospheric particulate matter over Northeast Asia. *Atmospheric Environment, 91*, 40–51.

Lin, J., Pan, D., Davis, S. J., Zhang, Q., He, K., Wang, C., Streets, D. G., Wuebbles, D. J., & Guan, D. (2014). China's international trade and air pollution in the United States. *Proceedings of the National Academy of Sciences of the United States of America, 111*(5), 1736–1741.

Lo, K. (2015). How authoritarian is the environmental governance of China? *Environmental Science & Policy, 54*, 152–159.

McHale, M. R., Pickett, S. T. A., Barbosa, O., Bunn, D. N., Cadenasso, M. L., Childers, D. L., Gartin, M., Hess, G. R., Iwaniec, D., McPhearson, T., Peterson, N., Poole, A., Rivers, L., III, Shutters, S. T., & Zhou, W. (2015). The new global urban realm: Complex, connected, diffuse, and diverse social-ecological systems. *Sustainability, 7*(5), 5211–5240.

Meng, J., Liu, J., Xu, Y., & Tao, S. (2015). Tracing primary PM2.5 emissions via Chinese supply chains. *Environmental Research Letters, 10*(5), 1–12.

Mol, A. P. J. (2009). Urban environmental governance innovations in China. *Current Opinion in Environmental Sustainability, 1*(1), 96–100.

Nam, S. (2002). Ecological interdependence and environmental governance in Northeast Asia: Politics vs. cooperation. In P. G. Harris (Ed.), *International environmental cooperation: Politics and diplomacy in Pacific Asia*. Boulder: University of Colorado Press.

Park, J. (2004). International and intersectoral R&D spillovers in the OECD and East Asian economies. *Economic Inquiry, 42*(4), 739–757.

Park, S.-U., Choe, A., Lee, E.-H., Park, M.-S., & Song, X. (2010). The Asian Dust Aerosol Model 2 (ADAM2) with the use of normalized difference vegetation index (NDVI) obtained from the Spot4/vegetation data. *Theoretical and Applied Climatology, 101*(1), 191–208.

Perkins, R., & Neumayer, E. (2008). Fostering environment-efficiency through transnational linkages? Trajectories of CO2 and SO2, 1980–2000. *Environment and Planning A, 40*(12), 2970–2989.

Ravetti, C., Jin, Y., Quan, M., Shiqiu, Z., & Swanson, T. (2014). A dragon eating its own tail: Public information about pollution in China. *Center for International Environmental Studies, Research Paper, 27*, 1–41.

Roberts, D. (2015) Opinion: How the U.S. embassy tweeted to clear Beijing's air, Wired, 6 Mar. http://www.wired.com/2015/03/opinion-us-embassy-beijing-tweeted-clear-air/.

Romer, P. (1990). Endogenous technological change. *Journal of Political Economy, 98*(5), S71–S102.

Schmookler, J. (1966). *Invention and economic growth*. Cambridge, MA: Harvard University Press.

Shapiro, M. A. (2012). Environmental legislation in East Asia: Rationale and significance. In Z. Zhu (Ed.), *New dynamics in east Asian politics*. New York: Continuum.

Shapiro, M.A. (2014a) International collaboration and green technology generation: Assessing the East Asian environmental regime. EAI Fellows Program Working Paper Series No. 46, pp. 1–23.

Shapiro, M. A. (2014b). Regionalism's challenge to the pollution haven hypothesis: A study of Northeast Asia and China. *The Pacific Review, 27*(1), 27–47.

Shapiro, M. A., & Gottschall, K. (2011). Northeast Asian environmentalism: Policies as a function of ENGOs. *Asian Politics and Policy, 3*(4), 551–567.

Shapiro, M. A., & Nugent, J. B. (2012). Institutions and the sources of innovation: The determinants and effects of international R&D collaboration. *International Journal of Public Policy, 8*(4–6), 230–250.

Sohn, I. (2012). Toward normative fragmentation: An East Asian financial architecture in the post-global crisis world. *Review of International Political Economy, 19*(4), 586–608.

Stoerk, T. (2015) Statistical corruption in Beijing's air quality data has likely ended in 2012. Grantham Research Institute on Climate Change and the Environment, Working Paper No. 194.

Sun, Y., Zhuang, G., Huang, K., Li, J., Wang, Q., Wang, Y., Lin, Y., Fu, J. S., Zhang, W., Tang, A., & Zhao, X. (2010). Asian dust over northern China and its impact on the downstream aerosol chemistry in 2004. *Journal of Geophysical Research-Atmospheres, 115*(D7), 1–16.

Thomson-Reuters. (2015) Web of science. Viewed 10 Oct 2015, http://wokinfo.com/.

Tsai, F., Tu, J.-Y., Hsu, S.-C., & Chen, W.-N. (2014). Case study of the Asian dust and pollutant event in spring 2006: Source, transport, and contribution to Taiwan. *Science of the Total Evironment, 478*, 163–174.

Tsunekawa, K. (2005). Why so many maps there? Japan and regional cooperation. In T. J. Pempel (Ed.), *Remapping East Asia: The construction of a region*. Ithaca: Cornell University Press.

Urpelainen, J. (2011). Technology investment, bargaining, and international environmental agreements. *International Environmental Agreements: Politics, Law and Economics, 12*(2), 145–163.

USPTO. (2009). Environmentally Sound Technologies (EST) concordance. Viewed 29 Apr 2016, http://www.uspto.gov/web/patents/classification/international/est_concordance.htm.

USPTO. (2015). United States Patent and Trademark Office Database. Viewed 10 Oct 2015, http://patft.uspto.gov/netahtml/PTO/search-adv.htm.

Victor, D. G. (2011). *Global warming gridlock*. Cambridge, UK: Cambridge University Press.

Wagner, C.S., Bornmann, L., Leydesdorff, L. (2014) Recent developments in China-US cooperation in science. Paper presented to the Conference on China's International S&T Relations, Arizona State University, April.

Wagner, C. S., Park, H. W., & Leydesdorff, L. (2015). The continuing growth of global cooperation networks in research: A conundrum for national governments. *PloS One, 10*(7), e0131816.

World Bank. (2015). *World development indicators, 2015*. Washginton, DC: International Bank for Reconstruction and Development/The World Bank.

Wu, J., Xiang, W.-N., & Zhao, J. (2014). Urban ecology in China: Historical developments and future directions. *Landscape and Urban Planning, 125*, 222–233.

Yoo, I.T., & Kim, I. (2015). Free trade agreements for the environment? Regional economic integration and environmental cooperation in East Asia. In *International environmental agreements: Politics, law and economics* (pp. 1–18).

Yoshimatsu, H. (2014). *Comparing institution-building in East Asia: Power politics, governance, and critical junctures*. New York: Palgrave Macmillan.

Young, O. R. (1990). Global environmental change and international governance. *Millennium: Journal of International Studies, 19*(3), 337–346.

Zhao, S., & Yang, B. (2007). A survey and analysis on the environment awareness of Lanzhou citizen [in Chinese]. *Journal of BUPT (Social Sciences Edition), 9*(5), 14–18.

Zheng, S., & Kahn, M. E. (2013). Understanding China's urban pollution dynamics. *Journal of Economic Literature, 51*(3), 731–772.

Zheng, S., Sun, C., Qi, Y., & Kahn, M. E. (2014). The evolving geography of China's industrial production: Implications for pollution dynamics and urban quality of life. *Journal of Economic Surveys, 28*(4), 709–724.

Zi, F., Yang, Z., Zhang, F., Tian, H., Wang, G., Wu, J., Liu, Y., Fang, G., Xiang, J., & Fang, R. (2012). Report of environment satisfaction and life satisfaction in ten cities in China [in Chinese]. *Journal of Beijing Forestry University, 11*(4), 1–7.

Chapter 14
Zaps and Taps: Solar Storms, Electricity and Water Supply Disasters, and Governance

Robert James Wasson

14.1 Introduction

Electricity distribution for water supply and sewage processing (Wilkinson 2011) (Fig. 14.1) is vulnerable to solar storms. This chapter aims to draw attention to this issue for those involved in water supply risk governance, to motivate more research and action to mitigate the likely catastrophe of an extreme solar storm. The main topics of the chapter are: the solar storm threat including the 'worst case', the vulnerability of power grids and water resources dependent upon electricity, current trends in power grid development in relation to the solar storm threat, mitigation strategies and governance issues, with particular emphasis on cross-border challenges.

14.2 Electricity for Water Management

Energy is needed for water resource management, and most comes from the generation of electricity that is distributed through high voltage power lines. In the USA about 3% of electricity production (4157×10^9 kWh) is for water treatment and disposal of wastewater, about 8% for cooking, cleaning and water heating, and about 1% for pumping and transport of water and wastewater (Novotny 2012). This amounts to about 500×10^9 kWh annually. In six Australian cities, that contain nearly half of the country's population, about 0.2% of the total energy used is for water management (Kenway et al. 2008). Globally there is considerable variability in the amount of electricity used in different parts of the water management system.

R.J. Wasson (✉)
Institute of Water Policy, Lee Kuan Yew School of Public Policy, National University of Singapore, Singapore, Singapore
e-mail: spprjw@nus.edu.sg

M.A. Miller et al. (eds.), *Crossing Borders*, DOI 10.1007/978-981-10-6126-4_14

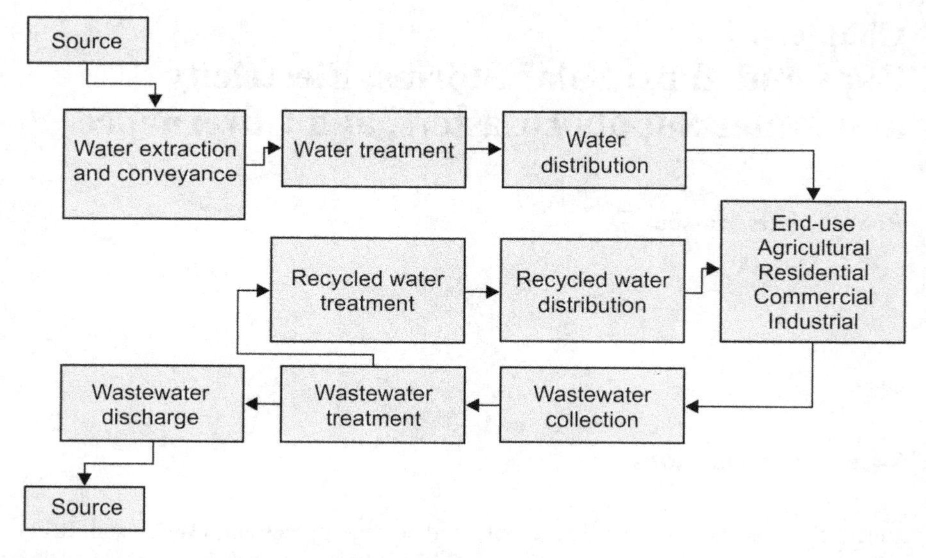

Fig. 14.1 Most of this system is vulnerable to failure of electricity supply

For example, water treatment can be achieved using between 0.03 and 7 kWh m^{-3} (Novotny 2012)—different by a factor of 233. Of the total energy used for water management in Bangkok, 20% is used for drinking water treatment, in contrast with 45% in Tokyo because of different water quality standards (Dhakal et al. 2015). But according to the same authors, 55% of the energy in Tokyo is used for water transport and distribution versus 80% in Bangkok, a share that is even higher in Delhi at 83.5%. From these few examples it is plain that energy is used in different amounts and in different ways depending upon many local factors. Therefore vulnerability to disruption is spatially heterogeneous.

14.3 Solar Storms and Some Examples of Their Impacts

The potential impacts of solar storms are a consequence of the combined effects of the threat, the vulnerability of electricity grids and the dependence of water supply on electricity. There are three components to solar storms: solar flares (SFs) that last for 1–2 h, solar proton events (SPEs) that last for days, and coronal mass ejections (CMEs) that also last for days (Marusek 2007). SFs are magnetically driven explosions on the surface of the Sun that produce electromagnetic radiation in the form of X-rays, extreme ultraviolet (UV) rays, gamma radiation, and radio wave bursts. SFs interfere with satellite communications, radar, and shortwave radio, and also affect the orbits of satellites. M-class SFs cause radio blackouts in the polar regions, while more powerful X-class SFs can trigger worldwide radio blackouts. SPEs consist of high-energy solar cosmic rays that disorient satellites, damage spacecraft electronics including solar panels, irradiate highflying aircraft, produce fading in shortwave radio signals, deplete ozone, and have human health effects (including, in extreme cases, heart attacks). Large SPEs are followed 96% of the time by a CME that is a mass of gas and charged plasma with an embedded magnetic field blasted from the Sun. When a CME reaches Earth it massively disturbs the magnetic field in a geomagnetic storm. Charged particles and electrons in the ionosphere induce powerful electrical currents in the surface of the planet, known as geoelectric-induced currents (GICs), that have spatially variable impacts depending in part on the conductivity of Earth's crust. CMEs cause satellite tracking errors and payload deployment problems, radar errors, radio propagation errors, compass realignments, oil and gas pipeline corrosion and failure, communication landline and equipment damage, electric shocks and fires, and human health impacts. Both SPEs and CMEs will hereafter be referred to as SPEs (Marusek 2007).

In addition to the severe solar storms, there is another space weather phenomenon that deserves attention: sudden impulses. These occur during geomagnetically quiescent periods and are caused by abrupt increases in the solar wind dynamic pressure that increases the northward-directed magnetic field and create GICs that can damage power grids at low latitudes (Carter et al. 2015).

The most readily observed manifestations of geomagnetic storms are aurora, known as the northern and southern lights. They result from the precipitation of charged particles from solar storms into the upper atmosphere. The resulting ionisation and excitation of atmospheric constituents emit light of various colours producing spectacular displays, mostly at high latitudes but also at much lower latitudes during extreme solar storms (Eather 1980).

The 1859 super-magnetic storm, known as the Carrington Event, erupted from Earth's poles to the equator but did not affect electrical supply, as it was not in wide use at the time. It caused telegraph systems to fail, with some machines bursting into flames and telegraph operators rendered unconscious (Clark 2007; Lloyd's 2013). The aurora associated with this event was seen as far south as 18° geomagnetic latitude (corrected for movement of the geomagnetic poles) near Panama, a long way south at about 8.5°geographic latitude (Boteler et al. 1998). With regard to electrical power systems, in 1972 AT&T (American Telephone and Telegraph Corporation) redesigned a trans-Atlantic power cable after a major solar storm stopped telephone communications. Also in the same year a 230,000-volt transformer exploded in British Columbia as a result of a solar storm, and in 1980 a similar failure occurred in Canada at St. James Bay, a replacement for which failed the following year for the same reason (Omatola and Okeme 2012). In 1989 a solar storm induced a geoelectric field that coupled with the Hydro-Québec electric power grid in Canada. The grid collapsed after the protective relays were compromised and about nine million people lost power (Bolduc 2002). Also in 1989, the Salem Pressurized Water Nuclear Reactor in New Jersey was affected when an induced current in the electrical transmission line damaged a step-up transformer (Anon 2014; Lloyd's 2013). During the same event large transformers were damaged in the UK and about 200 significant anomalies occurred in electricity grids across North America, with power interruptions as far south as California (Aon Benfield 2013). In 2003 a large solar storm caused system failure in the Swedish electrical grid by shutting down transformers (National Academy of Sciences 2004; Organisation for Economic Co-operation and Development/International Futures Programme [OECD/IFP] 2011). During the same event damage also occurred to the grid in North America including a capacitor trip and transformer heating, shutting down water and sewage pumps in New York City and spewing millions of gallons of sewage; in Detroit and Cleveland raw sewage polluted drinking water sources (Hines et al. 2009). The same event in South Africa led to 12 transformers being removed from service, a surprise at the time because the largest effects were expected at high latitudes (Lloyd's 2013). But as pointed out by Pulkkinen et al. (2010), geomagnetically induced currents (GICs) are a 'truly global phenomenon' even though the largest magnitudes of GICs are expected at high latitudes. For example, Pulkkinen et al. (2012) suggest that the magnitude of GICs is greater for the 100-year events (i.e. those with an annual probability of 0.01) at geomagnetic latitudes higher than about 40°, although there is enhancement near the geomagnetic equator in the equatorial electrojet. An elecrojet is a band of intense electrical current tens of kilometres above Earth's surface in the ionosphere, near the North and South Poles (the auroral electrojets) and also near the geomagnetic equator (Akasofu 2002; Carter et al. 2015; Lühr et al. 2004).

These examples suggest that in Asia the greatest effect is likely to be in northern China (Liu et al. 2014a), northern Japan and parts of Central Asia, with a smaller effect in Southeast Asia (Malaysia, Thailand, Lao, Cambodia, Vietnam, Indonesia) and India near the geomagnetic equator (Carter et al. 2015; Lühr et al. 2004). The most susceptible areas outside Asia are in Russia, most of Europe including the UK, and most of North America, with impacts in southern Australia, New Zealand and southernmost Africa, and smaller effects near the geomagnetic equator in central Africa and South America (e.g. in Brazil and Uruguay; Barbosa et al. 2015; Caraballo et al. 2013; Carter et al. 2015).

The disruptions to electricity supply described above may seem comparable with the impacts of meteorological storms, floods, earthquakes and heat waves and therefore well within modern society's capability to cope, albeit with some discomfort. An example of a non-solar disruption is provided by the massive power outage in the Midwestern USA in the summer of 2003 that was caused by the shutdown of one generating plant in Ohio: a case of a cascading failure as load was shifted to already overheated power lines that became hotter and shut down (Mitchell 2009). The largest impact of an apparent failure to manage an electrical grid occurred in India when a power outage affected 370 million people in 2012. But management and over-drawing of power may not have been the only cause of the outage. A solar storm is also implicated as the solar proton flux increased just before the outage, causing a trip in the grid (Mukherjee 2015; Mukherjee, personal communication, December Mukherjee 2015); this was possibly the result of enhancement in the equatorial electrojet (cf. Carter et al. 2015).

Perhaps the most disturbing impact of a long power outage is the failure of supply of cooling water for nuclear reactors (Kappenman 2010). Depending upon the magnitude of the impact of a SPE on power grids, hundreds of nuclear power reactors could melt down as their cooling water is depleted. Explosions and breaches of containment vessels would then spread radioactive material into the surrounding areas and further afield if wind speeds are sufficient; this would result in a cascading disaster across borders. The Chernobyl and Fukushima disasters are noteworthy examples of what could happen, but to many more reactors. While this is a realistic possibility, and the blogosphere is full of apocalyptic pronouncements, there appears to have been little serious analysis of the problem.

Society has become used to short-term power outages of a few days, but much longer outages as a consequence of a solar storm are not part of most country's planning (CRO Forum 2011). While researchers have focused on the physics of solar storms and impacts on electricity grids, there is almost no research on the links between electrical failure from solar storms and water supply disturbance. This is possibly because the identification of the 'downstream' cascading impacts of solar storms, including on water supply, requires simulation modeling that entails modeling of the AC power flow through the grid and detailed transformer specifications. Some countries may have undertaken simulations, but the results are not in the public domain. However, it is not difficult to imagine the impact on water supply of a power outage lasting for months, particularly in cities where householders and businesses have little or no water storage capacity. That such a scenario is not fanciful is clear, because even if spare transformers are available it can take weeks to months for transportation and installation (Bartley 2002). The construction of new transformers may take 5–12 months for domestic suppliers and 6–16 months for international suppliers (Corbin 2012; Office of Energy Delivery and Electrical Reliability 2012; United States International Trade Commission 2012). To add to the problem, the construction of new high voltage (HV) or extra high voltage (EHV) transformers requires electricity!

14.4 Threats, Vulnerability and Risk

The threat of solar storms of the intensity of the Carrington Event is real: auroral sightings over the past 2000 years have shown that the Carrington Event was not unique (Lloyd's 2013). An estimate of the probability of such a storm would be a useful input to risk assessment, and has been recommended in many reports on the threat of SPEs (e.g. North American Electrical Reliability Corporation [NERC] 2012). Also, some attention to the 'worst case' threat is provided because it has figured in much of the discussion of risk assessment (e.g. Langbein 2014). But an understanding of the threat will be a long way short of what is required, because the risk is a function of both the threat and the vulnerability of electricity supply and its effect on water supply and treatment.

In what follows the frequency, probability, and possible 'worst cases' of SPEs are considered, along with the factors involved in infrastructure vulnerability and estimates of total risk to electricity availability.

The physics of SPEs is insufficiently understood to estimate frequency from first principles. Therefore empirical methods are required, and scholars have used many different data types and analytical procedures, producing a wide variety of estimates. The 1859 Carrington Event was the largest SPE in at least 450 years according to Shea et al. (2006), and in 155 years according to Lakhina and Tsurutani (2016), with a likely return period of 150 years (an annual probability of 0.007) based on auroral sightings from as early as 480 BCE (Stothers 1979). The Québec event has a return period of 50 years (0.02 annual probability) based on the same information. Using power-law modeling of the Dst distribution (the disturbance storm time that is a measure of the geomagnetic disturbance level globally in units of nanoTesla, nT), Riley (2012) estimated the decadal probability of a Carrington scale event at 12% (an annual probability of 0.0127 assuming that the storms are independent of one another). He got the same result using CME speeds, but a value of only 1.1% using the nitrate in ice cores that has been interpreted to be the result of solar storms, a conclusion doubted by others (for a discussion see Lakhina and Tsurutani 2016). Kataoka (2013) extended Riley's work by using a longer record of magnetometer measurements in Japan, and concluded that the probability of a solar storm of the magnitude of the Carrington Event is 4–6% within the next decade. Love (2012) used the Dst distribution to estimate the decadal probability of a Carrington-scale event at 6%, although he used a Dst value for the event of −1760 nT, a value considered too high by a factor of two (Riley 2012) thereby reducing the probability estimate. Love (2012) calculated the 68.3% confidence limits for his estimate as 0.16–1.4% over the next decade, and the 95.45% confidence limits as 0–23%. Barnard et al. (2011) also used the nitrate record to calculate a rate for a major SEP of 5.2 per century (an annual probability of 0.052) for the period from 1700 to 1970 CE, and only 2.6 per century (0.026 annual probability) for the space era, a decline that may be real or may be the result of low-number statistics.

The longest record used in the analyses summarised above is from nitrate in ice cores, the veracity of which as a record of SPEs is in doubt (see Lakhina and Tsurutani 2016). However, long records are needed for accurate estimates of the probability of rare events. Marusek (2007) used calculated proton fluences (the total fluxes during SPEs) of solar storms >30 MeV cm^{-2} (F_{30}); that is, 30 million electron volts per square centimeter, referred to as F30 from satellite observations, nitrate spikes in ice cores, long records of the cosmogenic radionuclides (see Beer et al. 2012, for an account of these chemicals) ^{10}Be, ^{26}Al, ^{41}Ca, ^{81}Kr in moon rocks, ^{14}C in tree rings, and ^{10}Be in ice cores to conclude that solar storms a million times greater than the Carrington Event are possible on a over a period of a million years (annual probability of 0.000001). The maximum proton fluence based on a power law model of the upper tail of these data is of the order of 10^{16} cm^{-2}, an extreme value that is implausible according to the results of Townsend et al. (2006) who estimated the maximum at 18.8×10^{9} cm^{-2}. This lower value is consistent with the results of analyses by Usoskin and Kovaltsov (2012) and Kovaltsov and Usoskin (2014) that relied on the cosmogenic nuclides ^{10}Be and ^{14}C in terrestrial archives (ice cores and tree rings) and seven cosmogenic nuclides in moon rocks. They found two events during the past 11,400 years, in 780 CE and 1460 CE (an event that has also been ascribed to a supernova explosion rather than an SPE), which they claimed were extreme SPEs. From these data they estimated a conservative upper limit of F_{30} of 5×10^{10} cm^{-2}, with annual occurrence probabilities of extreme SPEs of 10^{-2} to 10^{-4} $year^{-1}$, and perhaps controversially no evidence for a very strong SPE at the time of the Carrington Event. Vasyliunas (2011) adopted a different approach to estimating the maximum possible SPE by using the physical theory embodied in the Dessler-Parker-Sckopke theorem that relates the disturbance magnetic field, created by a SPE, at Earth to the total kinetic energy of plasma in the magnetosphere. The maximum plausible Dst is –2500 nT according to this analysis, about three times the strength of the Carrington Event.

There are large uncertainties attached to all of the estimates of occurrence probability and the possible 'worst cases', and the Carrington Event as an extreme is in doubt. Love (2012) made the following telling statement about these large uncertainties: 'the 10-yr recurrence probability for a Carrington event is somewhere between vanishingly unlikely and surprisingly likely', a conclusion echoed by Lakhina and Tsurutani (2016). And the Carrington Event may not have been particularly severe. Despite the large uncertainties in the peer-reviewed scientific literature, it is noteworthy that Lloyd's (2013), one of the largest insurance companies, is prepared to identify recurrence intervals with error bands but without comment on their accuracy, its printed disclaimers notwithstanding.

The most probable extreme SPE, with an annual probability of 10^{-2}, has been derived from the long records of cosmogenic nuclides. This could be the 'worst case' but details of its likely impact on Earth are not available because the key measures of geoeffectiveness are not available; that is, the magnitude and spatial distribution of GICs. A SPE in 2012, which missed Earth, had a minimum Dst between −1150 and –600 nT, possibly 1.4 times greater than the Carrington Event (Liu et al. 2014b) producing GICs as large or larger than the largest observed GICs, although a parsimonious interpretation is that it was similar to the Carrington Event in strength. If the Carrington Event were to strike Earth today, the estimated cost in the USA alone would be US $2 trillion and a recovery time of 4–10 years (National Research Council 2008) or, with some attention to uncertainty, a cost of between US $0.6 and 2.6 trillion (Lloyd's 2013). The cost of the 2012 SME, had it hit Earth, would have been comparable to the estimate above, causing massive damage to electrical grids, water supplies, and other key facilities, but perhaps not to the extent of sending us back to 'a post apocalyptic Stone Age' as suggested by Anthony (2014) because not all of the planet would have been affected.

Because of the high quality of the data collected during the 2012 event, and the availability of advanced modeling capabilities (Ngwira et al. 2013), Daniel Baker from the University of Colorado observed: 'We would like space weather users, operators of systems, and policy makers (to) adopt this event immediately and do war game scenarios with it' (Byrd 2013). He went further by suggesting that the 2012 event should be adopted as the 'worst case' space weather scenario that should be used in modeling the effects on electricity grids. Even though this is unlikely to have been the worst SME during the past 11,400 years, as seen earlier, it is probably the best-observed extreme event. Baker's suggestion for a 'worst case' is therefore a compromise between the most extreme but poorly known event, and a less extreme but much better known event.

The threat to power grids is not only a function of the frequency and magnitude of SPEs, but also of the orientation of the SPE with respect to Earth; geomagnetic latitude (because GICs are stronger at high latitudes but not insignificant at lower latitudes); ground conductivity that can nonlinearly amplify GICs; and distance from the coast (because seawater is more conductive than rock and soil and the excess current can flow into grounded transformers on nearby land) (Alekseev et al. 2015; Lloyd's 2013; Pulkkinen et al. 2007). Therefore the threat is spatially heterogeneous, and each SPE will produce GICs of different magnitudes in different places. The vulnerability of power grids to GICs is also a function of the characteristics of the grid.

14.5 Vulnerability of Power Grids and Water Supply

Electric power transmission systems have generating plants connected by transmission lines in which voltages are controlled and high voltage is reduced for distribution at substations. Geomagnetic disturbances produce magnetic field variations that drive electric currents in the conducting ground that causes electrical currents (GICs) in conducting structures such as along transmission lines and through transformers into the ground. The magnitude of GICs is modulated by ground conductivity, as already discussed, a quantity that varies within and between countries by a factor of about 10 (International Telecommunications Union 1992) or about a factor of 55 when calculated differently (Alekseev et al. 2015). Damage to power grids by GICs consists of damage to bulk power systems, particularly to HV and EHV transformers, and also the loss of reactive power support (NERC 2012), the power that maintains the reliability of supply. The total vulnerability increases with the length of transmission lines (Lloyd's 2013): it reaches a maximum value in a few hundred kilometres in individual lines but continues to rise over much longer distances if the system length (i.e. all of the transmission lines) is taken into account (Zheng et al. 2014). The topology of the power network produces different GICs at the substations and in the transmission lines, with the largest GIC at the edges of the network and in the middle of individual transmission lines according to the deliberately simplified analysis by Zheng et al. (2014). Transformers can be overheated (but see Vergetis 2016), relays tripped, and/or they can fail completely from voltage instability. A significant loss of reactive power support, along with an increased demand for reactive power, is the largest source of transformer vulnerability. Based on past responses to GICs, transformers with high water and dissolved gas contents, and those nearing the end of their life span, are most vulnerable. Newer designs of transformers are less vulnerable and single-phase transformers are more vulnerable than three-phase transformers. Also the number and electrical resistance of transformers and transmission lines affect the magnitude of a GIC (Vergetis 2016, and references therein).

The 'perfect storm' of vulnerability for power grids could of course be produced by the coincidence of an extreme SPE with other sources of power grid failure. Birds, lightning, earthquakes, over-drawing of power, failure of old infrastructure, overheating in heat waves, collapse of transmission lines in ice storms, and instability caused by dead ends in the network (Menck et al. 2014) could coincide with an SPE. Non-SPE failures are planned for, usually using the n–1 criterion: that is, the losses of a single critical component (transformer, transmission line) without causing network overload or unstable operation. But this deterministic criterion is being replaced by a probabilistic approach that takes into account multiple failures (Heylen and Van Hertem 2014), an approach that will be essential for mitigation of the impacts of SPEs. Some countries (e.g. Australia and New Zealand) have already adopted the new approach, and so could be less vulnerable to SPEs depending upon how well they have assessed their vulnerability.

As already noted, reliance on electricity for water supply and treatment is spatially variable with some cities and countries using much more electricity per unit of water than others. Those that use most power for water treatment are most vulnerable to power outages. But if electricity supply is completely switched off by a SPE these differences will be unimportant, unless an adjoining country can still supply water because it has not been badly affected by a SPE (possibly because of differences in ground conductivity and/or the installation of power grid protection), and its electricity use for water supply is low leaving enough to provide water to its neighbour. Neighbours may also be able to supply electricity but not water. But the likelihood of such scenarios will probably depend more upon politics than technology. Other factors may also be important, such as the extent to which gravity flows allow transport of water within and between countries thereby avoiding the need for pumping; the availability of backup diesel generators, although they are likely to be hostage to fuel supply and the need for electricity to refill fuel storage tanks; the availability of alternative power sources such as solar panels and wind turbines that are separate from the grid and protected from SPEs; and the time taken to replace damaged HV and EHV transformers.

14.6 Risk

The current approach to assessing the risk for a power system of a SPE is to combine information in a simulation model of a plausible threat, often a Carrington-scale event (which may be considered similar to the 2012 event), with information about ground conductivity and the grid. Lloyd's (2013) applied this method to find that the GIC amplitudes in North America were highest in the Midwest of the USA extending into Canada. With information about transformer locations and designs (see Vergetis 2016 for a new view on transformer vulnerability), more detail can be achieved that shows large spatial variations. Storms other than a Carrington-scale event have also been used in simulations. The 2003 'Halloween storm' with a Dst of about –400 nT (Asia Insurance Review 2014) was used by Barbosa et al. (2015) to simulate effects in the low latitude Brazilian transmission lines, showing that events about half as strong as the Carrington Event are potentially damaging.

The North American Electric Reliability Corporation (NERC 2012), Kappenman (2010) and most recently the Institute of Electrical and Electronics Engineers (IEEE 2015) provide detailed accounts of simulation modeling, although they do not take into account network topology and are oriented to the extreme SPEs experienced at high latitudes. Gaunt (2014) makes the case for a systems model that includes space physics, network analysis, transformer engineering, network reliability and decision support, tailored particularly to low latitudes where fewer storms reach damaging levels and awareness of GICs is less well developed. Gaunt has raised the issue of decision support to find the best solutions for a complex system, although he doesn't mention operator error. In the US operator error accounts for 8% of blackouts (Hines et al. 2009), a figure that is likely to rise if space weather forecasts are to be used more to change the operations of power grids.

In addition, the absence of scenario analyses that rigorously include phenomena other than electricity, and water in particular, is a serious limitation on the design of mitigation and governance (OECD/IFP 2011). This absence suggests that solar storms are not well enough understood among government and private sector planners to be included in risk assessments, or they are viewed as having such a low probability that they can be discounted in the face of other risks such as weather and equipment ageing. There is also an issue of incentives for investment in expensive protection devices that may not be needed for a long time.

14.7 Trends in Electricity Grid Development: New Sources of Vulnerability?

In the interests of efficiency, reliability, cost reduction, and the social benefits of providing electricity to more people, cross-border power grid integration has occurred between many countries. NORDPOOL, for example, connects the Nordic Countries to The Baltic States, the UK, and Germany (Glachant and Lévêque 2009) in a region prone to extreme SPEs. Other cross-border grid networks have been established with varying degrees of interconnection, and include the Central American Power market (SIEPAC), the North American power grid, the Greater Mekong Sub-Region (GMS), the Southern African Power Pool (involving 12 nations) and the West African Power Pool (Singh et al. 2015; OECD/IFP 2011). Future potential and limitations of further integration are discussed by Economic Consulting Associates (2010).

In South Asia network links exist between Nepal and India, India and Bhutan, India and Sri Lanka, India and Bangladesh, Pakistan and Iran, and Afghanistan and several Central Asian countries, while an agreement between Pakistan and India is under discussion (Singh et al. 2015). Apart from the reasons given above for network integration, Singh et al. (2015) claim that lessening the role of the State in providing electricity to achieve affordable and reliable electricity is also an objective. But it is not clear if this is an objective of the authors of this World Bank report or an objective of the countries of South Asia. Certainly India has opened its power market to the private sector more than other countries in South Asia. But as we will see below, this trend may need to take account of solar storms.

In Southeast Asia an ambitious plan is underway to link the power grids of all ASEAN countries (Andrews-Speed 2016; ASEAN Power Grid Consultative Committee 2015). Eleven cross-border links already exist, 10 more are in progress and a further 17 are planned. The estimated cost saving from interconnection is US $1873 million in 2009 present value.

Another development that deserves attention is the move to 'smart grids' that connect different sources of electricity generation and involve interaction between users and the grid by means of sensors linked through the Internet. The likely benefits of 'smart grids' are reduced peak demand, tailored energy use, linkage to renewable sources of power (often a long way from users), automatic rerouting of electricity from disabled network components and routing of power to key facilities such as hospitals and emergency services during disruptions (ASEAN Power Grid Consultative Committee 2015). However 'smart grids' appear to enhance the vulnerability to GICs of power systems by extending transmission lines to connect to renewable generating sources and by relying on sensors that are vulnerable to satellite failure from SPEs. Once again, a lack of a comprehensive scenario analysis raises serious doubts about the ability of 'smart grids' or any other kind of grid to withstand GICs, although OECD/IFP (2011) notes that the modular components of 'smart grids' will be less vulnerable to GICs than large centralised networks.

14.8 Concluding Remarks: Mitigation and Governance

Mitigation can be achieved by planning, engineered hardening, better operational response and reform of governance (OECD/IFP 2011; NERC 2012). Planning might involve all or some of the following: scenario analysis and simulation of the effects on power networks of the 2012 SPE as the best known 'worst case'; simulations of cascading failures of key facilities and services that rely on electricity, including water resources and sewerage systems; simulations of both national and cross-border networks with sufficient spatial specificity and assessments of social vulnerability for operational purposes; cross-border agreements about how to communicate warnings of impending SPEs and when and how to act; cross-border agreements about the allocation of new or standby transformers in the event of a major loss of this equipment; and insurance against losses, business discontinuity, and the limitations imposed by territorial limitations of insurance (Aon Benfield 2013; Lloyd's 2013).

The simulations and scenario assessments should include the entire power network in a system dynamics framework along the lines suggested by Gaunt (2014). That is, scenario analysis could be in the form of dynamic systems models rather than large and often unreliable so-called deterministic models (Sterman 2000). They should also make use of modern thinking about how to overcome the vulnerability of networks (e.g. Barabási 2002; Little 2002; Lorenz et al. 2009). Helbing (2013) calls for a major overhaul of risk assessment and management, pointing to the absence in current approaches of coincidences of multiple threats and vulnerabilities (e.g. the 'perfect storm' of a SPE and other failure modes), the absence of feedback loops in analyses, linear rather than nonlinear thinking, downplaying of human errors and negligence in assessments, and insufficient attention to personal and government incentive structures in assessment of risk. He particularly calls for a reversal of the trend to dilution of responsibility in governments and corporations, so that those responsible for a failure are held responsible. But in a complex system with many feedbacks, responsibility can be a slippery concept.

Hardening of infrastructure is clearly necessary but is neither technically feasible nor economically possible for entire power networks, a problem that will be acute in poor countries (OECD/IFP 2011). Therefore, decisions have to be made about which critical facilities will be protected by the installation of neutral blocking devices and transmission line capacitors. Again, national and cross-border agreements will be necessary to ensure maximum protection of hospitals, water distribution and treatment, nuclear power plants, and emergency services. If possible, electricity supply to these critical facilities should not rely on power from neighbouring countries.

Well-executed operational plans and procedures are cheaper than hardening, but cannot fully replace hardening as a mitigation strategy. Operational effectiveness relies upon space weather monitoring (Pulkkinen et al. 2010), warnings (and therefore effective communications), quick and effective reactions to warnings (with sufficient training and flexibility given to operators to enable agile responses in the face of changing circumstances, rather than just following a rulebook); and cross-border coordination of operational plans and procedures. Warnings currently rely upon satellite observations that need to be maintained internationally rather than relying solely upon the USA to provide the hardware (OECD/IFP 2011).

All of these mitigation strategies seem sensible and achievable given sufficient knowledge, motivation, planning, and resources. But they may fail because of the inherent complexity of power networks that include cross-border connections and markets and 'smart grids'. Helbing (2013) argues, for example, that strongly connected networks that have produced highly interdependent systems are too difficult to understand and control top down, and may fail globally if perturbed by a SPE or other threat. Newell et al. (2011) argue that policy is too often designed by taking a narrow compartmentalised view, dominated by one worldview because of the bounded rationality of humans. This can be seen, for example, in the paper by Singh et al. (2015) who adopt a narrow economistic view of the benefits of cross-border power networks without paying attention to other issues. Such an approach almost always leads to policies that have unintended and often disastrous consequences (see Sterman 2000, for some iconic examples). Helbing (2013) goes further to suggest that bottom-up systems are likely to be much more resilient, and mentions the example of 'smart grids' as a solution to large-scale failure of networked power. But even 'smart grids' need high voltage transmission lines and transformers to connect them to part of the generating system, and so will not be entirely immune from SPEs and other sources of transmission line and transformer failure. A scheme analogous to smart electricity grids is the idea of water smart grids (Water Innovations Alliance 2012) that would be localised and optimised by sensors communicating with water and sewerage utilities. But such a scheme would not be entirely safe from SPEs as communications that rely upon satellites could shut down, and power supplies could be disturbed if any part of the local system needed to be connected to the larger grid.

So all power networks are to varying degrees vulnerable to SPEs and GICs, and many if not most water resource and sewerage systems are vulnerable to consequential electricity failure. If, however, reliance is placed entirely on the current trend for market mechanisms (and therefore the private sector) to build and manage new power networks, both within and between countries, the governance perspective may be too narrow to include network hardening and operating procedures to deal with GICs. Moreover, there may be little incentive for the private sector to make the necessary investments. Also, by giving priority to efficiency the opportunity may be lost to have on standby high voltage transformers and backup power systems to build redundancy into networks (see Newell et al. 2011 for examples of these issues in the context of climate change).

Still, localisation and modularisation of both electricity and water supply has many advantages. It will reduce the spatial extent of disruption by any cause, enable quicker recovery, and also reduce the need for difficult agreements between neighbouring states for cross-border network operation during GICs. It may also maximise cooperation between stakeholders because the network density of these small-scale networks is sufficiently small to avoid the erosion of cooperation as network density increases (Helbing 2013), thereby enabling self-organisation and agile responses locally to the threat of a SPE.

Whichever route is taken to rethink and reform governance of power networks, and their dependent functions such as water supply, to take account of SPEs, it is likely that reversing the trend to large centralised networks of the kind being designed and implemented in Asia will be essential, and could be achieved by including 'smart grids'. While this trend is underway in wealthy countries, it needs to be accelerated. Such reforms will require leadership from governments and cooperation by the private sector. It is also strongly recommended that narrow worldviews, such as economic efficiency, be balanced with other considerations by using a system dynamics approach.

Acknowledgements I thank my wife Merrilyn for inadvertently alerting me to the threat of solar storms, Michael Douglass and Michelle Miller for the invitation to write this chapter and their patience with my slowness, Karl Kim for encouraging my approach, and Sarah Starkweather for excellent editing. David Nott is thanked for advice on probability calculations. The project benefited from the financial support of a Singapore Ministry of Education Academic Research Fund Tier 2 grant entitled 'Governing Compound Disasters in Urbanising Asia' (MOE2014-T2–1-017).

References

Akasofu, S. I. (2002). *Exploring the secrets of the Aurora*. New York: Kluwer Academic Publications.

Alekseev, D., Kuvshinov, A., & Palshin, N. (2015). Compilation of 3D global conductivity model of the earth for space weather applications. *Earth, Planets and Space, 67*(7), 108–119.

Andrews-Speed, P. (2016). Connecting ASEAN through the power grid: Next steps. Policy Brief 11, Energy Studies Institute, National University of Singapore, Singapore, viewed 30 Apr 2016, http://esi.nus.edu.sg/docs/default-source/doc/esi-policy-brief-11---connecting-asean-through-the-power-grid-next-steps.pdf?sfvrsn=0.

Anon. (2014). Solar storm effects on nuclear and electrical installations. http://mragheb.com/NPRE%20402%20ME%20405%20Nuclear%20Power%20Engineering/Solar%20Storms%20Effects%20on%20Nuclear%20and%20Electrical%20Installations.pdf.

Anthony, S. (2014). The solar storm of 2012 that almost sent us back to a post-apocalyptic Stone Age. Extreme Technology, viewed 25 Apr 2016, http://www.extremetech.com/extreme/186805-the-solar-storm-of-2012-that-almost-sent-us-back-to-a-post-apocalyptic-stone-age.

ASEAN Power Grid Consultative Committee (2015). Energy regulatory commission ERC forum: ASEAN Power Grid, Road to Multilateral Power Trading. Retrieved 30 Apr 2016, http://www.energyforum2015.com/download/Session1-1present.pdf.

Barabási, A. L. (2002). *Linked: The new science of networks*. New York: Perseus Publishing.

Barbosa, C. S., Hartmann, G. A., & Pinhheiro, K. J. (2015). Numerical modeling of geomagnetically induced currents in a Brazilian transmission line. *Advances in Space Research, 55*(4), 1168–1179.

Barnard, L., Lockwood, M., Hapgood, M. A., Owens, M. J., Davis, C. J., & Steinhilber, F. (2011). Predicting space climate change. *Geophysical Research Letters, 38*(16). doi:10.1029/2011GL048489.

Bartley, W.H. (2002). Life cycle management of utility transformer assets. (http://www.imia.com/wp-content/uploads/2013/05/EP14-2003-LifeCycleManagement-Utility-Transformers.pdf; retrieved 1/09/2016).

Beer, J., McCracken, K., & von Steiger, R. (2012). *Cosmogenic radionuclides*. Berlin: Springer.

Benfield, A. (2013). *Geomagnetic storms*. Sydney: Aon Benfield Analytics.

Bolduc, L. (2002). GIC observations and studies in the Hydro-Québec power system. *Journal of Atmospheric and Solar – Terrestrial Physics, 64*(16), 1793–1802.

Boteler, D. H., Pirjola, R. J., & Nevanlinna, H. (1998). The effects of geomagnetic disturbances on electrical systems at the Earth's surface. *Advances in Space Research, 22*(1), 17–27.

Byrd, D. (2013). Proposed step to help society prepare for a solar storm disaster. EarthSky, viewed 25 Apr 2016, http://earthsky.org/space/can-the-july-2012-solar-storm-help-society-prepare-for-disaster.

Caraballo, R., Bettucci, L. S., & Tancredi, G. (2013). Geomagnetically induced currents in the Uruguayan high-voltage power grid. *Geophysics Journal International, 195*(2), 844–853.

Carter, B. A., Yizengaw, E., Pradipta, R., Halford, A. J., Norman, R., & Zhang, K. (2015). Interplanetary shocks and the resulting geomagnetically induced currents at the equator. *Geophysical Research Letters, 42*(16), 6554–6559.

Clark, S. (2007). *The sun kings: The unexpected tragedy of Richard Carrington and the tale of how modern astronomy began*. Princeton: Princeton University Press.

Corbin, R.B. (2012). The challenges of replacing a failed transformer. Fast Track Power, 26 Feb, viewed 18 Apr 2016, http://fasttrackpower.com/the-challenges-of-replacing-a-failed-transformer/.

CRO Forum (2011). *Power blackout risks: Risk management options*. Emerging Risk Initiative—Position Paper, CRO Forum, Amstelveen, the Netherlands.

Dhakal, S., Shrestha, S., Shrestha, A., Kansal, A., Kaneko, S. (2015). Towards a better water-energy-carbon nexus in cities. APN Global Change Perspectives Policy Brief No. LCD-01, Asia-Pacific Network for Global Change Research, Kobe.

Eather, R. H. (1980). *Majestic lights: The aurora in science, history and the arts*. Washington, DC: The American Geophysical Union.

Economic Consulting Associates. (2010). *The potential of regional power sector integration*. London: Economic Consulting Associates. retrieved 30 April 2016, http://www.esmap.org/sites/esmap.org/files/BN004-10_REISP-CD_The%20Potential%20of%20Regional%20Power%20Sector%20Integration-Literature%20Review.pdf.

Gaunt, C. T. (2014). Reducing uncertainties—responses for electricity utilities to severe solar storms. *Journal Space Weather Space Climate, 4*(A01), 1–7. doi:10.1051/swsc/2013058.

Glachant, J. M., & Lévêque, F. (2009). *Electricity reform in Europe: Towards a single market*. Cheltenham: Edward Elgar Publishing.

Helbing, D. (2013). Globally networked risks and how to respond. *Nature, 497*, 51–59.

Heylen, E., Van Hertem, D. (2014). Importance and difficulties of comparing reliability criteria and the assessment of reliability. IEEE Young Researchers Symposium, Gent, 24–25 Apr, retrieved 27 Apr 2016, https://lirias.kuleuven.be/bitstream/123456789/451702/1/YRS2014_EH_DVH.pdf.

Hines, P., Balasubramaniam, K., & Sanchez, E. C. (2009). Cascading failures in power grids. *IEEE Potentials, 28*(5), 24–30.

Institute of Electrical and Electronics Engineers (IEEE). (2015). *Guide for establishing power transformer capability while under geomagnetic disturbances*. Piscataway: IEEE Standard C57.163–2015, Institute of Electrical and Electronics Engineers Standards Association.

International Telecommunications Union (1992). World Atlas of ground conductivities. Retrieved 26 Apr 2016, https://www.itu.int/dms_pubrec/itu-r/rec/p/R-REC-P.832-2-199907-S!!PDF-E.pdf.

Kappenman, J. (2010). Geomagnetic storms and their impacts on the U.S. Power Grid. Meta-R-319, Metatech Corporation, retrieved 27 Apr 2016, http://fas.org/irp/eprint/geomag.pdf.

Kataoka, R. (2013). Probability of occurrence of extreme solar storms. *Space Weather, 11*(5), 214–218.

Kenway, S.J., Priestley, A., Cook, S., Seo, S., Inman, M., Gregory, A., Hall, M. (2008). Energy use in the provision and consumption of urban water in Australia and New Zealand, water for a healthy Country Flagship Report Series, Commonwealth Scientific and Industrial Research Organisation (CSIRO) and Water Services Association of Australia, Canberra, Australia.

Kovaltsov, G. A., & Usoskin, I. G. (2014). Occurrence probability of large solar energetic particle events: Assessment from data on cosmogenic radionuclides in lunar rocks. *Solar Physics, 289*(1), 211–220.

Lakhina, G. S., & Tsurutani, B. T. (2016). Geomagnetic storms: Historical perspective to modern view. *Geoscience Letters, 3*(5), 1–11. doi:10.1186/s40562-016-0037-4.

Langbein, F. (2014). Natural catastrophe: A solar superstorm—what if? Asia Insurance Review, October, viewed 26 Apr 2016, http://www.asiainsurancereview.com/Magazine/ReadMagazineArticle?aid=35552.

Little, R. G. (2002). Controlling cascading failure: Understanding the vulnerabilities of interconnected infrastructures. *Journal of Urban Technology, 9*(1), 109–123.

Liu, Y. D., Li, Y., & Pirjola, R. (2014a). Observations and modeling of GIC in the Chinese large-scale high-voltage power networks. *Journal of Space Weather and Space Climate, 4*(A03), 1–7. doi:10.1051/swsc/2013057.

Liu, Y. D., Luhmann, J. G., Kajdic, P., Kilpua, E. K. J., Lugaz, N., Nitta, N. V., Mösti, C., Lavraud, B., Bale, S. D., Farrugia, C. J., & Galvin, A. B. (2014b). Observations of an extreme storm in interplanetary space caused by successive coronal mass ejections. *Nature Communications, 5*, 1–8. doi:10.1038/ncomms4481.

Lloyd's. (2013). *Solar storm risk to the north American electric grid*. London: Lloyd's.

Lorenz, J., Battiston, S., & Schweitzer, F. (2009). Systemic risk in a unifying framework for cascading processes on networks. *European Physical Journal B, 71*, 441–460.

Love, J. J. (2012). Credible occurrence probabilities for extreme geophysical events: Earthquakes, volcanic eruptions, magnetic storms. *Geophysical Research Letters, 39*(10), 1–6. doi:10.1029/2012GLO51431.

Lühr, H., Maus, S., & Rother, M. (2004). Noon-time equatorial electrojet: Its spatial features as determined by the CHAMP satellite. *Journal of Geophysical Research, 109*(A1), 01306. doi: 10.1029/2002JA009656.

Marusek, J.A. (2007). Solar storm threat analysis. Unpublished white paper. *Impact*. Retrieved 25 Apr 2016, http://projectcamelot.org/Solar_Storm_Threat_Analysis_James_Marusek_Impact_2007.pdf.

Menck, P. J., Heitzig, J., Kurths, J., & Schellnhuber, H. J. (2014). How dead ends undermine power grid stability. *Nature Communications, 5*, 1–8. doi:10.1038/ncomms4969.

Mitchell, M. (2009). *Complexity: A guided tour*. New York: Oxford University Press.

Mukherjee. (2015). Northern India power grid failure due to extraterrestrial changes. *Earth Science and Climatic Change, 6*, 261–262.

National Academy of Sciences. (2004). *Severe space weather storms October 19–November 7, 2003*. Washington, DC: Government Printing Office.

National Research Council. (2008). *Severe space weather events: Understanding societal and economic impacts. A workshop Repor*. Washington, DC: The National Academies Press.

Newell, B., Marsh, D. M., & Sharma, D. (2011). Enhancing the resilience of the Australian national electricity market: Taking a systems approach in policy development. *Ecology and Society, 16*(2), 15.

Ngwira, C. M., Pulkkinen, A., Mays, M. L., Kuznetsova, M. M., Galvin, A. B., Simunac, K., Baker, D. N., Li, X., Zheng, Y., & Glocer, A. (2013). Simulation of the 23 July 2012 extreme space weather event: What if this extremely rare CME was earth directed? *Space Weather, 11*(12), 671–679.

North American Electric Reliability Corporation. (2012). *Special reliability assessment. Interim report. Effects of geomagnetic disturbances on the bulk power system*. Atlanta: NERC. 137 pp.

Novotny, V. (2012). Water and energy link in the cities of the future-achieving net zero carbon and pollution emissions footprint. In V. Lazarova, K. H. Choo, & P. Cornel (Eds.), *Water-energy interactions in water reuse*. London: IWA Publishing.

Office of Energy Delivery and Electric Reliability. (2012). *Large power transformers and the U.S. electric grid*. Washington, DC: US Department of Energy. http://energy.gov/sites/prod/files/Large%20Power%20Transformer%20Study%20-%20June%202012_0.pdf.

Omatola, K. M., & Okeme, I. C. (2012). Impacts of solar storms on energy and communications technologies. *Archives of Applied Science Research, 4*(4), 1825–1832.

Organisation for Economic Co-operation and Development/International Futures Programme [OECD/IFP] (2011). Geomagnetic Storms, OECD/IFP Futures Project on 'Future Global Shocks', retrieved 30 Apr 2016, http://www.oecd.org/gov/risk/46891645.pdf.

Pulkkinen, A., Bernabeu, E., Eichner, J., Beggan, C., & Thomson, A. W. P. (2012). Generation of 100-year geomagnetically induced current scenarios. *Space Weather, 10*(4), 1–19. doi:10.1029/2011SW000750.

Pulkkinen, A., Hesse, M., Habib, S., Van der Zel, L., Damsky, B., Policelli, F., Fugate, D., Jacobs, W., & Creamer, E. (2010). Solar shield: Forecasting and mitigating space weather effects on high-voltage power transmission systems. *Natural Hazards, 53*(2), 333–345.

Pulkkinen, A., Pirjola, R., & Viljanen, A. (2007). Determination of ground conductivity and system parameters for optimal modeling of geomagnetically induced current flow in technological systems. *Earth, Planets and Space, 59*(9), 999–1006.

Riley, P. (2012). On the probability of occurrence of extreme space weather events. *Space Weather, 10*(2), 1–12. doi:10.1029/2011SW000734.

Shea, M. A., Smart, D. F., McCracken, K. G., Dreschoff, G. A. A. M., & Spence, H. E. (2006). Solar proton events for 450 years: The Carrington event in perspective. *Advances in Space Research, 38*(2), 232–238.

Singh, A., Jamash, T., Nepal, R., & Toman, M. (2015). *Cross-border electricity cooperation in South Asia, Policy Research Working Paper 7328*. Washington, DC: World Bank Group.

Sterman, J. D. (2000). *Business dynamics: SystemsTthinking and modeling for a complex world*. Boston: Irwin/McGraw-Hill.

Stothers, R. (1979). Magnetic Cepheids. *The Astrophysical Journal, 234*, 257–261.

Townsend, L. W., Stephens, D. L., Jr., Zapp, E. N., Hoff, J. L., Moussa, H. M., Miller, T. M., Campbell, C. E., & Nichols, T. F. (2006). The Carrington event: Possible doses to crews in space from a comparable event. *Advances in Space Research, 38*(2), 226–231.

United States International Trade Commission. (2012). *Large power transformers in Korea: Investigation no. 731-TA-1189 (preliminary).*, Publication 4346. Washington, DC: USITC.

Usoskin, I. G., & Kovalstsov, G. A. (2012). Occurrence of extreme solar particle events: Assessment from historical proxy data. *The Astronomical Journal, 757*(1), 1–6.

Vasyliunas, V. M. (2011). The largest imaginable magnetic storm. *Journal of Atmospheric and Solar-Terrestrial Physics, 73*(11–12), 1444–1446.

Vergetis, B. (2016). Dire, dire hair on fire: read this for the real risks GIC poses to the grid. SmartGridNews, 7 Mar, viewed, 30 April 2016, http://www.smartgridnews.com/story/dire-dire-hair-fire-read-real-risks-gic-poses-grid/2016-03-07.

Water Innovations Alliance (2012). White Paper: The Water Smart Grid Initiative. Retrieved 29 Apr 2016, http://www.waterinnovations.org/PDF/WP_water_smart_grid.pdf.

Wilkinson, R. (2011). The water-energy nexus: Methodologies, challenges and opportunities. In D. S. Kenney & R. Wilkinson (Eds.), *The water-energy nexus in the American west* (pp. 3–17). Cheltenham: Edward Elgar Publishing.

Zheng, K., Boteler, D., Pirjola, R. J., Liu, L. G., Becker, R., Marti, L., Boutilier, S., & Guillon, S. (2014). Effects of system characteristics on geomagnetically induced currents. *IEEE Transactions on Power Delivery, 29*(2), 890–898.

Erratum to: Recognising Global Interdependence Through Disasters

Anthony Reid

Erratum to:
Chapter 2 in: M.A. Miller et al. (eds.), *Crossing Borders*, https://doi.org/10.1007/978-981-10-6126-4_2

The original version of this chapter was inadvertently published with incorrect impacts of natural disasters affect percentage on Asia region in Table 2.1 (Page no. 36). The correct percentage is 88.7%.

The online version of the original chapter can be found under https://doi.org/10.1007/978-981-10-6126-4_2

M.A. Miller et al. (eds.), *Crossing Borders*, DOI 10.1007/978-981-10-6126-4_15

Index

M.A. Miller et al. (eds.), *Crossing Borders*, DOI 10.1007/978-981-10-6126-4

CPSIA information can be obtained
at www.ICGtesting.com
Printed in the USA
LVHW081422190720
661079LV00018B/967